FLYING THE FLAG

Also by Hans-Liudger Dienel

INGENIEURE ZWISCHEN HOCHSCHULE UND INDUSTRIE:
Kältetechnik in Deutschland und Amerika, 1870 bis 1930

Also by Peter Lyth

INFLATION AND THE MERCHANT ECONOMY: The Hamburg
Mittelstand 1914–1924

Flying the Flag

European Commercial Air Transport since 1945

Edited by

Hans-Liudger Dienel
Managing Director
Centre for Technology and Society
Technical University
Berlin

and

Peter Lyth
Fellow, Research Institute
Deutsches Museum
Munich

First published in Great Britain 1998 by
MACMILLAN PRESS LTD
Houndmills, Basingstoke, Hampshire RG21 6XS and London
Companies and representatives throughout the world

A catalogue record for this book is available from the British Library.

ISBN 0–333–67354–9

 First published in the United States of America 1998 by
ST. MARTIN'S PRESS, INC.,
Scholarly and Reference Division,
175 Fifth Avenue, New York, N.Y. 10010

ISBN 0–312–16168–9

Library of Congress Cataloging-in-Publication Data
Flying the flag : European commercial air transport since 1945 /
edited by Hans Liudger Dienel, Peter Lyth.
p. cm.
Includes bibliographical references and index.
ISBN 0–312–16168–9 (cloth)
1. Aeronautics, Commercial—Europe—History.   2. Airlines—Europe–
–History.   I. Dienel, Hans-Liudger, 1961–   II. Lyth, Peter J.
HE9842.A35F58   1998
387.7'094'09045—DC21   98–17922
                                                                            CIP

This book is printed on paper suitable for recycling and made from fully managed and
sustained forest sources.

10   9   8   7   6   5   4   3   2   1
07   06   05   04   03   02   01   00   99   98

Printed and bound in Great Britain by
Antony Rowe Ltd, Chippenham, Wiltshire

Contents

List of Plates

Preface

The idea for this book grew out of a chance conversation between the editors at Munich's Deutsches Museum in 1995. At the time we were both working on airline histories and had become interested in the distinguishing features of Europe's flag-carriers, as well as the deep and historical divide between the air transport industries in Europe and America.. We were also struck by the fact that although the international airline industry was at a critical juncture in its history, with deregulation and strategic alliances making headlines in the press, economic and transport historians had paid it astonishingly little attention. This was particularly the case with Europe's airlines, about which no *comparative* history had ever been written.

In the course of the next year we gathered together a number of airline historians from different countries to talk about Europe's flag-carriers and to plan this volume. In addition to the Europeans we invited Professor Roger Bilstein to monitor our progress and write a concluding analysis from the American perspective. Our discussions took place in formal workshop sessions at the Deutsches Museum and occasionally in a neighbouring beer garden.

Perhaps one of the reasons why a comparative history of European airlines has never been produced is that studies across international borders are not only hard work but also logistically complex. Different scholars, writing in different languages, have to agree on common themes and goals, and be brought together within a uniform but fruitful working environment. If we have come anywhere close to succeeding in this task, then our colleagues deserve the credit and we would like to salute their good-humoured readiness to write – and then rewrite – their chapters at the command of a despotic English editor.

Our work would not have been possible without generous support from the Flughafen Frankfurt Stiftung. This foundation also contributed towards the funding of additional editorial meetings in Berlin, as well as a combined presentation by all the contributors to the London meeting of the Society for the History of Technology (SHOT) in 1996. We are very grateful and we would like to offer our thanks to the Stiftung's managing director Wolfgang Scherer.

Many people have helped in the book's preparation and we would like to take this opportunity to acknowledge their assistance and thank them. Bruce Seely, Helmuth Trischler and Burghard Ciesla offered advice and ideas, Sören Marotz prepared the figure in Chapter 1 and the index, and Kathy Franz and Stefan Zeilinger took the minutes of our preliminary discussions.

Preface

Marc Dierikx made valuable comments on an early draft of the Introduction. The Deutsches Museum has been especially helpful in providing accommodation and a meeting room for the workshops, and we are also grateful to Eva Mayring, the head of the Museum's Special Collections and to Albert Limmer. Amongst the airline archives, some have been rather more welcoming than others, as one would expect. We would like to thank them all and to mention Lufthansa, where the archivist Werner Bittner and the head of the picture collection Gerd Rebenich have been generous with their assistance.

Hans-Liudger Dienel Berlin

Peter Lyth Munich

May 1998

Notes on the Contributors

Roger Bilstein is Professor of Aviation and Aerospace History at the University of Houston. Among his many works are *Flight in America: from the Wrights to the Astronauts* (Baltimore: Johns Hopkins, 1994) and *The American Aerospace Industry: from Workshop to Global Enterprise* (New York, 1996.)

Hans-Liudger Dienel is Managing Director of the Centre for Technology and Society at the Technical University Berlin, Germany. He has edited, with Helmuth Trischler, *Geschichte der Zukunft des Verkehrs. Verkehrskonzepte von der frühen Neuzeit bis zum 21. Jahrhundert* (Frankfurt, 1997) and is currently writing a book on the history of commercial aviation and competing transport modes in Germany between 1945 and 1995.

Marc Dierikx is an independent scholar based in the Netherlands. His latest book is *Fokker. a Transatlantic Biography* (Washington, DC, 1997).

Joanna Filipcyzk teaches economic history at the University of Kraków in Poland.

Peter Lyth is a Fellow at the Forschungsinstitut of the Deutsches Museum in Munich, Germany, and an associate of the Business History Unit at the London School of Economics. He has published numerous articles on civil aviation and is currently writing on British Airways.

Amilcare Mantegazza is an economic historian at the University of Milan. He is the author of *L'Atm di Milano 1861–1972: un secolo di trasporto urbano tra finalità pubbliche e vincoli di bilancio* (Milan, 1993) and is now writing a book on Alitalia.

Nicolas Neiertz is professeur agrégé at the Conservatoire National des Arts et Métiers in Paris. His latest book is *Les Aéroports de Paris*. (Paris, 1997).

Abbreviations

AEA	Association of European Airlines
APEX	Advanced Purchase Excursion fare
ATLB	Air Transport Licensing Board (UK)
BA	British Airways
BAA	British Airports Authority
BEA	British European Airways
BOAC	British Overseas Airways Corporation
BCal	British Caledonian Airways
BSAA	British South American Airways
CAA	Civil Aviation Authority (UK)
CAB	Civil Aeronautics Board (US)
CRS	Computer Reservation System
CSA	CeskoSlovenske Aerolinie (CZ)
EC/EU	European Community/European Union
ECAC	European Civil Aviation Conference
IATA	International Air Transport Association
ICAO	International Civil Aviation Organisation
KLM	Koninklijke Luchtvaart Maatschappij (NL)
LOT	Polskie Linie Lotnicze (PL)
OPEC	Organization of Petroleum Exporting Countries
Pan Am	Pan American Airways
SABENA	Société Anonyme Belge d'Exploitation de la Navigation Aérienne
SAS	Scandinavian Airline System
SST	Supersonic Transport (aircraft)
STOL	Short Takeoff and Landing (aircraft)
TWA	Trans World Airlines
UTA	Union de Transport Aérien (France)

1 Introduction
Peter Lyth and Hans-Liudger Dienel

'The basic trouble remains that the world has too many airlines, most of them inefficient, undercapitalised and unprofitable.'

The Economist, 1960.[1]

In the late 1950s the major airlines of France, West Germany, Belgium and Italy got together to form what became known as Air Union. The aim was to coordinate operating schedules and quotas between the different countries and eventually, with the development of the European Economic Community, to create a fully-fledged European airline. There was a strong belief that only such an integrated undertaking would have the strength and resources to stand up to competition from the major American carriers. Instead of several national airlines, Europe would have a single operator the equal of Pan American Airways.

Air Union might well have been a very logical answer to Europe's civil aviation needs and such an enterprise had actually been proposed during the Second World War by people who felt that Hitler's 'unification' of European air transport under Lufthansa, should be turned to advantage in the creation of a unified postwar airline.[2] However, patriotic exclusivity triumphed and the Air Union idea collapsed in the 1960s when it became clear that the differences between the participants over aims and methods, and their respective shares in the enterprise, were irreconcilable.[3] Europe's national airlines went their own way and prospered in the boom which accompanied economic growth and the expansion of international tourism. What is interesting about the Air Union experiment is that it highlights one of the defining characteristics of the European air transport industry. Despite having a land mass of comparable dimensions to North America, Europe developed a very different airline system to that of the United States, based on a large number of states and sovereignties. And this has proved a persistent source of weakness in its attempts to keep up with the Americans in the years since the US industry first surged ahead in the 1930s. On the other hand the European system has served other, non-commercial, aims and given rise to one of the enduring features of twentieth-century transport history – the national flag-carrier. This book is about these flag-carriers, how they were created, how they have survived – in some cases for over 70 years – and what sort of future they may have now.

The following introduction begins with a chronological history of European civil aviation divided into sections according to the industry's main regulatory and technological milestones.

1

1   THE EARLY HISTORY OF EUROPEAN AIRLINES, 1920–45

Airlines originated in Europe. It was shortly after the end of the First World War that commercial air transport companies began offering rudimentary passenger services in Britain, France, Germany and Holland. A few of those early pioneers are still operating under their original names. Most of them evolved, through a process of government-induced amalgamation, into flag-carriers and the chosen instruments of national airline policies. They have all been state-owned or subsidized at one stage or another, and for long spells of their history have more or less relied on the taxpayer to keep them flying. This is not to say that European airlines have been entirely non-commercial in their operations, rather that their flag-carrying role has traditionally been decisive, and that many of their services and managerial decisions have been influenced by the need to maintain national prestige or pursue a political objective.[4]

As with the airlines, so the first aircraft capable of carrying fare-paying passengers were also built in Europe. The 1920s represents an embryonic stage in aeronautical engineering which was dominated by European firms like Junkers and Fokker, Farman and Handley Page. Junkers is especially noteworthy because it was this German company which produced the world's first metal-skinned monoplanes, the forerunners of the modern airliner.[5]

European airlines and aircraft have individual histories. The people of Belgium, Holland and Switzerland, as much as those of Britain, France and Germany, have developed a strong degree of familiarity, and even affection, for their flag-carrier. Moreover the flag-carrier's *style* has retained an institutional rather than corporate appearance and they have often been seen by their customers as a part of government. European passengers may want their flag-carrier to give good service and value for money, but they expect it to continue flying even if it does not stand up to the test of these criteria. Europe's airlines have become embedded in the tapestry of nation states that make up the continent, and this is one of the reasons why they have survived for so long in such a notoriously unprofitable business.[6]

Perhaps the easiest way to describe European flag-carriers is to contrast their history with that of the world's leading air transport power, the United States. American airline history took a very different course to Europe's. Passenger services were introduced there later, but when they arrived, towards the end of the 1920s, their development was faster and driven by demand from a much larger consumer market. Unlike most European carriers, American airlines were privately owned, although they did receive significant financial support from the federal government, usually in the form of mail subsidies. In the 1930s the basis of an extensive national air network was created, with the first transcontinental air routes set up by TWA and United Airlines. At the same time, by contrast, the Europeans were forging long, tenuous air links to their colonies in Africa and Asia, providing services for an elite of government officials and intrepid businessmen. Where the American airline industry drew its creative strength from the sheer size of its domestic market, the British, French, Dutch and Belgian flag-carriers gained impetus from the political desire to forge closer bonds with imperial possessions. And with similiar political aims, the Germans struck out for South America and the Far East.[7] The major contrast between the two systems was that Europe's airlines were born out of international operations while the industry in the United States was built on domestic foundations. Europe's air transport has remained an international operation and unlike those nations with great territorial land masses like the United States, Australia, Canada and the Soviet Union, it would never have flourished on domestic demand alone.

By the outbreak of the Second World War America already had a dynamic airline industry, but its potential to dominate international air transport was not realized until the end of the conflict. It was the experience of the war which baptised the American public in a culture of aircraft and air travel, and which led to the consolidation of America's lead. The war spurred on the American aircraft builders Douglas, Lockheed and Boeing in a way that could not be matched in Europe and, as a result, aircraft from the United States dominated the world's airline fleets in 1945. Furthermore American airlines now represented the best operational practice in the industry and it was against the standard of American airlines that the performance of European carriers was judged.

2 EUROPEAN AIRLINES IN THE POSTWAR ERA, 1945–65

It is said that the First World War taught people to fly, and the Second gave them the aircraft to do it economically. The international air transport industry in 1945 could have taken off into the clear blue sky of free competition, with every nation, in theory at least, fighting for a share of the business. The result of this would have been immediate and total American hegemony over civil aviation everywhere outside the communist bloc; the United States had the aircraft, it had the airlines and it had the passengers. This did not happen because nobody except the Americans themselves, and perhaps the Dutch, was prepared to give free competition, or *open skies*, a chance. The Europeans had not nursed their flag-carriers through the interwar years with expensive subsidies, to see them shot down now in competition with the Americans. Instead, and starting with the international conference held in Chicago at the end of 1944, they set about spinning a web of rules and regulations, fare-fixing pacts and bilateral agreements that held the industry in a 'firm corset' for the next 30 years.[8] This regulated system worked well.

The older flag-carriers survived, while new ones appeared to carry the flags of new nations and some, like Lufthansa, which were not so new. The underlying reason for the success of regulation, apart from the fact that everyone (including the Americans) benefited from it, was the rapid growth in demand for international air travel during the 1950s and 1960s.[9] There was enough business for all the airlines so long as everone played by the rules. *Open skies* would have destroyed this comfortable arrangement.

Europe's dedication to international regulation is clear from its loyalty to the International Air Traffic Association (IATA), the airlines' fare-fixing cartel, and from its easy adherence to the bilateral agreements signed by their governments, which covered every aspect of air transport between countries.[10] The regulation of European air travel was reinforced by the habit of pooling capacity and revenue between the airlines. Airline pooling was a peculiarly European practice and typical of an industry which was collaborative rather than competitive in nature. Pools restricted operations on nearly every European route to the services of the two designated flag-carriers and divided the revenue on a prearranged basis. It was a mutually rewarding system although occasionally it could be the source of dispute if one of the parties felt it was not getting a fair deal. In the mid-1960s, for example, the Italians became so exasperated by the generous traffic rights that the British enjoyed at Rome that they threatened to renege on the Anglo-Italian pool agreement if it was not reformulated in a manner more favourable to Alitalia.[11] To some degree the pooling system can be seen as an allocational cartel in which the participants agreed on the way that air travel markets would be divided, much as IATA agreed on the air travel prices that would be charged. On the other hand, the case for pools in Europe was fairly strong and became even stronger with the introduction of high-capacity jets in the 1960s.[12]

Pools were, not surprisingly, condemned on the other side of the Atlantic, yet the Americans could hardly claim the moral high ground. Bilateral agreements benefited them too. Indeed many of the deals reached between the American and European governments after the Second World War favoured the US by allowing their flag-carriers, Pan American Airways (Pan Am) and Trans World Airways (TWA), extensive *onward rights* between European capitals, while denying the same concession to Europe's airlines in the American hinterland. For smaller nations like the Netherlands, this was a major handicap. The Royal Dutch Airlines (KLM) gained a large share of international traffic in the years after the war, but had very little to bargain with when seeking rights from the United States since KLM's territory and population was small, even by European standards, and the American Civil Aeronautics Board (CAB) would only sign an agreement if it yielded the same benefits to its own carriers as to the Europeans.

After steady growth in the 1950s, two developments in the early 1960s transformed European air transport. First there was the realignment of route networks that accompanied the retreat from Empire, a factor of great importance to the British, French and Dutch. And secondly there was the far-reaching conversion of airline fleets from propeller-driven to jet aircraft. At the end of the war, Air France, the British Overseas Airways Corporation (BOAC) and KLM were all committed to rebuilding the pre-war air routes to their respective colonies in Africa, Asia and the Caribbean, and a considerable amount of their aircraft capacity was devoted to this task. Within a decade, however, these countries had loosened their ties with former colonies, or withdrawn from them altogether, and were transferring their attention to a more important route – the North Atlantic. The potential for revenue and profits represented by the American passenger market lured all the European airlines and made their North Atlantic services a priority in their long-haul operations. The introduction of long-range piston-engine aircraft like the Douglas DC-7C and the Lockheed Super Constellation in the mid-1950s, combined with the decline of ocean-going passenger shipping, turned the Atlantic into *the* profit-maker for European flag carriers.[13]

The move to jet aircraft, which, after the false start of the Comet 1, began in earnest with the introduction of the Boeing 707 by Pan Am in 1958, complimented the European route realignment towards the Atlantic. The major weakness of the Comet 1 had been its lack of range, which prevented BOAC from using it on the Atlantic, the one route where the pioneer jet airliner would have shown a significant competitive advantage over American piston-engined aircraft. The Comet's history is perhaps typical of the aircraft procurement problems which troubled both the British and the French in the 1950s. These two countries were not prepared to buy American equipment off the shelf like the Dutch, Germans and Italians.[14] Instead they sought to revitalize their own aircraft industries with civil airliner programmes. Both had a modest degree of success – for the British, the Vickers Viscount, for the French, the Sud-Aviation Caravelle – but in general it was an expensive and unrewarding enterprise, and in the end it only proved that no single European country could build aircraft on a scale to equal the American planemakers. In 1965 a major report into the British aerospace industry pointed out that aircraft manufacturing was subject to strong economies of scale and that unless the huge design and development costs of a new aircraft could be spread over a long production run, commercial failure was virtually inevitable.[15] This was an observation which applied equally to the whole European aircraft industry and it was in recognition of this cold logic of the civil aircraft industry the European Airbus programme was launched in the late 1960s.

One of the most striking things about European civil aviation is the degree to which its fortunes have been bound up with those of the aircraft industry. And this applies both to the period up to the mid-1960s when Britain and France were pursuing costly individual aircraft programmes, and to the period since then when the Airbus has carried the standard of European aeronautical know-how. It is a fact, and the source of much criticism from the United States, that far more public money has been spent on the European aircraft industry in the form of launch aid and grants for design and development, than has been received by all the European flag-carriers in the form of operating subsidies. This theme will be treated in more detail in the following chapters, highlighting the degree to which European governments and policy-makers have seen airline operation and aircraft manufacturing as two sides of the same coin; the one compelled to act as a captive customer for the other. However it is worth noting at this stage that American complaints about the degree of subsidy to the European aircraft industry are apt to ring rather hollow since the American manufacturers have themselves received huge amounts of government support in the form of critical defence contracts. The military aircraft that result from these Pentagon specifications have often led to the development of successful civil types, the classic example being the Boeing B-52 bomber and KC-135 tanker which led directly to the Boeing 707 airliner.

3 THE CHALLENGE TO REGULATION, 1965–80

In the 1960s the institutional arrangements that regulated international civil aviation began to come under strain. The IATA-controlled fare system that had successfully balanced high costs with high prices was questioned, as the sheer fact of air travel's popularization exerted a downward pressure on fare levels. IATA itself had already responded to this pressure with the judicious introduction of scheduled fare reductions like Tourist Class (1952/3) and Economy Class (1958/9), but this was not enough and European flag-carriers began to find themselves facing competition from a growing number of independent airlines, which operated outside the IATA framework and used low fares to find new markets. The appearance of these non-scheduled airlines, offering charter flights to holiday destinations, was an indication of the dynamism of the European market. Demand had begun to outstrip supply, which had hitherto reflected the traditional tasks of the flag-carriers rather than the true size of market. Further capacity was needed and although the introduction of big jets like the Boeing 707 and Douglas DC-8 provided extra seats, there was room for new airlines as well.

Charter airlines were a European speciality. They appeared at a time when international tourism was expanding in Europe and many Europeans – particularly the sea-bound British – were taking their first holiday abroad. It was this leisure air travel which provided the impetus for new carriers to start charter operations in Britain, Holland, Germany and Scandanavia. A new class of consumer was flying and air travel itself was becoming more price-elastic as the market was widened and democratized. While cheap air travel did not suit the flag-carriers with their high cost structures, it did appeal to shipping companies, which were anxious to diversify out of their declining passenger business. It was new capital from these shipping interests that launched independent European airlines, such as British United Airways, Hapag-Lloyd and Martinair Holland, and gave them the means with which to buy jet aircraft and compete with the flag-carriers.

As the airline industry refocused itself onto the holidaymaker, the nature of its business changed. As late as 1960 business and leisure travel formed roughly equal parts of the market, whereas by 1980 the ratio was 33 per cent business and 67 per cent leisure.[16] This had serious consequences for the flag-carriers. Until 1960 most passengers flew on IATA-controlled scheduled services, thereafter there was a steady haemorrhage to non-scheduled charter flights, where seats were sold either to members of an affinity group, or put together with hotel accommodation and sold as an Inclusive Tour (IT). These non-scheduled services became a distinguishing feature of European air transport and by the 1980s composed around 60 per cent of its total passenger market.[17] As long as the international market expanded, and it averaged about 10 per cent per annum during the 1960s, the flag-carriers could hold their own. But in the 1970s demand faltered with the recession of 1973–4 and competition with independent airlines became intense. The European market became polarized between the flag-carriers with their full timetables and non-scheduled airlines concentrating on IT in the peak season. Moreover the flag-carriers served routes in a spoke network, based on a capital hub airport such as London Heathrow and determined by bilateral agreement between governments. The non-scheduled airlines developed routes on a linear structure, often originating and terminating at provincial airports, e.g. Luton (UK) to Malaga (Spain). Furthermore, because they had lower operating costs than the flag-carriers, they were able to offer lower fares and the flag-carriers' attempts to match them with promotional or off-peak fares within the IATA framework were generally unsuccessful.[18]

For the European flag-carriers, the expanding IT market was an unwelcome phenomenon. Their response was to set up charter subsidiaries of their own to take care of the holiday market while working simultaneously towards the creation of lower rates *within* the IATA system, such as advance booking charters (ABC) and APEX-fares. However there is little doubt that the Europeans' affection for charter flights delayed the introduction of scheduled price reductions, in effect setting up a parallel system for cheap flights, and while the Americans moved towards a complete deregulation of their domestic market in the 1970s, the Europeans closed ranks in their adherence to the existing regulatory philosophy.

As is often the case in the airline industry, economic change coincided with new technology and just as the 1960s had brought jets, so the 1970s brought 'wide-bodies'. This time, however, the extra capacity caused by the introduction of the Boeing 747 *jumbo jet* and its smaller rivals, the Douglas DC-10 and Lockheed L-1011 Tristar, was more difficult to absorb. The industry's strong growth pattern faltered in the 1970s. Fare discounting increased as scheduled airlines sought to revive the market and beat off competition from the new independents. The so-called grey market which this practice created had the effect of blurring the distinction between scheduled and non-scheduled carriers, a process which was accelerated when non-scheduled airlines began to operate regular services – the first and most famous being Freddie Laker's Skytrain on the North Atlantic in 1977.[19]

Laker's appearance coincided with a more fundamental threat to the stability of the European flag-carriers. In the United States the huge domestic air travel market was about to be deregulated. This in itself did not weaken Europe's international protection under IATA and the bilateral system, but it did unleash a new doctrine of liberalization, and an increasing acceptance of its principles, which was to cross the Atlantic in the 1980s and transform the European air market in a process that is still in train today. America was vital for the Europeans. It was the richest source of passenger traffic and the market with the greatest commercial potential; as the flag-carriers had grown in size, so the importance of access to the American market had become critical to their profitability.

Table 1.1   Staff Productivity of US and European Airlines, 1978

		Staff	Tonne Kilometers (millions)	Tonne Kilometers per staff member	Passengers per staff member	Staff per aircraft
Eastern	US	35 899	4 151	116 000	1 099	156
American	US	40 134	5 356	133 000	762	158
TWA	US	36 549	4 786	131 000	665	156
United	US	52 065	7 008	135 000	657	156
Lufthansa	Eur	29 400	3 028	103 000	460	320
Alitalia	Eur	17 040	1 586	93 000	374	279
Pan American	US	26 964	4 899	182 000	358	355
Air France	Eur	32 173	3 423	106 000	333	314
KLM	Eur	17 812	1 957	110 000	231	326
British Airways	Eur	54 645	4 213	77 000	308	264

Note: Airlines with predominantly long-haul aircraft have a larger number of employees per aircraft, e.g. Pan American
Source: *House of Lords Select Committee on European Air Fares*, 1981, 185–7, European Air Fares, Air Transport Users Committee, Civil Aviation Authority, 1978.

The problem was that by any yardstick of airline performance, Europe was far behind the Americans. European cost levels were around twice those of American domestic carriers and European airlines usually had a higher proportion of staff in non-flying functions like maintenance and administration. In 1978, the year that the US Congress passed legislation to deregulate American air transport, a British study found staff productivity among US airlines to be around 50 per cent higher than in Europe. (See Table 1.1).[20]

In view of this weakness, the reaction of the Europeans to American deregulation was understandably defensive. The regulatory system, which had become very much a defender of state-owned flag-carriers, was increasingly under attack from the United States as an obstacle to widening the air travel market. While the Americans were driven by a ideological desire for market liberalization and greater competition, particularly strong in the CAB under Alfred Kahn's chairmanship, the Europeans were concerned to protect their relatively inefficient flag-carriers from US competition. It was an unsustainable position.

4 EUROPE AND DEREGULATION, 1980–97

Until the 1980s scheduled European airlines operated under a homogeneous system of regulation – the 'firm corset' whose classic features were single airline designation, tight capacity agreements and IATA-fixed fares.

What has happened to European air transport since the early 1980s has really been the belated enforcement of the Treaty of Rome's provisions on free competition between member states. Articles 85 and 86 of the Treaty had forbidden agreements between firms which limited competition or allowed the misuse of monopoly power in the Common Market, but these rules had never been enforced for air transport and restrictive practices such as pool agreements had been allowed to flourish.[21] In the 1980s cautious moves were made to create a more liberal and competitive European market, although these did not take place uniformly across the continent and some countries were more enthusiastic than others. Progress was hampered by the unanimity procedure in reaching agreements within the European Community so that member states which were anxious to protect state-owned flag-carriers were able to slow down the pace of deregulation. In the early stages probably as much was achieved at the bilateral level between member states, as by the European Union (EU) Commission in Brussels.[22] The EU's concern with developing a common air transport policy across international boundaries within Europe was sharpened by the 1985 Nouvelles Frontières case at the European Court, which confirmed the Commission's power to act on fares. Thereafter the Council of Ministers reluctantly moved towards liberalization, starting with the modest Stage One package in 1987. This was followed by Stage Two in 1989 which reduced the power of capacity-sharing agreements and introduced the principle of double disapproval on fares, i.e. *both* countries involved in a route have to object to a new fare set by an airline. Stage Three came in 1993 and launched a phased introduction of a fully deregulated market similiar to that prevailing in the United States. Since the spring of 1997 the process has been technically complete and full *cabotage* rights are now assured to all airlines within the EU as well as a nominally unregulated fare regime.[23]

Four general observations can be made about what has happened to European civil aviation since the early 1980s. Firstly the process of liberalization coincided, not only with an intensification of economic integration on the road to the creation of the single market, but also with an ideological shift towards greater reliance on market forces and a diminished role for state-owned industries in the transport sector.[24] Secondly the evidence that Europe's airlines were fundamentally less efficient than American carriers was now broadly accepted, and there was a growing sense that the traditional system of protection could not continue in the face of mounting consumer protest about the high cost of air travel in Europe compared to the deregulated American market. Thirdly the trend towards deregulation and the efforts of the Commission and individual countries to put this into practice, was limited to international services *within* Europe. On the North Atlantic, by contrast, a tough protectionist stance was retained towards the Americans, particularly towards powerful trunk carriers such as American Airlines, Delta and United, which had now begun transatlantic services of their own. Fourthly, and perhaps most interestingly, deregulation in Europe has not really hastened the reduction of air fares. This may be partly because the downward trend of average air fares in Europe since the 1950s, as shown in Figure 1.1, levelled out in the late 1970s and actually rose in the early 1980s.

While the legal side of European deregulation has proceeded at the ponderous pace dictated by the EU's legislative machinery, the operational side of airline liberalization has been influenced by practical considerations of which one of the most important has been European air traffic control (ATC) and airport development. ATC has always been more complicated in Europe than in the US because air space has remained a national responsibility and aircraft still have to deal with around fifty different control centres, using different equipment and computer systems, as well as a fair degree of what can best be termed technological protectionism on the part of some countries. By the late 1980s Europe's ATC system was stretched to breaking point, so it only needed a summer strike by French ATC staff to bring the continent to an aeronautical standstill.[25] ATC arrangements are another example of how national boundaries, rather than operational needs, have determined the shape of the European industry and made it more costly than the American.

US $

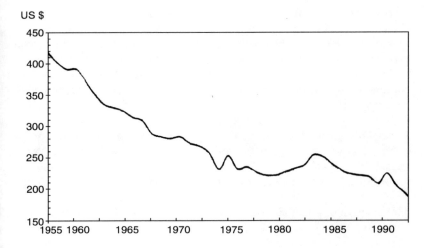

Figure 1.1 Average European fare for a 1000 km trip (constant 1993 US$ and to 1993 prices)
Source: Association of European Airlines, *Yearbook 1994* (Brussels: *AEA*, pp. 17–19)

Meanwhile Europe's airports have become increasingly congested with the relentless growth in passenger traffic, and this is despite continuous terminal expansion and the construction of some new airports. Again the problem is historical. Europe's main airports are national hubs, nearly always based at the capital, reflecting the prestige requirements of the 1930s rather than the commercial realities of the 1990s; for example Amsterdam, Brussels and Paris airports are barely 300 kilometers apart. On the other hand the fact that they have always been hubs, and served as headquarters for the individual flag-carriers, has proved an advantage in the era of *hub-and-spoke* airline networks – one of the most striking phenomena to accompany deregulation in the United States.[26] Hub and spoke networks are centred on a hub airport where a large number of incoming flights can be connected to outgoing flights within a short period of time. This massive and synchronized channelling of traffic through a single hub airport may create heavy peaks and be stressful for passengers, but it lowers an airline's operating costs and makes its operations more efficient than is possible with traditional networks of direct point-to-point flights.[27] European flag-carriers operated a type of hub-and-spoke network radiating out from their capital cities, although the spokes were determined by bilateral agreement with other countries rather than in response to operational needs, and the hubs did not offer the full connectivity of the Amercian carriers because of the regulatory restrictions imposed by pooling, that is airlines in Europe flew to other airlines' hubs rather than to provincial centres.

One of the consequences of the hub & spoke network growth in the United States has been the dominance of major airports by single airlines, so that these hubs became essentially closed to newcomers. Passengers tend to prefer the incumbent airline because of the greater connectivity it offers: it has more destinations. In Europe some airports are also becoming monopolized by flag-carriers, which depend on the capital hub as vital passenger collection and delivery points within their networks. British Airways (BA) owes much of its international market dominance to its entrenched position at Heathrow – one of the busiest airports in the world – and this privilege is a historical legacy of its days as a state-owned flag-carrier in the postwar years. In general, airline dominance at European airports has derived from the control of gates and landing slots by the oldest-serving airline – usually the flag-carrier. These so-called *grandfather rights* have enabled incumbent carriers to shut out potential new entrants from the most desirable slots, or at least force their services into less profitable times of day. The battle at Heathrow between BA and newcomer Virgin Atlantic in the early 1990s is a good example of this.[28] As the value of slots at Europe's major airports has grown, there has been sustained pressure by newcomers to have the method of slot allocation changed to the American auction system. This would contribute towards the Americanization of European air transport, by placing a market value on landing rights which in the past have been allocated by governments according to political criteria, i.e. they were anxious to secure reciprocal treatment from other countries.

5   LIMITS TO SURVIVAL : DOES THE EUROPEAN FLAG-CARRIER HAVE A FUTURE?

There is no doubt that in the fifty years between 1945 and 1995 the European flag-carriers have been less efficient and less profitable than American airlines. The reasons for this are many and varied, and will be dealt with in the pages that follow, but one or two general points should be mentioned at this stage because they have a strong bearing on the question of where these airlines might go from here.

It should be remembered that the imperative to be commercially efficient has been weaker in Europe than in the United States. Undertakings in which governments have invested a lot of tax-payers' money and which literally carry the flag, are rarely allowed to crash, and this political safety-net has reduced the pressure to cut costs and raise profits. In addition, civil aviation policy in most European countries has included a variety of objectives besides the achievement of profitability, the most common being the maintainence of air routes for political and strategic reasons. In their domestic operations European airlines have, until very recently, been underdeveloped in comparison with the United States and a large part of the reason for this has been the tenacious survival of Europe's railway network. Long after America's passenger railroad services went into decline, domestic travel in Europe still meant rail travel. Indeed one of the contrasts between the two systems is that America's airlines replaced the railroads, whereras Europe's airlines replaced the steamship companies. And since Europe's national railway systems are usually state-owned, with large labour forces, few governments have been willing to expose them to competition from state-owned airlines. Unlike the Americans, the Europeans have undertaken massive investment programmes in their rail networks, improving their speed and comfort, and in the case of France and Germany launching major *new* projects in the field of high-speed trains (TGV and ICE), which have enabled them to win back business travellers from the airlines. Not surprisingly therefore it has been economically as well as politically difficult to justify threatening this investment by allowing the airlines to compete too strongly for the railways' passengers. In short, Europe has protected its rail networks from the airlines. The outlook for the European flag-carriers in the future, on short-haul routes at least, promises a high degree of competition from the railways. Although trains will never be able to compete with aircraft in travelling speed, on European routes they give little away in overall journey time because distances are shorter than in the United States and, in an age of increasing traffic congestion, city-centre rail termini will always have the advantage over airports situated beyond the suburbs.

A further factor in the coming battle in Europe between the airlines and a reinvigorated surface transport system, is likely to centre on the highly-charged question of the environment. The future of the flag-carriers may depend on how *green* they become. Traditionally air travel has been seen as less polluting than the railways, the latter, after all, requires long ribbons of land on which to build its track. However modern high-speed electric trains are not only cleaner than their predecessors, but may also be less damaging to the environment than aircraft. Airports in Europe are likely to remain a particularly charged issue in the future. Not only will the aircraft need to be substantially quieter to remain tolerable to the residents in the neighbourhoods around airports, but there will be increasing resistance to the construction of new airports – and even the extension of existing facilities. How Europe's airlines respond to the challenge of airport siting and the improvement of interconnectivity with rapid-transit surface systems, may be critical to their survival in the next century.

Perhaps the most remarkable thing about Europe's flag-carriers, as the twentieth century draws to a close, is that they are all still flying. This, after all, cannot be said of America's most famous airline, Pan American Airways, which failed in 1991. But do they have a future? Can the combination of national politics and passenger affection which has kept them airborne for the last 70 years, keep them aloft? Or will the process of globalisation which is already sweeping the industry in the 1990s, finally swallow up the old and familiar names? Jan Carlzon, who was for a long time the head of SAS (Scandinavian Airline System), predicted in 1993 that European deregulation would lead to a process of consolidation and merger amongst European airlines from which only four or five very large carriers would survive.[29] So far this has not happened, although the process of global alliance formation has continued apace. Certainly the lesson of deregulation from America seems to be that the future belongs to a handful of very large airlines, but is the American example always useful for interpreting the European situation?

Whether or not globalization will ultimately be obstructed by European politicians anxious to protect vulnerable flag-carriers, or even by American politicians who are unhappy about the foreign ownership of their airlines, it seems likely that airlines will retain some element of their flag-carrying role. In Europe the administrative and policy-making apparatus of the European Union is acquiring increasing responsibility for the continent's airline industry, but on the other hand national governments can usually manipulate European directives to suit their interests if they feel their airline is threatened. But what are their interests? Although European airline nationalism may not have broken down as much as some deregulation supporters may claim, the trend towards the privatization of state-owned airlines suggests that many European countries wish to be relieved of the burden of subsidizing the airline industry. And if the death of the flag-carrier turns out to be a consequence of that process, then there will be governments which feel it is time to let economic logic come before prestige and allow the old names to die. There was always a strong element of symbolism in the flag-carrying role of airlines, and as that symbolism fades, the need for a national airline may diminish with it. Moreover the decline of national differences, which globalization implies, goes to the heart of European integration, if globalization does reduce the ranks of flag-carriers to three or four European megacarriers, they are more likely to reflect broad cultural values than national identities.

Two probable consequences of globalization in Europe, both of which were witnessed in the United States after deregulation, are a greater degree of consumer freedom in the choice of air transport services, particularly regarding fares, and the appearance of new *regional* airlines, not in competition with the flag-carriers, but operating as a complement to them, feeding traffic into their hubs. Both developments are already under way and there is every indication that the trends they represent will continue. Accompanying the appearance of new regional carriers will be a greater use of commuter aircraft and the construction of new inner-city airports to handle them. The *consumerist* argument, which has traditionally had a more receptive audience in America than in Europe, may prove hard to resist in the next century, since even European politicians will equate lower air fares with popularity and be willing to sacrifice the producing minority of airline workers, for the good of the great travelling public.

6 ORGANIZATION OF THE CHAPTERS

In the chapters that follow, the airlines have been chosen to represent as many differing types of European flag-carrier as possible. Also each chapter, while being the work of an individual author, has been organized as far as possible according to the same structure and deals with the same themes, in order to make possible a meaningful comparision between the airlines. These are not simply case studies, but separate parts of an integrated comparative analysis.

The six airlines chosen each represent a different tradition in the history of the European flag-carrier. Air France is a state-owned enterprise, very much within the French étatist tradition, and an airline that has contributed significantly to the technological evolution of European civil aviation. British Airways has a long and distinguished history like the French carrier, but one in which there has been a greater number of course changes, most strikingly to private ownership in the late 1980s. Lufthansa was a latecomer to the postwar airline industry, although it was able to draw on valuable experience from its pre-war operations. The German airline is also unique, in the sense that it was for a long time the western half of a divided airline tradition, reflecting the political schism within Germany itself. The treatment of the East German airline Interflug provides a novel source of comparison with Western European airlines. KLM's is a story of a national airline 'punching above its weight'; the Netherlands is a much smaller country than France, Britain or Germany, and yet has managed to sustain one of the world's major international carriers for nearly 80 years. Alitalia, by contrast, is a smaller airline from a larger country. Like Lufthansa, the Italian carrier was built on the foundations of a prewar tradition, although with less confidence than the German company and with more involvement by foreigners. The chapter on LOT shows how Poland's flag-carrier functioned as a national airline within the confines of a centrally-planned economy and under the watchful eye of its overbearing Soviet neighbour. The comparision between Western and Eastern Europe, in the chapters on Germany and Poland, reveals that there was a degree of overlap between socialist and free-market airlines. It is interesting to see that both used *American* carriers as a model of how to run an airline, in addition to flying American aircraft whenever they were able. There seems little doubt that LOT, for example, would have prefered to use Boeing or Douglas aircraft, if it had been politically acceptable. In any case the use of American technology imposed certain standardized operating procedures, regardless of whether the airlines were based in London, Frankfurt or Warsaw.

Each chapter tackles a different airline and is divided into themes common to the other chapters. In this way it is hoped that a comparative sense of European airline operations and management can be gained by the reader. The five themes are:

    i. origins and policy, i.e. the foundation and purpose of the flag-carrier,

    ii. aircraft procurement, a vital factor for airline success as aircraft productivity increases dramatically with aircraft size and speed,

    iii. route networks, including airport siting and policy,

    iv. operational and financial performance,

    v. collaboration and competition, both with other airlines and with surface transport systems.

The concluding chapter is written from the American perspective and will allow the reader to draw out the contrasts between the airline industry in the United States and in Europe.

NOTES

1. 'Unfree as the Air', *The Economist*, 28.5.1960.
2. For example, Sir Osborne Mance, *International Air Transport*, Oxford 1943: 55.
3. 'Air Union: Dis-Union', *The Economist*, 24.10.1964, pp. 415–16.
4. See Betsy Gidwitz, *The Politics of International Air Transport*, Lexington, Mass.:, 1980.
5. The remarkable Junkers F–13 was introduced as early as 1919. See Charles Gibbs-Smith, *Aviation: An Historical Survey from its Origins to the end of World War II*, London, 1970, pp. 178–93.
6. European airlines on scheduled routes were profitable (after interest payments) in only 18 of the 38 years between 1955 and 1992. Association of European Airlines, *Yearbook 1994*, Brussels, p. 19.
7. Harold Stannard, 'Civil Aviation: An Historical Survey', *International Affairs*, vol. XXI, no. 4, Oct. 1945; 499.
8. The expression *firm corset* comes from the first Director-General of IATA; Sir William Hildred. Anthony Sampson, *Empires of the Sky: the Politics, Contests and Cartels of World Airlines*, New York, 1984, p. 92.
9. In 1950 world scheduled airline traffic, both domestic and international, amounted to 31 million passengers, by 1970 it had increased ten times to 311 million. Revenue grew during the same period from US $1 521 million to US $17 817 million. *ICAO, World Scheduled Airline, Traffic Statistics*.
10. For a history of IATA, see J.W.S. Brancker, *IATA and What it Does*, Leyden, 1977.
11. Ministry of Aviation, EB/14/04, Review of Anglo-Italian Air Services Agreement, 1964–65, Department of Transport Records, London.
12. The main advantages of pooling are avoidance of over-capacity, maintenance of high load factors, and better scheduling and frequency spread, Stephen

Wheatcroft, *The Economics of European Air Transport*, Manchester, 1956, pp. 254–60.

13.   In 1948 only 28 per cent of passengers crossed the Atlantic by air, in 1963 the figure was 78 per cent, Kenneth R. Sealy, *The Geography of Air Transport*, London, 1966, p. 68.

14.   Though more modest in scope than the British and French industries, the Dutch (Fokker) and Germans also maintained aircraft industries in the postwar years, in the German case, mostly in the GDR. See Hans-Liudger Dienel, ' "Das wahre Wirtschaftswunder" – Flugzeugproduktion, Fluggesellschaften und inner-deutscher Flugverkehr im West-Ost-Vergleich, 1955–1980', in Johannes Bähr und Dietmar Petzina, *Innovationsverhalten und Entscheidungsstrukturen*, Berlin, 1996, pp. 343–52.

15.   *Report of the Committee of Inquiry into the Aircraft Industry*, (the Plowden Report), HMSO, Cmnd 2853, December 1965, pp. 7–10.

16.   Stephen Wheatcroft, Air Transport – 60 Years On, *Journal of the Institute of Transport*, Nov. 1979, p. 404.

17.   Stephen Wheatcroft and Geoff Lipman, *Air Transport in a Competitive European market*, Economist Intelligence Unit, Travel & Tourism Report, 1986, 23–5, also Sean D. Barrett, *Flying High: Airline Prices and European Regulation*, London, 1987.

18.   Peter J. Lyth and Marc L.J. Dierikx, 'From Privilege to Popularity: the Growth of Leisure Air Travel', *Journal of Transport History*, vol. 15, no.2, September 1994, pp. 97–116.

19.   Christer Joensson, *International Aviation and the Politics of Regime Change*, London, 1987, p. 43.

20.   There is a large literature on American deregulation, see Anthony E. Brown, *The Politics of Airline Deregulation*, Knoxville, Tenn., 1987, or Steven A. Morrison and Clifford Winston, *The Evolution of the Airline Industry*, Washington DC, 1995. For the international perspective, see George Williams, *The Airline Industry and the Impact of Deregulation*, rev. edn, Aldershot, 1996.

21.   J. Erdmenger, *The European Community Transport Policy: Towards a Common Transport Policy*, Aldershot, 1983, pp. 16–22.

22.   For example Britain and the Netherlands signed a path-breaking deal in 1984, allowing greater market openings to new carriers and loosening capacity restrictions. S.D. Barrett, 'Deregulating European Aviation – A Case Study', *Transportation*, vol. 16, 1990, pp. 311–27.

23.   See D. Vincent and D. Stasinopoulos, 'The Aviation Policy of the European Community', *Journal of Transport Economics and Policy*, vol. 24, 1990, pp. 95–100; also Kenneth Button and Dennis Swann, 'Aviation Policy in Europe', in Kenneth Button (ed.), *Airline Deregulation: International Experiences*, London, 1991, pp. 85–112.

24.   K.J. Button and T.E. Keeler, 'The Regulation of Transport Markets', *Economic Journal*, vol. 103, 1993, pp. 1017–28.

25.   'Air traffic congestion: the sky's the limit', *The Economist*, 23.7.1988.

26.   See Joseph Berechman and Jaap de Wit, 'An Analysis of the Effects of European Aviation Deregulation on an Airline's Network Structure and Choice of a Primary West European Hub Airport', *Journal of Transport Economics and Policy*, vol. 30, no. 3, Sept. 1996, pp. 253

27.   See Melvin A. Brenner et al., *Airline Deregulation*, Westport Conn., 1985, pp. 76–82.

28.   See John Dodgson, Y. Katsoulacos and Richard Pryke, *Predatory Behaviour in Aviation*, Brussels, Commission of the European Communities, 1991.

29.   'Airline regulation: the war in the skies', *The Economist*, 14.8.1993, pp. 66–7.

2 Air France: an Elephant in an Evening Suit?

Nicolas Neiertz

Air France has been state-owned since 1945 and this has entailed a close supervision over its management. The Government has continuously influenced Air France's attitude towards air transport regulation, aircraft procurement, network development, its partnership with Paris airports, operating and financial performance, labour management, marketing strategy, and collaboration and competition with other airlines. Air France has been for a long time a prestige-driven airline, reacting slowly to air transport evolution, which explains why it has acquired nicknames such as 'elephant in an evening suit'. It has however tried in each of these fields to gain some autonomy in order that in the future it could become a truly commercial firm and to be able to face competition, deregulation, globalization and privatization.

1 POLICY

1.1 Origins, 1933–48

Before 1933, French civil air transport was divided among several subsidized airlines. Then, following a financial scandal involving Aéropostale, the state extended its control over the management of these subsidized airlines. On 30 August 1933, Aéropostale, Air Union, CIDNA, Air Orient and SGTA-Farman were merged to create Air France, a private firm of which the state owned 25 per cent. Operating subsidies represented 78 per cent of the airline income in 1933, 65 per cent in 1939. After the German occupation in June 1940, Air France ceased operations.[1]

After the Liberation in 1944, French air transport was centralized in the French Military Air Network, the RLAF. Max Hymans, an engineer and former Member of Parliament, was recruited as Air Transport Director in the new Civil Service set up by General de Gaulle. Hymans was also the Administrative Supervisor and Managing Director of Air France, which was recreated as a nationalized airline.[2] Air transport was one of the first industries to be nationalized during this period, along with electricity and the coal mines. When the RLAF disappeared on 2 January 1946, Hymans became the Civil Air Transport General Secretary of the Transport Ministry.

The state completely controlled the activities of the airline and had to guarantee the revenue for each aircraft, within the limits of funds voted by Parliament.[3] Air France was only a part of the Transport Ministry, comparable to the Paris airports. Under this close government supervision Air France began again from scratch; it had to deal with destroyed airports and bombed aircraft, commercial agreements lost, staff scattered and departments disorganized. There were only a few heterogeneous aircraft left, mainly pre-war French models (see Table 2.1), no more than a hundred employees in Paris and a few hundred more in the maintenance centres in Toulouse and Marseille-Marignane.

While the fleet was being rebuilt and new staff recruited, the first services began to French colonies in Africa and Asia. Air France was an active member of the International Air Transport Association (IATA) from 1945 onwards. France renewed its bilateral agreements with other countries one by one, following the model of the Bermuda agreement of 1946 between the USA and the United Kingdom. Air France also renewed agreements with foreign airlines for its commercial representation abroad, for mutual technical assistance at airports and on pooling agreements for some routes in Europe, as will be explained in section 5.1. The United States supplied the company with aircraft, raw materials, tools and spare parts. The French government helped with the reconstruction of the air network, an instrument for French influence in the world. Henri Desbruères, managing director of the airline, wrote in 1946 that 'the radiance of a country can be measured through the importance of its civil air transport',[4] which means that air transport is supposed to show the commercial and industrial dynamism of a country, to attract foreign tourists to it and to make profits from investment abroad, such as airports in French colonies or in allied countries in South America.

In 1947 the Communists left the French government. Thereafter in economic policy socialist influence declined and Air France became a limited company as it had been in 1933.[5] The state however increased its hold to 70 per cent – private interests represented 2 per cent of the capital, the rest being owned by public institutions. In contrast to the pre-war period this led to significant control by the Transport and Finance Ministries over the composition of Air France's board of directors and the management of the airline. It was accepted that Air France was, from then on, distinct from the state but nonetheless dependent on it.

1.2 Government aims since 1948

Air France was in theory a commercial firm, whose statutes required (in Article 12) that it 'cover by its own revenue its expenditure, equipment and buildings depreciation, and funds to face risks of any kind'. However, the state, that is Transport and Finance Ministries, as the major shareholder, controlled its management. The number of government representatives on the Air France Board was increased in 1966. This supervision led to several obligations on the airline, such as the maintenance of routes to French colonies (1945) or to West Berlin (1948), *in the public interest*.

Hymans stressed in 1958 the 'civilizing action' of the airline, which could not be reduced to its 'pure technical, financial and commercial aspects'. French economy was seeking a 'social economy', securing the development and exercise of 'natural human liberties'. It was 'comforting', said Hymans, to think that the French had developed firms that could help this evolution.[6] This social role of Air France entailed public service obligations, such as flying to isolated domestic destinations and to French colonies. Financial compensation was defined in conventions with the state, but it did not give Air France a right to monopoly. The state had no right to give Air France a monopoly and the French parliament was divided on this issue.[7] The government's policy on competition to Air France was to develop air transport to French colonies as much as possible, through whichever French airline would undertake the service.

Air France was pronounced the *chosen instrument* of the government, but other airlines also flew on internal or international networks. After 1950, all but two of the small airlines disappeared; in the 1960s these two carriers – UTA (Union de Transport Aéromaritime) and Air Inter (24 per cent owned by Air France) – became serious competitors to Air France. This competition made a revision of the operating subsidies system necessary. Since 1953, Air France's routes in competition with other French airlines have not been subsidized, in contrast to routes to French colonies which were undertaken in the *public interest*. The subsidies that were paid consisted of compensation by the state for aircraft procured from French industry, as well as help with airport buildings and pilot training.[8] Air transport also received a regulatory framework in the 1950s; the Transport Minister had to authorize fares, new routes and major investment. In 1955 Air France and the other French airlines came to an agreement concerning their respective fields of operation, international routes being reserved for Air France.[9] These agreements were approved by the government, but they did not prevent some competition, especially on the main routes to French colonies. Until the end of French colonialism, i.e. the early 1960s, Air France's network remained under protection. Former colonies created their own airlines. In 1963, the government laid down spheres of operation for Air France and UTA in Africa. Domestic or colonial routes without any competition represented 37 per cent of Air France's traffic in 1960 and 13 per cent in 1964, whereas the share of international routes in competition with other airlines was 38 per cent in 1958 and 93 per cent in 1966, yet in spite of this sudden competition, the airline did become profitable in 1965.[10]

Air France's role was transformed by mass tourism and in the 1960s it became a commercial firm and one of France's main sources of foreign exchange.[11] The major shareholder, the state, concentrated its attention on the financial results of the airline. An unwritten rule was that state-owned firms like Air France should be neither unprofitable, insofar as the Treasury would have to cover operational losses, nor profitable because of the trade unions' strong pressure to obtain wage increases. In 1966, Air France chairman Joseph Roos explained the three assignments of the company. As an international airline, it had to link France to the numerous places where it had political or economic interests, and to compete in international markets. As a commercial firm, it had to pay continuous attention to its costs, but as the national airline, it had to preserve the public service, from which it received 'the constraints and the greatness'.[12]

In the late 1960s, the French state allowed public firms some measure of autonomy and the relationship between Air France, and the state was renewed by a management contract in 1978.[13] This policy was continued by the socialist government in the 1980s. The airline undertook to become profitable again as soon as possible in spite of the oil shock, and the state increased the financial compensation for public obligation imposed on the airline, especially for aircraft procurement and airport siting. This management contract meant that the state considered the airline as independent. Air France was therefore allowed to define a long-term development policy, and to concentrate on 'natural routes for the French foreign trade' and to put an end to 'artificial and unprofitable routes, (...) that are too distant from Paris'.[14]

European deregulation accelerated this evolution towards a more commercial management. In May 1993 with the return of conservative government, the Air France Group was placed on the list of the 21 state-owned firms to be privatized. However, this required a return to profitability by Air France. The government stated that the notion of public service could not exempt a firm from becoming profitable, and that by increasing 'for the last time' the airline's capital (see section 4.3), the state would facilitate this return to profitability.[15] The state, however, as the major shareholder, was accused by the Air France pilots' union of having left the airline in a serious state of 'sclerosis'.[16] According to the pilots' union, Air France's management was inefficient. The refusal to contract-out some activities increased the share of fixed costs. The decision-makers having been the same persons for a long time, Air France had acquired a bureaucratic mode of functioning and organization had not been adapted to competition. There still were unprofitable routes imposed by the Foreign Office for diplomatic reasons. The split into two airlines, Air France and Air Inter, and two Parisian airports, Orly and Roissy-Charles-de-Gaulle, deprived Air France of its domestic market and made a *hub and spoke* system difficult.[17] The heterogeneous fleet was more of a shop window for the aircraft industry than an efficient instrument. The state had let the airline get into debt for several years. Every new government changed the chairman according to its political ideas, which prevented management-continuity and greater autonomy. This severe criticism, in the context of the strikes in 1995, was an indication of the deteriorated climate that existed between the airline, its staff and its principal shareholder.

1.3 Attitudes to regulation

Air France participated in IATA conferences from the outset and was very attached to this protection. It tried continuously to prevent substantial fare reductions imposed by the Americans, because of its relatively high operating costs. In 1963, the American government rejected the decision of an IATA conference in Chandler, Arizona, on fare increases on North Atlantic and Pacific routes. Air France was among the toughest opponents of the American position. At the time, low fares were considered by Air France executives with contempt to be 'niggardly'.

American deregulation in 1978 and the support it got in some European countries forced Air France to ask the government for help in fighting fare cuts. Still Air France had to make compromises. It was opposed to 'excessive deregulation', but it admitted that competition might be worthwhile if 'equality of chances' between the competitors was guaranteed by 'a minimum of organization'. It would therefore try to fulfil its role of great international airline and public service organization for its country.[18] Air France has conducted a delaying battle, arguing that air transport could not be considered a strictly commercial activity and that it needed to be regulated. In the European Union France has been leading the forces opposed to American-style deregulation which has been championed by the British, Dutch and Scandinavians. The European Commission stated in 1985 that there should be limits to the rights of governments to restrict competition. The European Council accepted a very limited liberalization in 1987 which covered fares, capacity limits, and greater freedom for small regional carriers, and multiple designation became possible. This forced Air France to adapt itself to greater competition on internal routes and routes to the French Caribbean, but this adaptation imposed by the European Union was more tolerated than really welcomed. Air France has tried to oppose, within IATA, any reduction in European fares, which are among the highest in the world.

2 AIRCRAFT PROCUREMENT

When Air France was re-established in 1945, it had no more than twenty functioning aircraft, of outdated type, and the French government ordered aircraft and spares from the Americans through the Marshall Plan. However the French aircraft industry had to be rebuilt. Most of it had been state-owned since the French Popular Front government of 1936. As a national-ized airline, Air France had to participate in this national effort and to buy French aircraft.

Short-haul Languedoc were ordered from Société Nationale de Construc-tion Aeronautique du Sud-Est (SNCA). They represented in 1949 one-third of the Air France fleet, of which the other two-thirds were of American origin (see Table 2.1). In order to stimulate the French industry, the state required in 1951 that Air France order aircraft from Breguet at a higher price than the market level, in return for public participation in its deprecia-tion costs. Air France complained of the inadequacy of the state's support for the renewal of the fleet. New aircraft were then ordered from American and British manufacturers. One of them, the British company de Havilland, sold its Comet jet airliner to UAT (later UTA), the main French competitor to Air France. Initially Air France did not want the Comet because its technicians were not satisfied with its characteristics, but it could not allow UAT to become the only French airline to use jets so it too bought three Comets, which were delivered in April 1953 – three months after the UAT ones. They were suspended in May 1954 after BOAC and South African Airlines suffered Comet crashes. Air France was not closely bound to de Havilland and it did not lose much money in this episode.[19]

Later, while Air France was buying the Lockheed Super Starliner L-1649, Boeing and Douglas began to sell their long-haul jets. The French manu-facturer SNCA reacted by introducing the Caravelle, a short-haul jet, in 1956. Since 1951, the state had financed a research programme for a new short-haul jet, for use on colonial routes to North Africa. Air France had to interrupt its fleet's renewal, so as to support SNCA again. After a long series of failures in the late 1940s, such as the Latécoère 631, the SE-200, SO-30, SO-95, SE-2010, SE-161, NC-701/702 and the Bréguet 763, Air France had become very cautious with French manufacturers. However it wanted to be the first airline to use jets for both long-haul and short-haul flights, and took part in the design of the new Caravelle.[20] In February 1956 Air France ordered ten Boeing 707 and 12 Caravelle aircraft, justifying this choice of a heterogenous fleet with the diversity of its network. Unlike the Bréguet 763, the Caravelle was successful enough to make state participation in its depreciation costs unnecessary. Boeing was preferred over Douglas for long-range aircraft for several reasons. Air France could not afford to order both types as did Pan American Airways, and the links between Air France and Douglas had been weakened since the early 1950s in favour of Lockheed for long-haul piston-engined aircraft (see Table 2.1). Moreover the Boeing 707 seemed to offer technical advantages. The 707 and the Caravelle were put into operation in 1959 and progressively replaced the older piston-engined aircraft. The fleet did not increase in number in the 1960s, but its seat capacity grew quickly, resulting in a fall in load factor.

Within a decade the Boeing 747 'jumbo jet' had followed the 707 and this time the French industry was not able to face it alone. Instead it joined with other European companies to form Airbus Industrie in 1969. European airlines also had to share the costs of this new generation, at a time when charter flights were causing more competition. Air France had to face not only heavy investment and overcapacity, but also a decrease in average revenue per passenger. The airline shortened the depreciation of its aircraft and extended cooperation with other airlines. Unfortunately the first oil shock in the 1970s led airlines to postpone their orders to Airbus Industrie and Air France had to support the maintenance costs of the aircraft alone until 1977. Moreover, the Boeing 707 and Caravelle, which still represented 46 per cent of the Air France fleet in 1975, became increasingly costly with the rising price of fuel.

The supersonic Concorde programme, launched in 1962, followed a political decision by the French and British governments to give priority to performance and prestige over profit.[21] Concorde was an engineer's dream but an airline manager's nightmare. It concerned the principal aircraft manufacturers of the two countries: Sud-Aviation and SNECMA for France, BAC and Bristol Siddeley Engines for Britain. After the failure of the Comet, the British wanted to repair their reputation. The French, under General de Gaulle's presidency, were also seeking a new 'flying flag', some kind of Super-Caravelle. The name Concorde was supposed to signify the beginning of a new era of Anglo-French friendship. Air France was not consulted until the first flight in 1969, although then it was associated with the design of the cockpit.

In January 1973, the American carriers Pan American, TWA and American Airlines, rejected ordering the Concorde. Their example was followed by all the other airlines except two: Air France and British Airways, who now got a supersonic monopoly, with the financial support of their governments. Concorde, at three times the cost of a Boeing 747, was conceived for an elite market, whereas the time was in favour of mass transport. In 1976, the Concorde fare was 50 per cent more expensive than an ordinary first class ticket. Supersonic flights over the North Atlantic were more practical westwards towards the USA than eastwards towards Europe because of the time-lag. The Concorde programme did not exceed 14 aircraft, shared between the two airlines, and has been a commercial failure, sustained nonetheless for reasons of public image and national prestige. Concorde was conceived in a particular age of aviation, when people still considered aircraft as something exceptional; unfortunately it was delivered in another age, when air transport had become a mass industry.

Air France took part in the Airbus programme from the beginning. To give birth to this programme, the European aircraft industry put an end to the rivalry between state-owned national firms.[22] The German MTU, MBB and Dornier companies, the British Hawker Siddeley, and the French SNECMA and SNIAS-Aérospatiale were associated in 1969 in a new company, with government financial support at the outset, called *Airbus Industrie*. The first managing director was a French engineer, Roger Beteille, assisted by a German, Dr Felix Kracht, but the language used was English. The factories were in Hamburg and Toulouse. The first aircraft was delivered in 1972, and the first order from an American airline came in 1979, when Airbus Industrie adapted itself to the American market by producing a family of different aircraft competing directly with Boeing: the Airbus A-310 in 1979, the A-320 in 1988, and the A-330 and A-340 since 1991. In the 1980s, Airbus Industrie overtook McDonnell Douglas as the main competitor to Boeing, the world's leading aircraft manufacturer. In 1996, their market shares for civil aircraft were 37 per cent for Airbus Industrie and 59 per cent for Boeing.[23]

Air France engineers were associated with the design of the Airbus A-300, from which a made-to-measure derivative, the B4–200, has had large commercial success. The relatively high operating costs of Air France led to the choice of economy in operation, whereas other airlines preferred high frequencies with DC-9 or Boeing 737 aircraft. This preference for large capacity over high frequencies caused Air France to lose market share because it was unable to adapt its supply to the traffic as quickly as other airlines. The cooperation of Air France engineers continued with the Airbus A-320 and A-340, for which Air France was the launching airline. These aircraft, like Caravelle in the 1950s, were good enough to make state participation in their depreciation costs unnecessary.

As it can be seen in Table 2.1, the buying of wide-bodied aircraft from both Boeing and Airbus meant a heterogenous fleet for Air France in the 1970s and the 1980s, for long-haul and short-haul operations, although it did allow replacement of aircraft from one manufacturer with those from the other. In 1985, the Boeing 727, a big fuel consumer, began to be replaced with the A–320. Slowly Air France increased its load factor to 1950s levels (see Table 2.2). Air France, adapting its supply to demand through frequencies and capacity, preferred to lose market shares on some routes than preserve insufficiently loaded flights.

In 1989, Air France managing director Bernard Attali launched a major programme of aircraft orders, so as to double the fleet in ten years, and to make it competitive with those of the American megacarriers. According to Attali, bilateral agreements between the USA and European countries were unbalanced; Europe was about to undergo an American invasion because European countries were lacking foresight and did not stand together enough. Attali's solutions were massive investment and close cooperation between European airlines. While Air France got into heavy debt, its load

Table 2.1   Air France: aircraft fleet

Aircraft	1939	1945	1950	1955	1960	1965	1970	1975	1980	1985	1990	1995
Dewoitine 338	23	9	0									
Bloch 220	17	0										
Wibault 283	15	0										
LéO 242 Hydros	12	0										
Lockheed 14,18,60		13	0									
Lockheed Constellation 249,749			19	18	11	0						
Douglas DC-3			33	38	31	26	0					
DC-4/C-54			28	19	27	23	6	1	0			
Sud-Est Languedoc 161			31	7	0							
Lockheed Super Constellation				17	21	7	0					
Vickers Viscount				12	8	0						
Breguet 763 Provence					12	12	6	3	0			
Sud-Est Caravelle					24	41	44	29	8	0		
Boeing 707					17	25	30	26	10	0		
Super Starliner					10	0						
Fokker F-27							13	0				
Boeing 727							16	20	25	29	12	0
Boeing 747							4	14	23	24	29	33
Boeing 707 Cargo							3	7	4	0		
Boeing 747 Cargo								1	4	6	8	11
Boeing 737								2	0	12	19	39
Airbus 300B								7	17	18	15	10
Concorde								1	7	7	7	6
Airbus 310										5	11	10
Airbus 320											19	25
Airbus 340												13
Boeing 767												8
Others	19	42[2]	7[3]	0	0	0	0	0	0	0	0	0
TOTAL	86	64	118	123	161	128	119	108	98	101	120	155[4]
% French[1]	100	82	26	15	22	36	39	34	32	29	43	41

1. Aircraft involving French manufacturer.
2. 5 Junkers 52, 5 Bloch 221, 2 Farman 2200, 2 LéO 246, 1 Amiot 354, 1 Wibault 282, 1 Potez 54 and 25 various (Caudron, Go'land, Sikorsky, etc.).
3. 3 de Havilland Dominie, 2 Junkers 52, 2 Consolidated Vultee Catalina.
4. Renting or leasing: 8 Boeing 747, 8 Boeing 747 Cargo, 18 Boeing 737, 7 Airbus 300B, 6 Airbus 320, 10 Airbus 340, 3 Boeing 767.
1995: 31 March 1996.
Source: Air France Annual Reports and Accounts, 1948 to March 1996.

factor collapsed again. The merger of Air France and UTA in 1992 worsened the diversity of the fleet. Since 1992, procurement has been more spaced out (Boeing 747) and modified in favour of small aircraft (Boeing 737) for the short-haul sector, or new long-haul aircraft (Boeing 767, Airbus 340). Yield management halted the fall in load factor. The renewal of the fleet led to a reduction in fuel consumption, while increased leasing made the management of the fleet easier, in a climate that remained uncertain.

As in Britain, the French state continuously intervened with the national airline, in order to support the national aircraft industry. The main aircraft programmes, such as Caravelle, Concorde and Airbus, were launched under close government supervision and with its financial support. The government aimed to develop colonial or post-colonial routes with appropriate aircraft and to challenge American leadership with local products. The French aircraft industry however has never been able to supply Air France with all the aircraft it needed and since 1945 the major part of its fleet has been American. In 1996, Air France chose to buy ten Boeing 777s despite government pressure to favour the Airbus A–340; this strengthened Air France's autonomy while worsening the diversity of its fleet.[24] Air France has always had too many different types, as did British Airways for a long time, and this has raised its maintenance costs in contrast to airlines with more homogeneous fleets like KLM, Alitalia and initially Lufthansa.

3 ROUTES

3.1 Network development

Air France route network was a reflection of the geographic scattering of the French colonial empire in the Caribbean, Africa, Indo-China and Polynesia, and by 1948 was the longest in the world (188 000 km), which gave the airline its slogan in the 1950s. These colonial routes were extended to few independent countries, such as Egypt or Brazil. There were American routes, to New York and Boston (1945), then Montreal (1950) and Chicago (1953), but they remained relatively few. The European routes represented only 17 per cent of the traffic in 1957. Prestigious routes were preferred to short-haul ones: Mexico, Caracas and Bogota were opened in 1951, whereas Scandinavia only in 1954. Air France abandoned services in competition with other airlines and concentrated on protected ones, such as those to the colonies, justifying this defensive strategy by its contribution to the economic development of those territories.[25] In Africa for instance, Air France stopped only in French colonies. Only a few routes were exposed to competition, such as Paris to Tokyo or Mexico. The airline stated in 1954 that its network would not continue to grow and that it had reached its 'definitive' extent, which means that it lacked a commercial strategy. On European routes, open to competition, the load factor was far lower than on colonial ones. But the colonial traffic did not exceed the international traffic, because of its low frequencies, caused by small demand for long-haul colonial flights.

French withdrawal from the colonies in the late 1950s put an end to this situation. In 1955, with of the end of the Indo-China war, colonial traffic decreased from one-third to one-fifth of Air France's total. Air France began to develop the European and American routes, which were at first unprofitable for a few years. *Tourist Class* (1952) and *Economy Class* (1958) reduced the fare levels. In 1958 Air France opened the first polar route from Europe to Tokyo via Anchorage, in partnership with Japan Airlines, which reduced the journey time by half. The Atlantic routes became profitable and Air France held fourth position on the North Atlantic routes in 1962. International traffic (not including routes to French colonies) now represented 75 per cent of the total traffic of the airline. It had to close one half of its destinations in the former colonies. Its share in world air traffic decreased from 4 per cent in 1950 to 2.75 per cent in 1970.

Competition became fierce on North Atlantic routes in the 1960s because of the first charter flights and the jets which halved the duration of the Paris–New York flight. Air France increased its destinations in the United States including Los Angeles (1960), Houston (1962), Washington (1964) and Philadelphia (1969). It also developed its freight activities, the first cargo Boeing 707 services starting in 1965. There were then about twenty cargo routes in Europe, the USA and North Africa combined, which represented only 9 per cent of Air France's income. Air France remained below TWA, Pan American and BOAC for the international traffic and was the fifth airline for freight traffic in 1965. Its own charter flights were not so profitable as non-IATA airlines, despite the creation of its subsidiary Air Charter International in 1968. During the French political crisis of May 1968, it was on these American routes that Air France declined most.

Air France tried then to develop East Asian and European routes; first to Tokyo through Siberia in 1970, then from Paris or other French cities to southern and eastern Europe (1967). But the French were less interested than other Europeans in such services, and in so far as France itself was a southern country, the French did not need air transport to go on holiday; French tour operators could not be compared to those of other European countries. Air France had its own touring subsidiary, SOTAIR, after 1968.

The wide-bodied Boeing 747 created overcapacity on the North Atlantic from 1970 and Air France's load factors fell. Air France closed some of its unprofitable routes on the North Atlantic and declined from fourth to seventh rank in three years. Instead it developed services in the French Caribbean, where it did not have to face competition, in addition to Africa and East Asia, where the first regular route from Western Europe to China was created in 1973. Its Asian network became almost as important as its North American one, in terms of traffic and revenue. The opening of a trans–Pacific–route, Tokyo-Papeete-Lima, in 1973, created a global network of 500 000 kilometres.

A slump hit the European and Asian markets in 1974 and Air France had to balance it through South America, the Caribbean, North Africa and the Middle East. The trans-Pacific route was cancelled in 1977. On another hand, the first Concorde routes were opened to South America (Paris–Caracas, Paris–Rio) and to the United States (Paris–Washington) in 1976, which reduced the flight time by one third and one half respectively If the airline had succeeded in opening routes to New York and Tokyo, Concorde might have become profitable. But the American authorities raised environmental objections and legal obstacles, which delayed the first Concorde flight to New York until 1977, while the Japanese route never materialized. By way of compensation, the Washington route was extended to Mexico in 1978. Concorde remained very marginal in Air France's activities and has been used mainly for its public image. This commercial failure led to the suspension in 1982 of all Concorde routes except the principal one: Paris–New York. This regular service became profitable for the first time in 1983, and has been complemented since 1985 by special flights. According to the airline, the maintaining of Concorde flights on reduced services proved successful and contributed to the promotion of Air France and French prestige, without altering the airline's financial results.[26]

In the late 1970s, long-haul flights to America, Africa and Asia provided less revenue per passenger-kilometre than short-haul ones in Europe, North Africa and Middle East; and short-haul flights represented one third of the traffic but one half of the revenue. However, long-haul load factors were usually higher than short-haul ones. This paradox results from the growing share of reduced fares for tourists on long-haul revenue. Air France said it would henceforth prioritize its 'most natural markets', which comprise mainly routes close to France.[27] Some 'artificial' and unprofitable routes, or routes that were too distant from Paris, would be cancelled.[28] But in general Air France faced the slump of 1974–9 without any 'severe amputations' to its network, which was extended to 600 000 km and 160 destinations in 1979. According to the annual report of 1980, it chose a half-way strategy between opposite extremes and modified the weight of each geographic sector in its network through frequencies and capacities, adapting itself to the economic streams all over the world. In 1978, North America represented only one fifth of its traffic and one eighth of its revenue, and it was overtaken by South America and the Caribbean. Some unprofitable routes were closed, such as the trans-Pacific in 1977, others were opened or re-opened, like French Guyana–Peru in 1977, or France–Vietnam in 1978. The freight network was also continuously developed and represented in 1980 one fifth of revenue. Contrary to the management contract of 1978 with the state, the airline was not completely autonomous or did not feel free to close any unprofitable route.

Because of previous measures, the second oil shock in 1979 was easily overcome although recovery benefited Air France less than other IATA airlines. A new commercial policy was put into operation in 1980: a simplified and less expensive first class ('business' class), 'holiday' fares on tourist routes (Caribbean), special fares for young people or students, families, touring groups (Europe), or for the slack season ('blue' flights). This policy was successful, especially in the Caribbean sector, but this service was seasonal and Air France transferred most of it to its subsidiary, Air Charter International.

During the 1980s, when deregulation extended from the US to the European market, Air France developed its network as much as possible. Its length increased from 600 000 km and 160 destinations in 70 countries in 1979, to a million kilometres and 195 destinations in 78 countries in 1989. The airline increased its destinations in the United States and in Europe, in order to strengthen its claim to be one of the foremost airlines in the world, before complete deregulation. Air France *followed* the trend more than it anticipated it. In the Caribbean sector, for example, opened to competition since 1987, it discounted fares according to the date, maintaining traffic but reducing revenue. This method had been used by Air Inter for internal flights since 1977. On the North Atlantic routes, it created the business class, 'Le Club' in 1983 and later discounted fares on all long-haul flights. In 1989, within the framework of Bernard Attali's expansion plan, Air France had the longest European network, 107 destinations, with time schedules that allowed a return journey in one day, starting from Paris or any French city. The trans-Pacific route was re-created that same year, while most of the Middle East sector was abandoned.

This ambitious strategy was suddenly interrupted in 1990. The Air France Group, that is Air France, UTA and Air Inter, had to reduce its 238 destinations to those which were the most profitable. European routes, especially those starting from French provincial cities, were cancelled or transferred to Air Inter. Air France and UTA merged in 1992, which increased the Air France network in North America, Africa and Asia. Direct flights were developed and unprofitable take-off times were dropped according to the operating costs. A more aggressive commercial policy was put into operation: special fares for frequent flyers, discounted fares purchased the days preceding some flights, and variation of the services provided to passengers during the flight, according to route. The idea of an extensive network was replaced with the idea of profitable routes and a 'hub and spoke' system based on Paris-Charles-de-Gaulle airport. The network in 1996 comprised 165 destinations in 96 countries, with Air Inter and Air France becoming complementary in Europe, in anticipation of complete deregulation in 1997.

Network management reflected weak commercial policy at Air France, where the idea was that the airline just had to wait for the passengers to come, and if they did not, it was useless to pay any attention to them. Slowly this mentality is disappearing.

3.2 Airports environment

As in Britain, but unlike Germany or Italy, French civil aviation has been characterized by an imbalance between Paris airports and those of other French cities.[29] Air France's first airport was at Le Bourget in 1933, then at Orly in 1953, and since 1974, at Roissy-Charles-de-Gaulle. Since 1945, these Parisian airports have been grouped together in a state-controlled organisation called Aéroport de Paris or ADP.

Strange as it may seem, this establishment first developed its airports under the supervision of the Transport Ministry rather than in cooperation with Air France. Air France hardly participated in their design and their construction although it took one third of the space at the airlines' disposal at Orly in 1961. Moreover, Air France and ADP became competitors for some activities, such as offering technical assistance to foreign airlines.

Air France and ADP developed a complex relationship, the latter trying to preserve its independence towards its principal client. Air France put pressure on the Transport Ministry to restrain airport taxes and behaved like the owner of the airport, requiring the most profitable sites among the hangars and the workshops. At Parisian airports, the airlines did not own the ground or the facilities, but had to rent them from ADP. During this period of strained relations, two former managing directors of ADP, Louis Lesieux and Pierre-Donatien Cot, held the same positions at Air France, but took an opposite stand.

For the building of Roissy-Charles-de-Gaulle (CDG), in the late 1960s, Air France was in charge of its own offices and workshops, and its sites inside the airport were agreed with ADP as soon as they were designed. But this cooperation remained limited, as can be seen through the example of surface transit transport. The opening of CDG in 1974 doubled the capacity of the Paris airports, but reduced the ease of transfer between Air France flights at CDG and internal ones at Orly. Neither ADP, nor the railways (SNCF) nor the Parisian public transport system (RATP) was willing to transport passengers between Orly and CDG, which forced Air France to create its own subsidiary for the purpose – SATTE. These services were later complemented by buses from RATP, but ADP took no interest in them. The first subway line at CDG, opened in 1976, did not service the passenger terminal at all and subway passengers had to use a shuttle to reach it. Until the 1980s, ADP paid no attention to what was happening outside its airports, and the SNCF and RATP did not have the capacity to service both Parisian suburbs and the airport. Special railways were necessary, but the state was not interested in such investment.[30]

The second terminal at CDG – CDG-2 – was opened in 1982, designed in close cooperation between Air France and ADP, and it can therefore be considered as a turning-point in their relationship. Air France wished to have a whole terminal at its disposal, and ordered it from ADP. The design respected the priorities set by the airline for simplicity and large capacity. CDG-2 had the shape of an '8', to which new modules could be added as traffic increased. But CDG was far from saturation-point, in contrast to most European international airports. The layout of CDG-2 reduced passenger walking distance and speeded-up luggage delivery. It could be laid out to make easier connections between long-haul and short-haul flights. Air France concentrated its take-off and landing schedules at specific times each day, in order to benefit from the American-style 'hub and spoke' system.

Since 1987, Air France and Air Inter have improved connections in CDG-2 between internal and international flights, removing the half-way baggage check-in. The merger of Air France and UTA in 1992 led to the replacement of 50 unprofitable direct routes from French provincial centres to the rest of Europe with connections in CDG-2. The traffic between these French centres and Europe did not decrease significantly, but the move reinforced the centralization of the Air France network on CDG-2. In 1994, ADP handled 60 per cent of French airport traffic; so far as Air France was concerned, the connection traffic between long and short-haul traffic at CDG-2 was far less important than it was for British Airways at Heathrow, Lufthansa at Frankfurt, or KLM at Schiphol. This was the reason why, in 1996, Air France tripled the connections capacity of CDG-2, by laying out schedules and airport facilities in close cooperation with ADP. This strategy was considered by the airline as the 'key for Air France recovery', since it increased the traffic while reducing operating costs: 'if Air France does not succeed in making a hub out of CDG-2', says ADP Commercial Director Alain Falque, 'Paris will only be the airport of the Parisian region'.[31] Taking KLM at Schiphol, Swissair at Zurich and Lufthansa at Frankfurt as models, this reform guaranteed a 'minimum connecting time' (MCT) of 45 minutes between the twelve Air Inter routes starting from CDG-2 and Air France's network. MCT was a decisive criterion for competition among airlines, as it determines the order in which flights appear on travel agencies' computers for reservations.

ADP has been developing independently of the French Transport Ministry. Because of the rising competition between European airports, it has been trying to adapt its operations to a more commercial strategy. This explains the connection with high speed TGV trains at CDG-2, and the creation of a business area right in the middle of CDG – 'Roissypole' – that is serviced by the subway. Air France placed its new registered office there in 1995, in order to benefit from the economic dynamism of the airport area, the so-called *aéroville*. ADP now seeks more autonomy in its relationship with Air France through this new strategy and through the deregulation of French civil aviation generally.

4 OPERATING AND FINANCIAL PERFORMANCE

4.1 The age of direct subsidies: 1945–65

Between 1945 and 1965, Air France remained unprofitable.[32] Until the end of 1948, direct subsidies could not be distinguished either by type of aircraft or by route. Subsidies simply covered the annual deficits. Detailed accounts did not exist before 1 January 1949 and capital was paid up before 1950, at which point a general meeting of the shareholders was possible. From 1945 to 1948, investment was financed by government funds voted each year by parliament. Afterwards, the 1948 status made long-term loans possible, partly within the Marshall Plan framework. Between January 1948 and June 1949, for instance, Air France bought American spare parts for US$ 6 million of which 1 million was financed through the Marshall Plan. In June 1949, French manufacturers recommenced their own spare parts production. Investment was financed by a government fund for reconstruction and equipment. The amount of these-long term government loans increased sixfold between 1950 and 1956. However despite this investment, Hymans said in 1953 that the airline did not have 'all the required equipment for increasing frequencies on routes where the traffic would need it'.[33] The rapid expansion of traffic led however to massive investment, mainly aircraft purchases. Financial charges and depreciation costs represented 10 per cent of expenditure in 1953–4.

Air France's persistent losses were caused by high operating costs, though Max Hymans, its chairman between 1948 and 1961, tried to minimize this characteristic, stating in 1956 that the 'lack' of an important European market was a 'serious handicap' for European airlines and deprived them of the 'solid basis' that American airlines enjoyed in their own domestic market.[34] The European population was larger than the American but its standard of living was lower and much more heterogeneous in the various countries and regions. Europe suffered moreover from the economic partition between nations and from the 'Iron Curtain' that did not exist in the United States. National protectionism limited air transport development, in spite of ICAO and IATA talks.[35]

The arrival of jet aircraft forced the airlines to aim at mass transport in order to be profitable and Hymans worried about the accompanying acceleration of competition. Until 1959, according to him, technical performance had had right of way over management, as if the course of the air transport industry had been determined by aircraft manufacturers, often running on

military orders. Hymans called for an end to this 'unreasonable' race for
speed and 'apparent profit', thus revealing his anxiety about the new scale of
air transport implied by the switch to jet aircraft.[36]

In the 1950s, Air France was undercapitalized and financed its investment
essentially through loans: while the capital remained steady and the equipment value was multiplied by six, long-term debt increased eightfold and
financial charges ninefold. This debt, combined with Economy Class, special
fares, charter flights on the North Atlantic, unprofitable cargo services, and
the growth of labour costs, explained the deficits. In 1962, the state increased
Air France's capital fivefold through a transfer from the Treasury, which led
to a reduction of the debt rate, and therefore of the interest charges in
expenditure. Staff reduction, computerization of staff management, concentration of administration in a single building in Paris and productivity
increases all combined to lead to the first operating profit in 1965. 'Air
France, as a national airline, can and has to live by itself', said chairman
Georges Galichon in 1969. 'There cannot be for a competitive firm like Air
France any expansion without profitability'.[37] Air France distributed dividends to its shareholders after 1966.

Initially Air France had to choose between two labour strategies: either to
use sub-contractors for as many activities as possible, like Swissair or Scandinavian Airline System (SAS), or to provide them itself. It chose the second
solution. In fact Air France provided numerous services complementary to
air transport, such as aircraft maintenance, food and beverage preparation,
hotel accommodation and tourist services. The first hotel subsidiary of Air
France was created in 1950. Then hotel subsidiaries were merged in 1970 in
the Méridien hotel chain. This choice from the beginning explained the
number of job categories existing at Air France: 350 in 1950, 200 in 1985.
Some of these services were provided on behalf of other airlines, such as
maintenance and runway operations, and passenger handling. A comparison
with staff management in other airlines can be drawn through the example of
French pilots. Air France employees were not civil servants, but they benefited from a similiar level of employment security and relatively high wages,
thanks to special status approved by the Transport Minister. The first strike
at Air France was caused by the introduction of this status in 1951, subsequently put into operation in 1953. Air France's navigators came initially
from the Army, then from a training centre dependent on the airline, until in
1958 the National School for Civil Aviation (ENAC) was created by the
Transport Ministry. Their military origins gave pilots a highly elevated
position in French society. Moreover their special status among the Air
France staff explains why it was so difficult to fire them later on when
Air France was trying to make profits. French pilots' wages were equal to
those of Pan American's pilots in 1945. Thus Air France functioned like a
part of the civil service, with strong financial support from the state and

powerful trade unions. The navigators' unions were soon more powerful than others because their strikes could completely paralyse the airline. French pilots were therefore overpaid and usually older than foreign pilots. Air France executives often came from the *Polytechnique* (engineers) or the National High School for Civil Service (ENA), in which they received an elitist *'esprit de corps'*.[38] This mentality explains how they managed the airline for such a long time; Air France's business was national prestige, it was of secondary importance to pay attention to revenue and expenditure.

The arrival of jet aircraft transformed working conditions, and caused the first major strike of navigators in 1960. The first jets were regarded as inconvenient and difficult to work in, according to air stewards. The composition of the crews and flight plans were modified, wages were increased and the number of navigators was expanded by 25 per cent. Air France was never able to oppose strikes by the navigators. The greatest increase in staff numbers took place amongst the navigators; other categories remained steady or decreased. The large number of navigators at Air France extended their power inside the airline. Labour expenditure represented about a third of the total, which was comparable to other airlines, but while recruitment almost stopped during the 1960s, expenditure increased 40 per cent in 1963 thanks to a new pension system for the navigators and by the return to France of Air France employees from the former colonies. For instance 1000 employees returning that year from Algeria were helped by the airline to find accommodation. Labour expenditure was at its highest level, 43 per cent, in 1968; then the arrival of wide-bodied jets in 1970 caused the same kind of strike as in 1960 when the first jets were introduced.

'Marketing' has always been rather unfamiliar to the French mentality, and it is still untranslatable into the French language. In the early years, Air France had to face shortages of every kind and did not pay attention to commercial strategy. Priority was on national prestige and high technical performance ('comfort that flies fast'), rather than on commercial competition on fares. The latter were strictly defined by IATA, and the airlines competed with each other only on service and reputation. Air France commercial policy was based on its worldwide presence: 'in all skies' or 'the longest network in the world'. It was a public service more than a commercial firm. The first market studies were mentionned in Air France's Annual Report for 1955. Air transport was conceived of as being for an elite; the airline just had to wait for the passengers to come to it. Service on board had to give an impression of luxury, through gastronomic meals and expensive French wines. In the 1950s the Paris–New York flight, in Super Constellations, offered private cabins with couchettes, the *Parisien spécial*, to its richest passengers. Air France Caravelles and Boeing 707s had French castle names, like *Château de Versailles* or *Château de Chenonçeaux*. Advertising posters was created by famous French designers, such as Villemot, Savignac and

Mathieu. Air France was itself an advertisement for the luxurious French *art de vivre*, or way of life – its aircraft were like flying Eiffel Towers.

Table 2.2   Air France: financial performance

	Revenue	Surplus/deficit	Profitability rate	Productivity	Capital
	millions FF	millions FF	(revenue expenditure)	(revenue staff)	millions FF
1938	–	–	40%	–	–
1946	–	–	75%	–	10 000
1950	26 447	−1930	93%	1.811	10 000
1955	51 985	−3260	94%	3.138	10 000
1960	1 254	−61	95%	0.054	100
1965	1 864	34	102%	0.078	500
1970	3 269	12	100%	0.118	1 000
1975	6 633	−354	95%	0.217	1 761.2
1980	14 856	−143	99%	0.444	1 818
1985	30 402	729	102%	0.854	1 975.75
1990	34 433	–	98%[1]	0.897	3 156.57[2]
1995[2]	39 140	−2950	93%	1.072	5 744

1. April 1995 to March 1996.
2. AF Group.
Source: Air France Annual Reports and Accounts, 1948 to March 1996.
N.B.: 1 FF of 1990 (= 6 US $) = 8.33 FF of 1950, 0.14 new FF of 1960, 0.21 FF of 1970, 0.54 FF of 1980.

4.2   A profitable public firm: 1965–78

In 1967 a new chairman, Georges Galichon, and a new managing director, Pierre-Donatien Cot, were appointed at Air France. Galichon, in contrast to Max Hymans (1948–61) and Joseph Roos (1961–7), was a newcomer to the air transport business. He was a civil servant, but he stressed the need for profitability in Air France and for mass transport development. Cot was an engineer, influenced by American management methods, which he had introduced in his former position as head of ADP. At Air France, the increasing traffic in the late 1960s raised again the question of investment planning and financing. The airline was not able to finance from its own reserves more than a third of the required investment. But average revenue per seat/kilometre was decreasing because of charter flight development. Investment planning became uncertain. Galichon and Cot abandoned the original five-year planning system in favour of a 'gliding', computerized one, using the American Planning Programming Budgeting System. This reform had to 'give a first concrete answer to the questions raised in all French firms by quick changes imposed by the insertion of France in an industrial world where it has entered without being really prepared'.[39] Cot noted the economic backwardness of France compared with the United States. Air France's capital was doubled in 1970, which improved its plough-back rate to twice the average rate of the other six French state-owned firms in the transport and energy fields. Thus Air France achieved an investment policy closer to that of private firms than other state-owned enterprises.[40]

For three weeks during 1968 events paralysed Air France in common with the rest of the country. Monetary variations as well as wars in the Middle East and Vietnam also reduced Air France's revenue so that 1968 was a year of deficit for Air France, whereas it was one of the most profitable years for other airlines. On the other hand, charter flights were less important in France than in other countries, because of the low internal traffic and the short distance to the Mediterranean. Air France became profitable again in 1969, but strikes in 1971 and 1973 led to new deficits.

The recession began after 1970, whereas the heavy investment required by the arrival of wide-bodied aircraft assumed further growth. Because of the new equipment necessary to service the Boeing 747, airport fees increased quickly – by 70 per cent between 1967 and 1971. Air France decided to preserve its plough-back rate rather than reward its shareholders: the depreciation span for new aircraft was shortened to nine years in 1973 for the Caravelle, and 10 years for Boeing 707, 727 and 747; usually it was 12 years for short-haul and 14 years for long-haul aircraft in Europe.[41] Air France began also to use leasing, which represented a quarter of its aircraft in 1974. Finally, Concorde required further capitalization from the state and created more debt, so short-term debt exceeded long-term after 1973.

The 1973 oil shock hit Air France severely, in so far as all its fuel supply contracts had to be renewed at the same time that year, whereas other airlines were able to space them out over twelve months. Fuel expenditure increased sharply at Air France in January 1974. Moreover, benefits had been shared with the staff after 1965 through generous wages increases, and with the shareholders through dividends, which deprived the airline of the accumulated financial reserves available to other airlines. Because of IATA regulations, fares could not be revised before the end of 1974. This delay entailed a considerable loss for Air France. Meanwhile CDG airport and the first Airbus aircraft were put into operation that year, which led to additional expenditures, the loss of the connection market with French secondary airports, and overcapacity. Air France knew that using CDG would deprive it of connectivity to other French airports and was not compelled to move there. But Orly was close to saturation point and the state had promised financial compensation for the costs of moving to CDG. A further problem was that other European airlines had delayed their orders for Airbus aircraft and Air

France had to postpone the use of its own Airbuses, or underutilize them, because traffic agreements with other airlines had stipulated that they should all be put into operation at the same time.

Deficits reappeared and the investment programme was reduced. Meanwhile the first jets (Caravelle and Boeing 707) needed to be replaced. Concorde accounted for half of the 1976 deficit, and the state delayed that year financial compensation for Air France's public service obligations, increasing the debt and financial charges of the airline. From 1977, government funds covered most of the Concorde deficit, and compensated for the Parisian airport duplication factor and the Caravelle replacement. Thanks to these direct subsidies, the airline was returned to profitability, since long-term debt exceeded short-term and Air France now negotiated a management contract with the state.

The airline's organization was very centralized. From the late 1960s, the managing director, Pierre-Donatien Cot, introduced American management techniques, with the help of the consultants, McKinsey. He centralized the Air France organization through the creation of regional networks in 1972 and of long-haul and short-haul flight centres. This reform helped the airline to face the problems associated with the airport duplication between Orly and CDG in 1974. The transfer of Air France from Orly to CDG was progressive; in 1976, one-third of Air France passengers still landed at Orly, which explained the labour recruitment, which stopped after 1974. Air France maintained its staff and network in the late 1970s in order to preserve its development potential. This was also supported by the trade unions and the French diplomatic service; although this delayed Air France's financial recovery, it did prevent a rapid growth of expenditure when traffic picked up again after 1980. This policy was officially explained in 1978 by 'a preoccupation of social balance shared with the Government',[42] that is to spare a 'procession of human damage and of material wastes'.[43] It helped the airline overcome the 1979 oil shock more easily than other airlines. Recruitment started up again in 1982. The switch to two-member flight crews on Boeing 737s in 1981, however, caused another pilots's strike.

Air France's advertising had been intensified with the first jets. Under Galichon and Cot, the commercial function was reinforced in Air France's organization and the department of 'commercial development' became a 'general direction' in 1968. From this point onwards Air France aimed less at selling its product than at defining a product that could be sold. Its methods were revised to take into account each segment of the market – businessmen, tourists and cargo, and to react more quickly to demand through a decentralized organization and to measure more precisely its results. After 1966, computerization aimed at speeding up reservation and check-in formalities for passengers through the Alpha 3 programme (1969). This system integrated car rentals and hotel booking in 1972. Air France and its subsidiaries, such as Air Charter International and the tour operator SOTAIR, worked together with French tourist authorities to produce computerized market studies, advertising campaigns and the development of Air France's hotel subsidiary Méridien. The Air France Group was now selling not only air transport but whole journeys, although Concorde from 1976 continued the tradition of prestige, luxury, and technical performance. This tradition was famous among Air France passengers and when the airline tried to replace *haute cuisine* with picnics to Economy Class passengers in 1979, it was a complete failure and the experiment had to be abandoned three years later.

4.3 The age of management contracts: 1978–95

After 1969 the French state developed management contracts with the enterprises it owned, within the framework of public investment planning. These contracts planned government investment and subsidies for the following years, and aimed at a return to profitability as soon as possible. The first management contract with Air France was signed on 26 January 1978. It noted the growing segmentation of the air transport market between charter flights, business trips and freight, and set the airline the following objectives: 'to improve its productivity, to adapt its network to demand, to order the most appropriate aircraft for this network, to offer profitable and suitable services, thus to return to profit'.[44] This meant a real turning point for Air France towards a more commercial style of management. In 1979, the airline became profitable even without operating subsidies, and financed for the first time more than 80 per cent of its investment from its own reserves. A mixed fleet of passengers-and-freight aircraft, a commercial policy based upon the segmentation of demand, and the modernization of surface equipment led to a steady decrease in operating costs.

The 1979 oil shock was less violent for Air France than the first one, because this time fuel contracts had been spaced out and it was able to adjust its fares to the progressive increase in fuel expenditure. Moreover, the replacement of aircraft with more fuel-efficient ones led to a reduction in Air France's fuel costs between 1974 and 1980. The supply in seats/kilometres was adapted to decreasing demand. In contrast to other airlines, the deficits remained slight in 1980–82 (see Table 2.2) and debt was well controlled. The airline overcame this new crisis more easily than in 1974. Air France became profitable again in 1983 and reduced its debt. The renewal of the management contract was negotiated in 1984 with good conditions for the airline, whose capital was almost doubled during the 1980s. The computerization of numerous activities, such as the computer-aided design of maintenance workshops, freight management (Pélican) and computerized reservation systems (Amadeus), provided substantial savings. In stock management, for instance, computerization allowed a continuous supervision system and, for the first time, a reduction of stock volume. Computerized supervision of Airbus A-320 maintenance from 1987 simplified aircraft management.

With the first step of European deregulation in 1987, the state authorized an important development of Air France subsidiaries. Its share in Air Inter grew in 1987 from 26 per cent to 36.5 per cent, as it was argued that France was the only European country 'where the national airline may not service the internal market' (except for Nice and Corsica).[45] Air France had never been interested in domestic services, which were all unprofitable in the beginning. But nevertheless it would not accept that another international airline take over the internal network, all the more so since Air Inter's favoured status would not have permitted Air France to compete with it. For the first time, the accounts of the 'Air France Group', that is the airline and its subsidiaries, appeared in the airline's annual reports. The subsidiaries were all profitable except the Méridien hotel chain.

Bernard Attali became Air France's chairman in 1988 and launched an ambitious investment programme in order to cope with the growing competition from American carriers. The four leading US airlines owned about 500 aircraft each, while the biggest European airline operated no more than 200. Moreover, bilateral agreements appeared to be unequal. Whereas European airlines were allowed to fly to about twenty American cities, American airlines were servicing more than forty European cities, and new competition was arising with East Asian airlines which were more dynamic and profitable. Air France became the major shareholder of its main French competitor, UTA, which allowed it to control almost all French traffic taking off from French airports. By becoming a minor shareholder of Touraine Air Transport (TAT), the fourth French airline, Air France was able to improve its *connectivity* between internal and international flights. At the end of 1989, the airline had as many aircraft on order as it had in use. Investment increased 45 per cent (mostly for industrial equipment) in a single year, financed from the company's own reserves. Debt represented one half of the capital in 1989, and one and a half in 1990. The Air France Group, according to Attali, was in 1989 'the first European air transport company and one of the very first in the world', ready to 'face the challenge of competition everywhere'.[46]

But while the airline launched this vast programme of expansion, a new cycle of the air transport business, together with a new increase in fuel costs accompanying the 1991 Gulf War, suddenly reversed the gains and Air France faced heavy deficits again. This crisis was 'the most serious air transport has ever known', pleaded Attali in 1992.[47] All the activities of the Air France Group were hit. In three years the airline's losses on international networks erased the profits made in the previous twenty.[48] Moreover, after 1990, demand in the French domestic market fell while other European airlines recovered. According to Attali, Air France's deficit represented 5 per cent of IATA airlines' total deficit in 1990, 12.5 per cent in 1992, and almost one half in 1993!

A plan for a financial recovery, *Cap 93*, was launched in 1991. It consisted of 4400 redundancies of which 1200 (1.85 per cent of the staff) were to take place within a year, a freeze on wages and a 2.3 per cent reduction in expenditure. The network was to be reduced and Air France and UTA wholly merged. Aircraft orders were to be renegotiated and capital increased through loans. This plan caused a major strike in October 1993 and was abandoned as a result. Bernard Attali resigned and was replaced by Christian Blanc, a civil servant and former chairman of RATP where he had faced powerful unions. His *Project for the Company* was submitted for staff agreement.[49] This second recovery plan consisted of a one-third increase in labour productivity over three years, 2100 dismissals (5 per cent of the staff), a wages and promotions freeze and longer working hours. The fleet was reduced without capacity loss through a better management of capacities after 1994. Organization was deconcentrated through the creation of eleven results centres for each field of activities. 12 000 employees became shareholders, representing 5 per cent of the airline's capital, and in return accepted wage reductions. The French government increased Air France's capital for the last time. The trade unions understood that the financial situation was serious and the survival of the airline was in question. The European Commission gave its authorization, provided that Air France would receive no further injections of French state funds after 1997. Aircraft orders were renegotiated with Boeing and Airbus Industrie, then simply frozen, and the Group sold some of its subsidiaries, such as the Méridien hotel chain.

Initial results were positive; revenue remained steady while expenditure decreased significantly. Between April and September 1995, the Air France Group was profitable, and for the period April 1995 to March 1996, the airline's deficit was limited to one-third that of the previous year. The capital increase reduced financial charges. Strikes in the Autumn 1995 did not damage this improvement and Christian Blanc predicted profits for 1997.[50] It remained a fact, however, that while KLM and Lufthansa were out of danger and British Airways was very profitable, Air France, like Iberia and Alitalia, was a *lame duck* in the world of international airlines.[51]

Whereas other airlines like British Airways were cutting staff, recruitment increased at Air France in the 1980s, especially for navigators, whose strength among the staff rose from 13 per cent in 1969 to 20 per cent in 1989. Air France justified this in its 1981 annual report, the first under the new socialist government, as its contribution to the 'national effort in favour of … employment'. This entailed an expensive training programme.

Computer training began in 1980 but was interrupted in 1990, as the airline was forced by its financial situation to reduce its staff. Air France (38 000 employees in 1991) and UTA (8000 employees) merged in 1992 and this led to numerous redundancies. The new staff comprised a total of 44 000. Meanwhile Air France negotiated a revision of the 1953 agreement with the unions. The wages of French pilots were 20 per cent higher than American pilots and 100 per cent higher than British, yet this difference was not justified by higher productivity.[52] Air France stewards, who were also overpaid, flew 535 hours a year, while Lufthansa flew 585 hours with average wages which were 27 per cent lower.[53]

Table 2.3   Air France: labour productivity

| | Traffic | | Load Factor | | Staff | Productivity | |
	millions pass.	millions pass.-km	millions freight tons-km	(traffic/ supply) seats-km	employ- ees	(traffic/ staff) pass.-km/ employee	(supply/ staff) seats-km/ employee
1945	–	146	–	–	5 015	29 112	–
1950	0.775	1 134	38	65 %	14 603	77 655	119 469
1955	1.8	2 282	61	71 %	16 564	137 768	194 040
1960	3.2	4 049	99	66 %	22 852	177 183	268 460
1965	4.0	6 347	153	56 %	23 893	265 642	474 361
1970	6.1	10 657	355	53 %	27 593	386 221	728 719
1975	8.0	17 925	753	58 %	30 550	586 743	1 011 625
1980	10.9	25 495	1569	64 %	33 453	762 114	1 190 803
1985	12.4	28 649	2408	68 %	35 598	804 792	1 183 518
1990	15.7	36 730	3252	69 %*	38 377	957 083	1 387 077
1995	14.9	51 712	4766	72.8 %	36 484	1 417 388	1 946 962

* Air France Group.
1995: April 1995–March 1996.
Source: Air France Annual Reports and Accounts, 1948–March 1996.

Further labour troubles at Air France in 1993 led to another recovery plan: redundancies were resumed, especially among non-flight crew. The results centres made directors responsible for the profitability of their departments. This reform was supposed to remove the bureaucratic decision-making, which had undermined the motivation of staff and earned Air France the nickname 'elephant in an evening suit'.[54]

In general, so far as performance is concerned, Air France followed the trend more than it anticipated it after 1978. The lack of an aggressive commercial policy was one of its main weaknesses. Recently, the airline has taken more account of the growing segmentation of the market, offering an enlarged variety of fares and services on board; larger seats and special waiting rooms in airports. But its Frequent Flyer Programme (FFP) remained, with about a million passengers in 1996, below the level of the Lufthansa and British Airways FFPs.[55] It has also developed new freight services, such as Air France-Express in 1984, Domicile-Express in 1986 and Mach-Plus in 1988, with the help of a computerized parcels management, as well as the Pélican programme, a trucking service, and an association with TAT for internal flights.

5 STRATEGY

5.1 Collaboration and competition with other airlines

With the end of French colonialism, Air France transformed its local networks in Africa and Indo-China into subsidiaries, such as Air Atlas, Air Algérie, Tunis Air, Air Afrique and Air Vietnam – national airlines in which Air France kept an interest (except for Air Algérie). In Africa, it kept the direct management of two major routes: Paris–Dakar and Paris–La Réunion. At Paris airports, it offered maintenance and handling agreements to numerous foreign airlines and took part in *pools* mainly with European and Asian airlines. It also participated in scheduling and fare agreements in New York with Eastern Airlines and United Airlines.

Then between 1957 and 1965, Air Union sought to create a technical and commercial group out of Air France, Lufthansa, Alitalia and Sabena, partly in order to cope with the new generation of jet aircraft.[56] At the beginning in 1957, the project was called 'Europair' and aimed to replace competition with capacity-sharing for the long-haul sector. It was enlarged in 1959 to include wider objectives and move closer to being a single firm, like SAS in Scandinavia. The project had a political dimension, in so far as it had been launched by the six countries that had founded the EEC in 1957, but it was initiated by the airlines, not by the governments. Negotiation was long and difficult, under French leadership. Air France and Sabena wanted to leave out their colonial networks from the association, KLM abandoned its association and Alitalia demanded a higher share than its actual traffic justified. The final agreement between the airlines in 1960 was 34 per cent for Air France, 30 per cent for Lufthansa, 26 per cent for Alitalia, 10 per cent for Sabena, but government agreement was necessary because the airlines were state owned. The governments turned out to be very reluctant, especially in France, to accept what they considered was a loss of national sovereignty. A diplomatic conference began in 1964, and in the following year the French President, General de Gaulle, who was opposed to any kind of European federalism, put an end to Air Union for fear of losing Air France's major position. National interests prevailed over European interests and the airlines remained 'flying flags'.

A second technical maintenance pool was launched in 1967 for Boeing 747 and DC-10 aircraft, composed of Air France, Alitalia, Lufthansa, Sabena and, after 1972, Iberia. This Montparnasse Comity[57] became the Atlas Group in 1969, devoted to pilot training and aircraft maintenance, and similiar in aims to the KSSU that grouped Swissair, KLM, SAS and UTA. Charges were equally shared, each member had specific tasks, and the whole Group ordered aircraft under the same conditions from Boeing and Douglas. Air France was in charge of the maintenance of Boeing 747 and of DC-10 engines. This pool was a real success and was extended to Airbus aircraft and partly to Concorde.

The 1973 oil shock compelled several airlines to delay their orders to Airbus Industrie in 1974, which meant that Air France had to postpone the use of its own Airbus aircraft because of bilateral agreements on capacity, and carry out its own maintenance. After 1978, American deregulation put an end to IATA agreements on North Atlantic routes.[58] In 1987–9 Air France strengthened its technical cooperation with Lufthansa for assistance at airports, on routes to West Berlin, on freight traffic, and with the CRS Amadeus. The latter grouped Air France, Lufthansa, SAS and Iberia, and was extended to other European airlines when it was put into operation in 1989. Air France chose Lufthansa and Amadeus, instead of British Airways, KLM and the Galileo CRS, because their markets were similar and because they had got used to working with the Germans. Between 1989 and 1995, Air France tried to create a 'family of allied European airlines'[59] taking stakes in Sabena and the Czech airline CSA, but it had to sell its shares because of opposition from the European Commission. Computerization however led to strong technical cooperation. The Traxon freight programme grouped Air France, Lufthansa, JAL, Cathay Pacific and others, in order to create a world-wide computer system.

Whereas British Airways, Lufthansa and KLM have formed partnerships with American airlines, Air France remained isolated, even ending some bilateral agreements with American airlines. Deficits made Air France unattractive to possible partners. Moreover the Americans were unwilling to participate in a foreign state-owned firm, especially if it belonged to a country such as France that was reluctant to deregulate. Air France did sign commercial and technical agreements in 1995 with Japan Airlines and Delta Airlines, and in 1996 with Continental.[60]

5.2 Collaboration and competition with other transport modes

There has been no competition between Air France and surface transport modes, since the airline's route network has been almost completely international. Except for Paris–Nice and Paris–Corsica, domestic routes belonged to Air Inter. Collaboration was more important.[61] A combined air and railway service existed on the London–Paris route during the 1950s, involving the French railways (SNCF) and two secondary airlines, the French Air Transport and British Skyways, but Air France did not take part in it. Road transport was considered as an auxiliary for the freight traffic and Air France has owned the truck subsidiaries, SODETAIR and SATTE, since 1974. Transfer between trucks and aircraft in CDG was computerized in 1989. In 1988 Air France and the Post Office created a joint subsidiary, SODEXI, for express parcel delivery, and the airline worked together with transit operators on Traxon (see above). In general Air France has never competed with sea or land transport on routes that did not require a rapid delivery service.[62]

The high speed TGV train, although a direct competitor for Air Inter, became in 1994 complementary to Air France, in so far as it serviced two major French airports, Paris-CDG and Lyon-Satolas. The rail links to the airports, considered in the past as 'the wolf in the sheep enclosure', were henceforth an important asset in Air France's competition with other airlines. The TGV station at CDG-2, designed and built by ADP and SNCF, has strengthened Air France's 'hub' there. It has led Federal Express, the express parcel delivery firm, to choose CDG as its European gateway.[63] The TGV, including the Paris–London Eurostar, was not just a train that went faster than others – it was a new transport mode whose character is more like air transport than rail. Air France uses it as a major element of its commercial strategy and has been selling combined tickets, for example Lille–CDG–New York. Such combined tickets raise the problem of a connection between Air France CRS Amadeus and the SNCF computer system Socrate. A connection between air transport and the Eurostar could also be put into operation, but the latter is a competitor to Air France and responsible for the loss of 20 per cent of its traffic on the Paris–London route between 1994 and 1995.[64] Although it was difficult to know precisely the economic consequences of the combination between the TGV and air transport, it was obvious that a European network for high speed trains in the future will force the airlines to define their commercial strategy in intermodal terms.[65]

5.3 Globalization, privatization and strategic alliances

Since it was created in 1933, Air France has never been a monopoly, in so far as some internal and African destinations were also serviced by other French airlines. In 1967 Air France gave up its internal network in favour of Air Inter, to which it was bound by financial participation and a convention. The following year, it created a specialized subsidiary for charter flights, Air Charter International, which held a leading position in the French market. The other main French charter airlines were Minerve, Corsair, Air Liberté and Point Air.

In 1987, British Airways signalled a wave of concentration in Europe, by buying British Caledonian. UTA, the major French competitor to Air France, tried to take control of Air Inter. After a two-year fight, UTA, Air Inter, Air France and its subsidiaries had become all one group, in order to face the increasing competition from foreign airlines. Air France became in 1990 the major shareholder of UTA and therefore of Air Inter. It tried in 1989 to become a minor shareholder of TAT, the fourth French airline, but the European Commission opposed this and British Airways bought TAT in 1993. Air France and UTA merged in 1992, and Air Inter developed its European network, becoming in 1996 'Air France Europe'. A real merger between Air France and Air Inter was impossible before 1997, because of insuperable differences in their management methods. Air France was considered by Air Inter to be a prestige-driven spendthrift, whereas Air Inter had a more cost-conscious management. Air France executives were supposed to be the 'lords' while those at Air Inter were considered 'stingy'.[66]

This unfinished concentration of the French airline industry raised the question of Air France's survival. Its disappearance or its purchase by a foreign airline in the case of privatization, would have been considered a loss of national sovereignty and difficult for the French government. On the other hand, the bilateral agreements of European airlines with American carriers were unbalanced.[67] There was no 'open sky' for European airlines in the USA, and these agreements did not prevent dumping. American mega-carriers profited easily from European deregulation, while European airlines were not able to work together. The Air France Group would therefore have to cope, after 1997, at once with American airlines, European and French competitors, as well as the high speed train network.

Has Air France become something more than an 'elephant in an evening suit'? It is still a 'chosen instrument' of the government, which has continually subsidized its deficits, but the most recent capital increase has been accepted by the European Commission provided that it will be the last one before privatization. The French state and trade unions have accepted management contracts aimed at a restoration of profit. Air France has kept a balance between French and American manufacturers in its aircraft procurement and has maintained a heterogeneous fleet. It has recently replaced the idea of an extensive network with concentration on the most profitable routes, and has developed a strong partnership with Paris airport so as to create a hub at CDG-2. A merger with the second and third French airlines (UTA in 1992, Air Inter in 1997) has strengthened this network. Marketing has become more effective and collaboration has been sought with foreign airlines (Lufthansa, Japan, Delta and Continental Airlines) and with the high speed train network that serves CDG-2. However despite these recent changes, the challenge of competition still has to be faced against the background of cultural obstacles in French society.

NOTES

1. Air France Annual Report and Accounts–Report on the former Air France company since January 1943, August 1948, p. 15; 'Air France, entreprise nationale de transports aériens', *Notes et Etudes documentaires*, no. 3849–3850, 30 December 1971, pp. 10–20; 'Air France et son histoire, 1933–1983', *Icare* (French Aviation Review), no. 106–7, October 1983, pp. 51–100.
2. 26 June 1945 Order.
3. 26 June 1945 Order, 20 September 1945 Convention.
4. 'Air France...', *Icare*, op. cit., p. 100.
5 16 June 1948 Act.
6 Air France Annual Report and Accounts, 1957, p. 6.
7. See the debates in the Assemblée Nationale, *Journal Officiel*, 1948, pp. 1734, 2333, 2510, 3096 and 4264.
8 Air France Annual Report and Accounts, 1954, p. 14.
9. 'Air France..', *Icare*, pp. 110–18.
10. Air France entreprise..., *Notes et Etudes documentaires*, p. 19.
11. Comité des transporteurs aériens français, 'Le transport aerien français'. Paris, CTAF, 1977, pp. 19–23.
12. Air France Annual Report and Accounts, 1965, p. 6 (Chairman's speech, 12 September 1966).
13. S. Nora, 'La gestion des entreprises publiques'. Paris, Documentation française, 1967.
14. Air France Annual Report and Accounts, 1978, p. 5.
15. 'Un long chemin pour devenir une entreprise comme les autres', *France Aviation* (Air France Journal), special issue for the 50th anniversary of the new airline, February 1995, p. 7.
16. G. Bordes-Pagès, 'Air France, cas exemplaire', *Le Monde*, 11 November 1995, p. 12.
17. P. Funel and J. Villiers, 'Le transport aérien français'. Paris, Documentation française, 1982, p. 84.
18. Air France Annual Report and Accounts, 1980, p. 9.
19. 'Air France...', *Icare*, op. cit., p. 109. Espérou, R. (former General Inspector for Civil Aviation) written interview, 21 May 1996.
20. Air France Annual Report and Accounts, 1955, p. 6. Esperou, R., op.cit.
21. E. Chadeau, 'Le rêve et la puissance, l'avion et son siècle'. Paris, Fayard, 1996, pp. 369–76.
22. E. Chadeau, op. cit., pp. 389–407; E. Chadeau, E. dir. 'Airbus, un succès industriel européen'. Paris, Institut d'Histoire de l'Industrie, 1995, pp. 3–15.
23. C. Jakubyszyn, 'Airbus Industrie s'affranchira en 1999 de la tutelle de ses fondateurs', *Le Monde*, 10 July 1996, p. 14; 'L'envol américain d'Airbus', *Le Monde*, 17 December 1996, p. 18.
24. C. Jakubyszyn, 'Le PDG d'Air France affirme son autorité en imposant l'achat de Boeing 777', *Le Monde*, 21 November 1996, p. 22.
25. Air France Annual Report and Accounts, 1954, p. 13.
26. Air France Annual Report and Accounts, 1984, p. 8.
27. Air France Annual Report and Accounts, 1977, p. 27.
28. Air France Annual Report and Accounts, 1978, p. 5.
29 G. Maoui, N. Neiertz dir., 'Entre ciel et terre' (History of Paris Airports). Paris, 1995, pp. 1–128, for this whole part 3.2.
30 M. Dacharry, 'Géographie du transport aérien'. Paris, 1981, p. 304.

31. V. Malingre, 'Une organisation en étoile', *Le Monde*, 26 January 1996, p. 17.
32. Air France Annual Reports and Accounts, 1948 to March 1996, for this whole part 4.1.
33. Air France Annual Report and Accounts, 1953, p. 5.
34. Air France Annual Report and Accounts, 1956, p. 5.
35. J. Naveau, 'L'Europe et le transport aérien'. Brussels, 1983, p. 110. J. Pavaux, 'L'économie du transport aérien, la concurrence impraticable'. Paris, 1984, p. 128.
36. Air France Annual Report and Accounts, 1959, p. 8.
37. Air France Annual Report and Accounts, 1968, p. 4 (Chairman G. Galichon speech, September 1969).
38. C. Stoffës dir., 'Transport aérien, liberalisme et déréglementation', Paris, 1987, pp. 11–18.
39. Air France Annual Report and Accounts, 1967, p. 59.
40. Air France Annual Report and Accounts, 1971, p. 38.
41. Air France Annual Report and Accounts, 1974, p. 43.
42. Air France Annual Report and Accounts, 1978, p. 4.
43. Air France Annual Report and Accounts, 1983, p. 6.
44. Air France Annual Report and Accounts, 1978, p. 5.
45. Air France Annual Report and Accounts, 1987, p. 12.
46. Air France Annual Report and Accounts, 1989, p. 52.
47. Air France Annual Report and Accounts, 1991, p. 4.
48. Air France Annual Report and Accounts, 1992, p. 2.
49. C. Blanc, 'Projet pour l'entreprise', 9 March 1994, Air France documents, pp. 13–47.
50. V. Malingre, 'La compagnie Air France devrait retrouver l'equilibre en 1997', *Le Monde*, 19 April 1996, p. 15.
51. V. Malingre, 'Le trafic aérien international devrait croître de 8% en 1996' *Le Monde*, 23 April 1996, p. 20.
52. P. Merlin, 'Les transports en France', Paris, Documentation française, 1994, p. 65.
53. These figures were given in 1994 by a German financial consultant, Berger, and have been contested by the unions. G. Bridier, 'Le redressement d'Air France se heurte aux oppositions syndicales', *Le Monde*, 27, September 1995, p. 16.
54. 'Pachyderme en smoking'. A. Routier, 'Air France, réveil douloureux', *L'Expansion*, 3 October 1991, pp. 122–32.
55. Air France Annual Report and Accounts, 1995–96, p. 12.
56. 'Air France et son histoire, 1933–1983', *Icare*, op. cit., p. 130–7. R. Esperou, op. cit. N. Spira dir., 'Le transport aérien', *Les Cahiers français*, May–June 1976, p. 12–15. J.-G. Marais and F. Simi, 'L'aviation commerciale', Paris, 1964, pp. 112–15. J. Naveau, op. cit., p. 176–86.
57. Montparnasse is the area of Paris where Air France had its registered office until 1995.
58. Air France Annual Reports and Accounts, 1980 (p. 14), 1981 (p. 8), 1982 (p. 16).
59. Air France Annual Report and Accounts, 1991, p. 7.
60. C. Jakubyszyn, 'Air France s'allie avec Delta Airlines et Continental Airlines', *Le Monde*, October 17th 1996, p. 15; B. Marck, 'Air France- Continental-Delta: une alliance pour un ménage à trois', *Aéroports Magazine*, no. 274, December 1996, pp. 16–19.
61. N. Neiertz, 'La coordination des transports en France de 1918 à nos jours', unpub. PhD thesis, Paris IV-Sorbonne University, 1995, pp. 44, 121, 164, 199,

229, 270, 338, 427, 512, 578 and 634. P. Funel and J. Villiers, op. cit., p. 74. See also: J. Pavaux dir. 'Les complémentarités train-avion en Europe'. Paris, Air Transport Institute, 1991. J. Varlet, 'L'interconnexion des réseaux de transport en Europe'. Paris, Air Transport Institute, 1992. N. Neiertz, 'Etude d' un mode d'interconnexion de réseaux: le cas d'Aéroports de Paris'. *Transports urbains*, no. 84 (July–Sept. 1994), pp. 15–22.

62. Air France Annual Report and Accounts, 1980, p. 24.

63. F. Lemaitre, 'FedEx installera sa plate-forme européenne à Roissy', *Le Monde*, 26 April 1996, p. 17.

64. D. Lonchambon, 'Train-avion: je t'aime moi non plus', *Le Journal d'ADP*, no 18, October 1995, pp. 6–8.

65. M. Saint-Yves, 'L'intermodalité air-fer, une nécessité pour l'Europe'. *Aviation International*, no. 995 (1 December 1989), pp. 32–4. J.-M. Thomson, 'Road, rail and air competition for passengers in Europe'. *Aeronautical Journal*, London (April 1978), pp. 139–47.

66. M. Gilson, 'Les seigneurs contre les radins', *Le Nouvel Observateur*, 1–7 October 1992, p. 72.

67. P. Lemaitre, 'L'Europe du transport aérien se délite face aux Etats-Unis', *Le Monde*, 14 June 1995. S. Kaufmann and V. Malingre, 'Les Etats-Unis veulent forcer l'ouverture du ciel français', *Le Monde*, 23 March 1996, p. 17.

3 Chosen Instruments: the Evolution of British Airways

Peter Lyth

When Lord King became chairman of British Airways in 1981 he took over the leadership of an airline with an accumulated history of over forty years and a tradition stretching back nearly sixty. It was an illustrious history, but one with its share of disasters and disappointments. British Airways and its predecessors had suffered from misguided policy, poor management and troublesome aircraft. However they had also built up one of the longest international route networks in the world, carried millions of passengers in comfort and safety, and pioneered the world's first passenger jet aircraft.

1 POLICY

1.1 Origins, 1924–44

British Airways is now a profitable private enterprise, but it has not always been either private or profitable. It began life as the state-owned British Overseas Airways Corporation (BOAC) in 1940 and to understand BOAC's creation one has to go further back, to the pioneering days of commercial air transport.

Britain's first national carrier, Imperial Airways, was created as a private monopoly in 1924 after five years of fruitless competition between the early airlines.[1] Subsidized on a regular basis by the government, Imperial Airways devoted itself to the task of building air mail and passenger links to India, Australia and South Africa. In the 1930s its famous Short flying boats became a regular feature of life around the outposts of the Empire but, against the background of growing national rivalry in Europe, it was seen to be failing in its prestige role.[2] In the 1938 Cadman Report it was criticised for neglecting the wider needs of national policy and despite earlier government encouragement for its development of colonial routes, it was now seen as imperative that better services be established between London and the principal capitals of Europe.[3] The need for a more visible British presence on the Continent had been partly met by a second international carrier, British Airways Ltd, which had begun mail and passenger services to destinations in northern Europe in 1936.

However, in the wake of the Cadman Report the decision was taken to combine the two airlines and form a nationalized air corporation – BOAC. It was recognized that Britain's *chosen instrument* could not succeed against foreign competition unless it was properly subsidized and supervised. And it was this need for supervision that led to the unusual decision by a Conservative government to bring the airline into public ownership.[4]

In August 1939 the British Overseas Airways Corporation Bill received the Royal assent and in April 1940 BOAC began operations. The timing could hardly have been worse. It was war, the skies were full of combat aircraft and there was neither space nor resources for commercial operations. No sooner had it been created than BOAC was placed under the control of the Air Ministry and the Royal Air Force (RAF). Normal business procedures were suspended and its task was reduced to ferrying men and material along whatever international routes were still safe from enemy action. By the end of 1940 the only European services that BOAC was flying from Britain were to Portugal and occasionally to Sweden.[5] During the rest of the war BOAC rendered valuable service as a quasi-military operation, particularly with its North Atlantic *Return Ferry Service* which facilitated a steady flow of aircraft from American factories to the war in Europe, and was, incidentally, the first regular all-year-round Atlantic service.

1.2   Government objectives, 1944–84

It was not until the autumn of 1944 that the wartime government was able to give serious consideration to the future of BOAC, and the Minister of Civil Aviation, Viscount Swinton, set about assembling a plan for its development. The plan, which was presented to parliament in March 1945, envisaged British civil aviation in the hands of several organizations with participation by surface transport interests. BOAC was to remain a state-owned enterprise but two new corporations, British European Airways (BEA) and British South American Airways (BSAA), would be added, with mixed ownership, of which the majority would be private.

The Swinton plan aimed to fulfil a number of criteria, which are worth recording because of their enduring influence on postwar policy. Firstly, it sought to provide vital, but inherently unprofitable, *social* services within Britain and to support them, not by direct subsidy, but by granting monopoly rights on the airline's profitable routes; in other words asking it to cross-subsidise. Secondly, it tried to ensure that the airlines flew British aircraft, and thirdly, that they operated without subsidy. This last objective, which was in contrast to the 1939 policy, can be seen in the stipulation that the participating companies be chosen not only for the contribution they could make in skill and experience, but also because they were 'prepared to invest their own money without any Government guarantee'.[6] The Swinton plan was superseded by the electoral victory of the Labour party in the summer of 1945, but its structural provisions reappeared in Labour's nationalization programme and the 1946 Civil Aviation Act. In the main debate on its bill, the government reaffirmed the need for subsidies and argued that nationalization would enable these to be employed more effectively. However it accepted Swinton's proposal for *three* airlines flying in a non-competitive relationship to each other. It justified this on the grounds that a multiplicity of instruments would enable the testing of different managerial styles, although the separation of long-haul (BOAC) from short-to-medium haul operations (BEA), left the latter with a handicap in higher unit costs and the burden of loss-making domestic services.[7] There were various proposals to unite the three air corporations from as early as 1946. These were rejected, but the government did decide to reduce the number from three to two by allowing BOAC to absorb BSAA following the unexplained disappearance of a BSAA Tudor aircraft over the Atlantic in early 1949.[8]

Neither BOAC nor BEA made any profits in the early years and between 1946 and 1952 BOAC received about £35 million in Treasury grants, while BEA got about £16 million in the years until 1955.[9] Thereafter British air transport operated without direct operating subsidies, although this independence of the British flag-carriers from government financial support was not matched by any greater clarity in government policy towards them. The central and unresolved question concerned whether the airlines were to function as purely commercial entities, or whether they were to undertake additional assignments in the national interest.

Unlike some other European flag-carriers and in contrast to the provisions of the Swinton plan, BOAC and BEA remained entirely state-owned enterprises. They also remained monopolies, although their monopoly privileges were increasingly eroded as the industry expanded. In 1952 the incoming Conservative government allowed applications for new services from independent airlines to be heard by the Air Transport Advisory Council (ATAC). It was a small gesture on the path towards competition, but few Conservatives wanted to dismantle or privatize the air corporations and the government assured BOAC and BEA that their *scheduled* routes would not be opened to competition. In 1960 a statutory change was made with the Civil Aviation [Licensing] Act. This threatened to deprive them of their exclusive rights to scheduled services, and it replaced the ATAC with the more robust Air Transport Licensing Board (ATLB). However, any dilution of BOAC and BEA's monopoly was limited, and new applicants for air routes had to prove that their projected service would not cause 'material diversion of traffic from any air transport service which is being, or is about to be provided.'[10] Only after 1969 did the corporations' existence as state-owned monopolies face serious challenge. The Edwards Report, which was published that year, recommended the creation of a second force airline in the private sector and a joint holdings board to oversee the future development of BOAC and BEA.[11] These proposals were put into effect in the 1971 Civil Aviation Act, which set up the Civil Aviation Authority (CAA) in place of the ATLB, cleared the way for British Caledonian Airways (BCal) to become the second force, and established the British Airways Board to oversee the management of the two airlines. Although Edwards had not envisaged BOAC and BEA losing their separate identities, a process of amalgamation was now in train. The multiplicity of instruments favoured in 1946 was now replaced with a large single flag-carrier, while hopes for more genuine competition were placed in BCal.

British Airways (BA) began life in a difficult decade for airlines. Moreover BA's management proved unable to reap the rewards of rationalization which should have flowed from the merger of BOAC with BEA. This theme will be returned to when we look at BA's operational and financial performance; so far as the government is concerned, it seems that, having finally blessed the marriage of BOAC and BEA, it was unwilling to force the new airline to make the logical adjustments to its structure. By 1975 it was clear that the Labour government's policy sought primarily to protect BA and the jobs it provided, and its position as a state monopoly was defended by the responsible Minister (Peter Shore) against both Freddie Laker's Skytrain and BCal which was prevented from expanding into BA's sphere of operations.[12] Only with the arrival of Margaret Thatcher's government in 1979 did the full integration of the airlines begin, with the inevitable staff redundancies which this involved. However the Conservative government's reputation for encouraging greater competition in the airline industry should be put into context. Certainly the CAA strove to create a more competitive environment, both within Britain and abroad, and it was with the support of the CAA that the Transport Secretary (Nicholas Ridley) achieved a partial liberalization of Britain's bilateral relations within Europe. However the CAA's 1984 report recommending more competition for BA, and the transfer of BA routes to BCal, was rejected by the government.[13] An alliance of BA and the Treasury blocked the recommendation on the grounds that nothing should be allowed to jeopardize BA's forthcoming privatization. When BA did eventually pass into private ownership in 1987, it did so as a giant in the industry, freed of its debts by a government write-off and ready to engulf its smaller competitors. Ironically, the government celebrated for its dedication to competition, obstructed it in the case of the airline industry in order to allow the creation of a profitable private monopoly.

2   AIRCRAFT PROCUREMENT

Although the Americans dominated the manufacture of transport aircraft during the Second World War, the British were determined to produce their

own civil aircraft for the postwar era and the government was determined
that BOAC would fly them. To meet this objective the Brabazon Programme
was launched in 1943, with the aim of developing designs for five new types
of airliner. As little progress was made during the remaining months of
the war, BOAC's fleet in 1945 was composed largely of obsolescent
types ranging from the Avro York to the Short Solent flying boat. As an
interim measure BOAC was allowed to replace most of these with American
equipment, but for its long-term requirements it was expected to introduce
the appropriate Brabazon type. By 1950 there was no doubt at BOAC
as to that aircraft's identity: the de Havilland Comet, the world's first
jet airliner.

2.1 The Comet

BOAC ordered eight Comet 1s in 1947 and began commercial services with
the aircraft from London to Johannesburg in May 1952. The jet was an
instant success with passengers and immediately profitable for the airline. By
the end of its first year in service the Comet's future seemed assured and de
Havilland received the ultimate accolade when Pan American Airways
placed an order for the bigger Comet 3.[14] Then disaster struck. In May
1953 a BOAC Comet broke up in a thunderstorm after leaving Calcutta.
Seven months later a BOAC Comet fell out of a clear sky after taking off
from Rome and all Comets were grounded for investigation. In March 1954
services were restarted, but only two weeks later a Comet operating a South
African Airways service crashed near Naples. A fatal pattern had been
established and the aircraft's certificate of airworthiness was withdrawn.
After a complete aircraft had been tested to destruction, the cause of the
crashes was found to be structural failure of the pressure cabin due to metal
fatigue. The lead which BOAC had established in jet aircraft operation had
vanished.

The Comet crashes not only damaged BOAC's reputation, they also
complicated its aircraft procurement process. The immediate consequence of the Comet's withdrawal was a substantial shortfall in capacity.[15]
To fill this gap the government authorized the purchase from the United
States of the piston-engined Douglas DC7-C. In 1955 BOAC renewed
its trust in the Comet with an order for 20 of the completely rebuilt
Mark 4s, but the following year it also received government approval to
buy the Boeing 707.[16] On the one hand it needed the American aircraft
to stay in business, on the other it was obliged to buy British whenever
possible. One solution to this procurement dilemma, already adopted by
BEA, was a form of compromise between piston-engines and pure jets:
the turboprop.

2.2 Turboprops: Viscount, Britannia, Vanguard

The turboprop engine was a technology to which the British were especially attached and two of the aircraft which emerged from the original Brabazon Programme were powered by turboprop engines: the Vickers-Armstrong Viscount with Rolls-Royce Darts and the Bristol Britannia with Bristol's own Proteus.

The Viscount, which was developed for BEA, was the most successful civil aircraft ever built in Britain. It was sold to airlines all over the world, including several in Europe, and thanks to its engines, operated (with American airlines) at cost levels considerably below close competitors like the Convair 340.[17] The success of the Viscount led to the understandable belief that turboprop engines should be used in the Viscount's replacement, and this was the genesis of the Vickers Vanguard, which first flew in 1959 and entered service with BEA in 1961. BEA was convinced that turboprops would have a cost advantage over jets on short-haul routes where the jet's speed advantage was likely to be minimal. What was not foreseen however was that turboprops lacked the passenger appeal that made jets immediately popular and the proof of this emerged in 1959 when Air France flew the short-range Caravelle into London, gaining an immediate competitive advantage over the British airline.[18]

The Vanguard's history at BEA was paralleled by the Bristol Britannia at BOAC. The Britannia was a first-generation turboprop like the Viscount, but much larger and unlike the Viscount it was plagued by problems with its engines. In the wake of the Comet crashes and BOAC's temporary loss of faith in jets, the Britannia was prioritized, but it was to be another three years before the smaller 102 version entered service with BOAC and by 1957 BOAC's focus had turned back to jets amid the global preparation for the Boeing 707 and Douglas DC-8. Because of the Britannia's late delivery it arrived at the same time as the stopgap DC-7Cs and BOAC now faced a surplus of capacity, aggravated further in 1958 when the Comet 4 was delivered. By the time that BOAC began flying its own 707s in 1961, the airline had 75 aircraft consisting of big American jets (Boeing 707s), small British jets (Comet 4s), British turboprops (Britannias), and American piston-engined aircraft (DC-7Cs) (see Table 3.1). This was too many types and it deprived the airline of the chance to gain economies through the use of a more standarized fleet.

2.3 British jets: VC10, Trident and BAC 1–11

In the course of the four-year interlude between the Comet crashes and Boeing's introduction of the 707, BOAC rejected a Vickers design for a large jet, the V-1000, and then, having accepted the need for such a jet when the

Table 3.1   BOAC, BEA and BA Fleet Statistics, 1953–96

(revenue-earning aircraft, includes leased aircraft and all freighter versions, not including helicopters)

As at March	1953	1957	1961	1965	1969	1973	1977	1981	1985	1989	1993	1996
			BOAC						BA			
Long Range												
Avro York	6	–	–	–	–	–	–	–	–	–	–	–
Handley Page Hermes 4	9	–	–	–	–	–	–	–	–	–	–	–
De Havilland Comet 1	8	–	–	–	–	–	–	–	–	–	–	–
Bristol Britannia 102 / 312		7	31	3	–	–	–	–	–	–	–	–
De Havilland Comet 4			19	17	–	–	–	–	–	–	–	–
BAC VC10/Super VC10				12	28	27	15	–	–	–	–	–
BAC/Sud Concorde*							5	7	7	7	7	7
Canadair Argonaut	22	8	–	–	–	–	–	–	–	–	–	–
Boeing Stratocruiser	10	16	–	–	–	–	–	–	–	–	–	–
Constellation L.049/749	12	9	–	–	–	–	–	–	–	–	–	–
Douglas DC-7C		6	10	–	–	–	–	–	–	–	–	–
Boeing 707			15	20	23	27	20	15	–	–	–	–
Boeing 747						12	25	27	30	47	56	63
Tristar L.1011/200/500								10	10	8	–	–
McDonnell Douglas DC-10										8	7	7
Boeing 767										4	20	24
Boeing 777												3
Total	67	46	75	52	51	66	65	59	47	74	90	104
Per cent British	34	26	67	61	55	41	31	12	15	9	8	7

Short to Medium range	BEA					BA						
DH 89 Rapide Islander	9	3	–	–	–	–	–	–	–	3	3	–
Vickers Viking	46	–	–	–	–	–	–	–	–	–	–	–
Bristol Freighter	1	–	–	–	–	–	–	–	–	–	–	–
Airspeed Ambassador	20	19	–	–	–	–	–	–	–	–	–	–
Vickers Viscount	4	35	62	39	33	15	15	8	–	–	–	–
De Havilland 114 Heron	–	2	2	2	–	–	–	–	–	–	–	–
Vickers Vanguard	–	–	20	19	15	–	–	–	–	–	–	–
De Havilland Comet 4B	–	–	–	6	13	7	–	–	–	–	–	–
Handley Page Herald	–	–	–	7	–	–	–	–	–	–	–	–
Argosy	–	–	3	4	–	–	–	–	–	–	–	–
Trident 1/2/3	–	–	–	3	33	60	61	52	26	34	–	–
BAC Super One-Eleven	–	–	–	14	10	21	25	26	–	6	–	–
HS 748	–	–	–	–	–	2	2	2	6	8	8	–
BAe ATP	–	–	–	–	–	–	–	–	–	8	14	14
DHC-7-100	–	–	–	–	–	–	–	–	–	–	–	5
DHC-8	–	–	–	–	–	–	–	–	–	–	–	6
Pionair (Douglas DC-3)	50	46	–	–	–	–	–	–	–	–	–	–
Tristar L.1011/1	–	–	–	–	–	9	9	–	9	–	–	–
Boeing 737	–	–	–	–	–	–	–	9	46	47	74	66
Boeing 757	–	–	–	–	–	–	–	25	24	36	42	44
Airbus A320	–	–	–	–	–	–	–	–	–	8	10	10
Total	130	106	87	94	108	124	112	122	111	150	151	145
Per cent British	61	57	93	100	100	93	92	75	29	34	13	10
Combined Total	–	–	–	–	–	190	177	181	158	224	241	249
Per cent British	–	–	–	–	–	75	69	52	25	25	11	8

* Anglo-French

Source: BOAC, BEA and BA Annual Report and Accounts.

delay with the Britannia showed that turboprops would be surpassed on long-range routes, opted for another Vickers design: the VC10. Bowing to government pressure to continue its support for British aircraft, BOAC ordered 35 VC10s in 1958 and two years later, ten more of the long-range Super VC10s. It was the largest expenditure on aircraft made by any airline in the history of British aviation.

The VC10 was an undoubtedly graceful aircraft, with its four Rolls-Royce Conway jets mounted at the rear and, unlike the Britannia, it was built on time exactly to BOAC's specification. The problem was in the specification itself. Because the American jets needed long runways, it was thought that they would be unsuitable for BOAC's African and Asian routes, whereas the clean-winged VC10 would be able to 'land on a dime'. Unfortunately by the time the VC10 entered service in April 1964, the Boeing 707's performance had been greatly improved and everyone had built longer runways. As *The Economist* noted in a cutting remark, 'concrete is cheap and aircraft are not'.[19] To make matters worse, the VC10 now turned out to be significantly more costly to operate than the 707.

While Vickers were building the VC10 for BOAC, a similiar miscalculation was being made at BEA, which was working out the specifications for its own *tailored* jet, the de Havilland DH.121, known as the Trident. The background to the Trident order was the simultaneous production of the Vanguard and the short-range version of the Comet 4: the 4b. With so many aircraft in the pipeline, BEA anticipated overcapacity and in early 1959 the decision was taken to downgrade the Trident to a smaller aircraft with 87 instead of 110 seats.[20] As a result of this change the Trident's Rolls-Royce Medway engine was replaced with the less powerful Spey, a step which ruined the 'stretch' potential of the aircraft and made it less attractive to foreign buyers.[21] The consequences of its reduction in size and power became clear when the Trident began operations in 1964. With jets in widespread use in Europe, it had to be pressed into service on the short but prestigious London–Paris service although for that route it was too small and had operating costs above those of the Vanguard.[22] Hawker Siddeley (into which de Havilland had now merged) began to stretch the Trident, but it was too late: the 150-seat Boeing 727, a very similiar tri-jet design but with greater range, capacity and take-off performance, was already collecting orders across Europe. By 1966 the 727's superiority was recognized by BEA and it requested government permission to buy the 727-200 as well as the smaller Boeing 737 twin-jet. The Labour government's refusal of that request was accompanied by the payment of substantial compensation to BEA for taking the Trident 3 and the BAC One-Eleven instead.[23]

The British Aircraft Corporation's BAC One-Eleven originated with the small Hunting company in 1960. Being a short-range twin-jet, it was in some respects an early example of a European airbus. Unlike the Trident it was not commissioned by BEA and was first adopted by the independent carrier British United Airways (BUA). The longer Super One-Eleven version was only accepted by BEA after its application to buy 737s had been rejected. It was a successful aircraft in so far as BAC sold over 200 of them and BA continued to fly them until the early 1990s, but it was overshadowed by its close rival the Douglas DC-9; despite its late start, the American aircraft sold four copies to every One-Eleven.

2.4 Concorde

The VC10 and Trident would have been the last major civil transport aircraft built in Britain, had it not been for the Anglo-French supersonic Concorde. This aircraft was not commissioned, or even sponsored by BOAC, but was rather an example of state-funded prestige technology. It was a project 'not so much entered into, as plunged into, in a mood of technical enthusiasm approaching euphoria, without serious thought of the market, the cost or the consequences'.[24]

Originating in a 1962 design and development agreement between the British and French governments, the Concorde consumed a vast amount of government money before it finally began scheduled services with BA in 1976. And even then it was to be another year and a half before BA was able to overcome American environmental objections and fly it to New York, the only route on which the aircraft had any chance of breaking even. The fact that in the 1990s it is still flying without rivals has given Concorde an exceptional status in the international airline industry, but it has to be recognized that its commercial survival has only been possible thanks to a total write-off of its development costs and that there has never been anything particularly extraordinary about it as a technological achievement. Its engines (Bristol Olympus) use a straight jet technology which was already dated when the aircraft was conceived. More critically the Concorde is small, and BOAC felt even in 1964 that its payload would not be big enough to carry its cost on the North Atlantic. In some ways the Concorde was similiar to the first Comet, in that it pushed aeronautical know-how too far. Had the manufacturers waited until supersonic aircraft could be made more economic, the British and French tax-payer might have been saved a lot of money. As it was the Americans had more or less decided by the mid-1960s that the speed of airliners had been raised to its economic limit (about 575 mph) and that future designs should concentrate on making aircraft bigger, not faster.[25] BA remained highly sceptical of Concorde's economics and in the end it was careful to ensure that the government provided a guarantee to cover the aircraft's expected operating losses, remaining convinced 'that the aircraft could not be operated successfully by any normal commercial standards'.[26]

2.5   American jets

With the exception of the Viscount, the use of British aircraft by BOAC and BEA created a financial burden the extent of which was probably unmatched in any other European airline. BOAC had a particularly unfortunate experience. BEA seems to have fared better with British equipment until the Trident, but thereafter was also unhappy with its aircraft. In its submission to a Committee of Inquiry in 1967, BEA warned that its financial prospects were endangered because the lower air fares which were acceptable to foreign airlines with American aircraft, could hardly be borne by BEA which was operating more costly British equipment.[27] Two years later, when the future of Britain's *airbus* project (the BAC Three-Eleven) looked in doubt, BEA's chairman warned that if the government did not support the development of the Three-Eleven it would have no alternative to buying American aircraft and 'by 1985 its whole fleet would be American'.[28] Prophetic words. (see Table 3.1)

BA's long relationship with Boeing had begun with the propeller-driven Stratocruiser and continued with the 707, but the major impetus came with the introduction of the wide-bodied Boeing 747 'jumbo jet' which entered service with BOAC in 1971. With the decision to buy the 707 BOAC had had some choice since it could have bought the Douglas DC-8, or managed with the Comet 4 until the VC10 was ready. With the Boeing 747 it had no choice whatever if it wanted to remain competitive in the 1970s.

For BEA the situation was less straightforward. In 1971 it had a main fleet composed of Tridents, Super One-Elevens and Vanguards. To replace the early Tridents two wide-bodied aircraft were under consideration: the Lockheed L-1011 TriStar and the European Airbus A300B.[29] To the annoyance of Britain's European partners, it bought the TriStar instead of the Airbus and introduced it in 1974 onto BA's European routes. In 1978 BA announced that it would also buy the Boeing 737 as well as the newly-conceived Boeing 757 as a replacement for the later model Tridents – in the latter case acting, with Eastern Airlines, as a launching customer for Boeing. BA's preference for 757, and renewed rejection of the Airbus, was explained on the grounds that the larger Airbus would have been incompatible with its new TriStar fleet. An additional factor was the use on the Boeing aircraft of the Rolls-Royce RB-211 engine; the Airbus was powered by General Electric engines.[30]

3   NETWORK STRUCTURE

3.1   Routes

The evolution of the British air route system was dictated in the pre-war period by colonial objectives. Imperial Airways extended its network from

England to Cairo, and then eastwards to India, Singapore and Australia, and south to Nairobi and Johannesburg. The final key element in the system was added after 1941 when regular crossings of the North Atlantic began. By 1946 BOAC had re-established the Imperial Airways network and begun a North Atlantic service. After the take-over of BSAA in 1949, routes to South America were added as well. The growth of its route mileage in the next twenty years was relentless: in March 1953 it was 136 563 kilometres, by March 1973 it was 567 745 kilometres.[31]

The large number of destinations served by BOAC was made possible initially by Britain's strong position in air rights negotiations. At the end of the Second World War Britain controlled a string of territories around the world which served as actual or potential air bases vital to the future development of international air transport. Later, network expansion was brought about by greater aircraft range and the consequent use of more direct flights. The growth in direct flights also meant a decrease in the relative importance of historical staging posts such as Rome and Cairo. The trend culminated in very long haul services such as the BOAC polar route through Anchorage, and after 1970, a direct service to Japan through Moscow. While in the 1950s BOAC's network was still composed of a succession of relatively short links, so that a trip from London to Johannesburg required stops in Rome, Khartoum and Nairobi, by 1970 Nairobi could be reached in a single flight.

BOAC's status as an exclusively long-haul carrier meant that it carried comparatively few passengers, in 1960 barely a fifth of the number carried by BEA. Its long, non-connective linear network was good for showing the flag, but less useful for maximizing passenger traffic. In contrast to other European airlines, BOAC managed to obtain landing rights at a large number of gateway cities on the North American continent – 15 by 1972 – yet despite the overwhelming importance of the American market it had difficulty maintaining even the share to which it was entitled because of capacity shortages.

In comparision with the North Atlantic, BOAC showed little enthusiasm for serving South America. Having inherited BSAA's route from West Africa to Rio de Janiero, Buenos Aires and Santiago, BOAC struggled in vain to make it profitable, and abandoned it in 1954 following the Comet crashes. It was resumed in 1960 and in April 1961 a west coast service from Nassau to Lima was added, but problems with traffic rights in Brazil and Argentina made it difficult to break even. In general South America was a political rather than commercial route, and for a long time the Foreign Office encouraged BOAC to maintain services there for prestige reasons. In 1964, having sought and been refused a government subsidy to operate the route, BOAC again withdrew from South America and subsequently the east coast route was awarded to the independent BUA.[32]

As problematic as its South American routes were BOAC's operations on the Pacific. It did not have the capacity to complete its circumnavigation of the globe with a Pacific link until the late 1950s. A transcontinental route from New York to San Francisco was opened with DC-7Cs in 1957 but it was not until August 1959 that the airline had both the means and approval from the US Civil Aeronautics Board to mount a westbound Pacific service through Honolulu and Tokyo to Hong Kong.[33] As with the South American routes, BOAC's Pacific service suffered from poor loads and was difficult to make profitable. The Pacific was an American pond and BOAC's interest in creating a round-the-world link was little more than another prestige objective, and a costly one at that, since no additional fifth freedom traffic could be carried westwards over American territory between New York and Honolulu.[34]

While BOAC laboured under the lingering burden of imperial pretension, BEA focused on bringing air travel to the masses and played an important role in the discovery of continental Europe by the many ordinary Britons who took their first holiday abroad in the 1950s and 1960s. In order to make its short sectors profitable, despite their lower yield, BEA had to fly intensively. Thus in 1957 it was the European carrier with the most flights per day (147), but the shortest average sector distance (438 km); by contrast BOAC had 46 daily flights but an average sector range of 1808 km.[35] In its first decade BEA built up a network with routes radiating out of London to points in a semicircle stretching from Oslo in the north, to Ankara in the east and Gibraltar in the south. New routes were steadily added, but the overall structure remained the same. A major addition in the 1950s was the internal German network, an anomaly of European civil aviation which arose as BEA, in the company of Air France and Pan American, established air services in the zones of occupation administered by the Western Allies. BEA also extended its links to connect with capitals behind the Iron Curtain, starting with Prague (1957), Belgrade (1957) and Warsaw (1958) and culminating in its Moscow service in 1959. By 1972 BEA was serving 76 destinations in Europe, North Africa and the Middle East, its expansion over a quarter of a century reflecting the trend within the air travel industry towards greater emphasis on holiday destinations; in 1947 all BEA's destinations were capital cities, in 1972 at least a third were tourist resorts, the major growth area being the Mediterranean.

The most lasting source of concern to BEA was its domestic network. Britain's internal air routes had grown up with the independent airlines of the 1930s and had been taken over by BEA in 1947. They consisted of trunk routes from London to Manchester, Edinburgh, Glasgow and Belfast, holiday routes to the Channel Islands and the Isle of Man, and so-called 'social' services in the Highlands and Islands of Scotland. Only the trunk routes had the potential to be profitable, but the 'social' services were politically essential in the absence of adequate surface transport in that region. They were inherently unprofitable, with very short sectors, low revenue rates and thin traffic. At places like Barra, in the Outer Hebrides, BEA offered an Australian-style outback service where aircraft had to land on the beach, tide allowing! [36]

Unlike BEA's international services, there was no rapid expansion of the domestic network and some minor routes were abandoned. Some of the trunks showed steady growth in volume, but generally the routes were profitless and BEA was required to provide them without a subsidy. The government held to the view, established with the Swinton plan, that as long as BEA enjoyed monopoly profits on its international routes it should bear the losses incurred on the 'social' services through cross-subsidization. This policy was condemned by successive inquiries, but Ministers felt that industries which were given monopolies should 'take the rough with the smooth' and be willing to perform uneconomic services for the public good in exchange for that privilege.[37] Unlike domestic air transport in the United States, local air services in Britain suffered from a second class image and there seems to have been an underlying assumption that profits from international operations should be used to support them and aircraft handed down to domestic services when they had outlived their usefulness on overseas routes, such as the Viscount and the Vanguard.

BA inherited the enormous network of BOAC and BEA in 1974, building on BOAC's strength in Commonwealth countries and its comparatively large number of North American gateways, as well as BEA's comprehensive structure in Europe. By 1984 it was already the largest international airline in the world in terms of passenger mileage – the basis for its claim that it was 'the world's favourite airline'. In 1994 its scheduled services encompassed 75 countries and 167 destinations, 78 of which were in Britain and Europe, and 22 in North America. In Britain it was running 90 shuttle services per day between London and Manchester, Belfast, Glasgow and Edinburgh, while in Europe it carried 14 million passengers with 700 daily flights on 58 routes. In Europe it also developed a major non-British hub at Paris, with 100 arrivals and departures per day at Charles de Gaulle airport, shared with its partners TAT and Deutsche BA.[38]

3.2 Airports

One of the distinctive characteristics of British air transport has been the strong centralizing influence of London. For a long time air transport in Britain meant air transport in and out of the capital, and neither BEA nor BOAC showed much interest in conducting operations from major provincial airports like Manchester or Glasgow. Not only were their operations heavily centralized at Heathrow airport, but Heathrow has also become one of the most valuable assets for British civil aviation. With the coming in the late 1940s of regular scheduled services on the North Atlantic, London moved from a peripheral to a pivotal position in European air transport; indeed so much international traffic goes through Heathrow that the advantage to the British flag-carrier which is based there is almost incalculable.

Heathrow, like most major European airports, was a postwar creation. After the war the Imperial Airways base at Croydon was considered too small to handle the larger aircraft which were now expected and operations were moved to Heathrow, which had been planned originally as an RAF transport airfield. The airport was opened to international traffic in May 1946, with one runway and a few hastily erected tents for waiting passengers. BOAC moved all its operations there immediately, with BEA following in the early 1950s.[39] It grew steadily with investment in terminal buildings and by the 1980s had acquired its present sprawling appearance. Like Gatwick, positioned to the south of London, and other major airports across Western Europe, it has mutated from a rather lacklustre government department – British airports were originally run by the Ministry of Civil Aviation – to a full-blown private business with over fifty per cent of its revenue coming from non-flying (i.e. retailing) sources. And this process has not been without its critics. Particularly since the privatization of the British Airports Authority (BAA) in 1987 it has been argued that an inefficient public monopoly has been transformed into a profit-driven private one. On the other hand, airports can remain profitable even when the air transport business is losing money, because they are able to spread their fixed costs further as aircraft and passenger numbers rise. As one expert has put it in a description that can readily be applied to Heathrow, 'the more congested, crowded and uncomfortable an airport, the more likely it is to be highly profitable'.[40]

With the merger of BOAC and BEA in 1973, the British flag-carrier's domination of available 'slots' at Heathrow continued to provide it with a valuable bonus. In 1994 BA was the largest operator at Heathrow, accounting for nearly 45 per cent of all passenger movements there. This domination of the airport has been challenged with only moderate success by British competitors like British Midland and Virgin Atlantic.

Other British independent airlines such as BCal mounted their challenge to BA from London's second airport, Gatwick, which, since the 1970s, has actually shown a faster growth rate in passengers than Heathrow.[41] Gatwick originated in the 1930s and benefited from its own station on the rail link between London and Brighton. However flooding and other problems held it back until the late 1950s when it was redeveloped as London's second airport. The key to Gatwick's development, and that of other airports close to London such as Luton and Stansted, has been the rise of non-scheduled airlines specializing in inclusive tour operations, for whom the convenience and lower charges of these airports has been a major advantage.

The centralizing force of London has been balanced to some extent since the 1970s by the growth of provincial airports such as Manchester and the greater freedom that they have been given to develop international services. Various sites have nonetheless been proposed, particularly in the 1970s, for a *third* London airport, to relieve the pressure on Heathrow. Cublington, near Oxford, Maplin Sands and the Isle of Sheppey were all considered and rejected, mainly in the face of sustained local objections. In the end the existing military airfield at Stansted was chosen for major extension and modernization, although it was considerably further from London than Heathrow or even Gatwick. However its lack of appeal to the flag-carriers and BA in particular, despite the opening of an attractive new terminal in 1991, has dimmed its prospects somewhat; it remains to be seen if the regional and holiday carriers which use it at present will be joined by larger international airlines as Heathrow's congestion worsens.

4   PERFORMANCE

The operating and financial performance of BOAC and BEA can be usefully analysed in three chronological sections encompassing the years from 1946 to 1973, and that of BA in two sections from 1974 to 1996.

4.1   Early days: 1946–53

BOAC got off to a difficult start. It had practically no economic aircraft and in 1946/47 recorded a loss of over £7 million on £11.5 million revenue

BEA was in similar difficulties with a lack of proper ground facilities at its Northolt base and obsolescent aircraft. The basic problems, which were to remain with the airlines for the next 30 years, were overmanning and uneconomic aircraft, and the high operating costs that they caused.

Expectations of the air corporations' employing capacity at the end of the war were seriously exaggerated and there seems to have been a pervasive attitude both in government and among the public at large that nationalized airlines would find a job for every demobilized member of the RAF. By 1946 BOAC employed close to 23 000 people. Gradually staff numbers were reduced, despite the absorption of the BSAA labour force in 1949, and by 1951 numbers were down to 16 000. But the incubus of overmanning, born out of BOAC's wartime origins, was to remain. At BEA the early problems centred on inefficient aircraft and low utilization. Its Douglas DC-3 and Vickers Viking aircraft, the latter an interim type based on the Wellington bomber, had difficulty competing with the more modern aircraft of other

Table 3.2 BOAC and BEA operations, 1947–53

Year ended March	1947	1949	1951	1953		1947	1949	1951	1953
	BOAC					BEA			
					millions				
Passengers	0.13	0.12	0.20	0.29		0.07	0.57	0.94	1.40
CTM	59	86	161	210		6	28	53	69
LTM	38	53	98	130		4	18	30	43
					%				
L/Factor %	65	62	60	62		65	63	58	63
					£millions				
Revenue	11.5	15.1	23.5	35.7		1.2	5.3	8.9	13.1
Surplus/				0.06					
Deficit	7.3	6.2	3.3			2.1	2.5	0.8	1.1
					£				
Cost/CTM	0.32	0.24	0.17	0.17		0.55	0.27	0.18	0.20
					thousands				
Staff	24.5	18.9	16.0	17.8		4.2	7.1	7.0	8.5
CTM/staff	2.4	4.6	10.1	12.0		1.4	4.0	7.5	8.1

KEY:

CTM	:	Total Capacity Ton Miles
LTM:	:	Total Load Ton Miles (incl. commercial freight)
L/F	:	Overall Load Factor
Revenue	:	Total Traffic Revenue
Operating Surplus/*Deficit*	:	Before interest on capital and taxation

Source: BOAC and BEA *Annual Reports*, Winston Bray, *The History of BOAC, 1939–1974*, London, 1973, pp. 454–57.

European carriers. They also seem to have been used less intensively. Certainly in comparison with American practice, BEA's aircraft utilization was very low and in a report he made after visiting a number of United States airlines in 1949, BEA's chief project engineer remarked that 'no-one over there could believe we could operate at such low utilizations as we currently achieve and stay in business'.[42]

Nowhere was BEA's cost-to-revenue ratio so disastrous as in Scotland. No commercial airline would have been able to sustain the losses BEA had on these social services, but its obligation to continue them was not negotiable. In 1948 the Minister of Civil Aviation instructed BEA that there were certain domestic routes 'which will be required for as long ahead as we can presently see. It is clearly most desirable that these should be provided by BEA itself.'[43] Knowledge of BEA's domestic services was reasonably widespread by 1950, but there was a strong impression that air travel within Britain was more expensive, and also more complicated, than first-class rail travel.[44] This was something that only time and advertising would change.

4.2   Flying British: 1954–63

In the 1950s, while BOAC struggled to recover from the impact of the Comet crashes, BEA introduced the Airspeed Ambassador (known as the Elizabethan) and the Vickers Viscount. BEA's profits began with the integration of the Viscount into its fleet in 1955 and lasted until 1961 when almost every airline in the world lost money. BOAC stumbled through the decade and ended it with record losses and a managerial crisis. Productivity, measured as capacity-ton-miles (CTM) per employee, did rise steadily at both airlines, with BOAC registering the greater increase thanks to the advantage of long-haul routes (see Table 3.3). But BOAC's costs per CTM were also 25 per cent higher than the industry average.[45]

Table 3.3   BOAC and BEA, operations, 1955–63

Year ended March	1955	1957	1959	1961	1963		1955	1957	1959	1961	1963
	BOAC						BEA				
					millions						
Passengers	0.28	0.39	0.49	0.89	0.97		1.87	2.46	2.82	3.99	4.91
CTM	205	268	377	667	931		98	139	181	237	327
LTM	127	165	202	322	372		63	89	109	155	196
					%						
L/Factor %	63	63	57	56	45		65	64	60	65	60
					£ millions						
Revenue	36.6	48.6	57.9	88.0	92.4		17.1	3.9	31.8	42.3	51.2
Surplus/	1.2	2.1	0.8	4.3			0.06	0.21	0.23	1.54	
Deficit					*4.7*						*0.3*
					£						
Cost/CTM	0.17	0.17	0.15	0.13	0.11		0.17	0.17	0.17	0.17	0.16
					thousands						
Staff	17.1	18.7	18.8	20.6	21.4		9.1	10.4	11.5	13.2	16.1
CTM/staff	12.6	14.4	19.7	33.5	42.2		10.7	13.3	15.7	17.9	20.3

KEY:

CTM	:	Total Capacity Ton Miles
LTM	:	Total Load Ton Miles (incl. commercial freight)
Revenue	:	Total Traffic Revenue
Operating surplus/*Deficit*	:	Before taxation and interest on capital.

Source: BOAC and BEA *Annual Reports*, Winston Bray, *The History of BOAC, 1939–1974*; pp. 454–7.

In 1956 Sir Miles Thomas resigned after eight years as BOAC's chairman, disillusioned by the Comet saga and the task of making profits with British aircraft. He was replaced as chairman by Gerard d'Erlanger who had been BEA chairman briefly in the 1940s and now accepted the BOAC post somewhat reluctantly and on a part-time basis only. The d'Erlanger years were characterized by sagging productivity and renewed losses in 1958. This was partly caused by the problems with the Britannia, but more fundamentally with the airline's costly and overstaffed engineering operation. Attempts to reorganize the engineering department and cut staff ran into opposition from the trade unions which expected privileged treatment from a nationalized industry. An overtime ban was threatened and in October a strike by engineering staff overshadowed the airline's introduction of the Comet 4. Some improvement was made in the turnaround time for aircraft maintenance and overhaul but it was slow progress, with reductions to the overblown staff levels coming only through so-called 'natural wasteage'.[46]

In 1959 BOAC faced severe criticism in the House of Commons when successive speakers claimed that the airline was poorly managed and that its earnings on capital were low. BOAC's aircraft were fast but were not worked hard enough, and offered fewer seats than other airlines. There was a general view that the airline needed more tourist class passengers and had to speed up 'its rate of conversion from a high quality service operator to a mover of mass traffic'.[47] Losses at BOAC mounted precipitously after Sir Matthew Slattery took over the chairmanship in 1960. Moreover the full cost of the aircraft procurement confusion of the mid-1950s now became clear: the short working lives of the Britannia and the DC-7C had been insufficiently accounted for in their rate of depreciation and by 1962 BOAC's accumulated debts, caused by writing down the book values of these two propeller types, plus the Comet 4, had reached £64 million.[48]

Another problem at BOAC, of comparable magnitude to its inefficient engineering department, was its investments in unrewarding subsidiary airlines, to which a considerable portion of its losses was attributable. BOAC's involvement in these airlines was supposed to ensure the flow of feeder traffic onto its own trunk routes as well as promoting British aircraft sales.[49] However any success in this endeavour was vastly outweighed by the cost. The airlines were generally inefficient, but were able to continue operations in the sure knowledge that they were politically important to the British government. BOAC managed to begin the process of untangling itself from this aftermath of Empire and divested itself of its holdings in a number of airlines, although in 1962 its list of subsidiaries still amounted to eleven. Among them was British West Indian Airways (BWIA). BWIA's annual losses were regularly in the region of £400 000, yet BOAC gained little by way of local credit for its involvement and it was unable to challenge the 'inflated organisation and extravagant policies' of BWIA for political reasons. Only when the airline was taken over by the Trinidad government did BOAC rid itself of this particular burden.[50]

Meanwhile BEA, under the robust leadership of Lord Douglas of Kirtleside, enjoyed a steady decade of growth. BEA's short stage lengths inevitably lowered its productivity since short hauls are more expensive to operate, per kilometre, than long hauls. On the other hand BEA had the advantage of European air rates, which were among the highest in the world and Europe's largest traffic source (London) as its main hub. By 1956 BEA was Europe's largest airline in terms of passenger traffic and Lord Douglas could justifiably claim that it was 'now one of the major airlines of the world'.[51]

Like BOAC, BEA suffered from high operating costs, but its need to reduce them was even greater because short-haul airlines have lower average yields per passenger and have to rely on narrower operating margins. The other reason why BEA needed to pay close attention to costs was the increasing competition it faced from low-cost charter carriers. By the end of the 1950s the importance for BEA of the fast-growing European holiday market was clear. Non-scheduled British airlines were already taking a substantial share of the traffic to holiday resorts in the Mediterranean. If BEA was to gain a share of that market in the future it had to develop cheaper forms of service, noted BEA's commercial director in 1958, and that meant lower costs.[52] Table 3.3 shows that BEA's unit costs fell only slowly. And like BOAC, a major area of concern was its aircraft maintenance and overhaul operation, but it also had generally low aircraft utilization and labour productivity. In 1959 BEA's total operating costs were about 45 per cent higher, per CTM, than American domestic airlines with comparable average stage lengths, despite the higher wage levels in the United States. BEA's lower aircraft utilization was partly the result of the great difference it faced between summer and winter traffic levels, while the Americans' higher productivity stemmed from their use of larger aircraft which produced a greater CTM-per-hour rate with the same crew strength.[53]

4.3 Finding the mass market: 1964–73

In 1964 there was a change of leadership at both BOAC and BEA. At BEA the retirement of Lord Douglas, after a chairmanship of 15 years, was crowned by the airline's most successful year, and he was replaced in a seamless transition by the long-serving chief executive Anthony Milward. In sharp contrast, the transformation at BOAC was painful and acrimonious. Sir Matthew Slattery had taken over a highly troubled airline, and had to face a financial situation not of his own making. With mounting losses in 1961 criticism of BOAC became so loud and persistent that the Minister (Julian Amery) for forced to order a confidential report on BOAC's management. The White Paper that was based upon its findings charged the airline

Table 3.4 BOAC and BEA, operations, 1965–73

Year ended March	BOAC 1965	1967	1969	1971	1973		BEA 1965	1967	1969	1971	1973
			BOAC			millions			BEA		
Passengers	1.23	1.50	1.58	1.89	2.81		6.12	7.32	7.73	8.66	9.57
CTM	1150	1347	1581	1996	2922		435	547	590	699	851
LTM	521	698	787	975	1404		254	324	349	378	461
Load factor	49	54	51	49	52	%	58	59	59	54	54
Revenue	114.4	137.4	169.6	195.5	267.7	£ millions	66.2	87.0	117.6	152.4	178.8
Surplus/	16.8	23.2	21.9	5.9	22.7		4.1	4.4	1.8	11.4	
Deficit											0.3
Cost/CTM	0.09	0.08	0.09	0.09	0.08	£	0.15	0.16	0.18	0.20	0.21
Staff	19.6	19.0	20.9	24.1	24.7	thousands	17.7	20.2	22.2	24.8	24.9
CTM/staff	57.3	70.3	76.6	85.1	117.8		24.6	27.1	26.6	28.1	34.1

KEY:

CTM   : Total Capacity Ton Miles
LTM   : Total Load Ton Miles (incl. commercial freight)
Revenue   : Total Traffic Revenue
Operating Surplus/Deficit : Before taxation and interest on capital

Source: BOAC and BEA *Annual Reports*; British Airways, *Annual Report and Accounts 1972–73*; pp. 70–1, 118; Winston Bray, *The History of BOAC, 1939–1974* London, 1974: 456–8.

with a whole range of failings from 'unduly optimistic' traffic forecasting, to 'ineffective financial control' and a persistently 'excessive' level of maintenance costs.[54] The criticism was so devastating that both Slattery and the managing director, Sir Basil Smallpiece, felt they had no option but to resign. Their departure in November 1963 was overshadowed in the press by President Kennedy's assassination, but was nonetheless such a shock that it precipitated a general exodus from the BOAC board over the next 12 months.

Slattery was replaced by Sir Giles Guthrie. His task seemed clear: it was to get the airline working harder. The productivity of the new jet aircraft was so high that airlines needed fewer pilots and flight crew. Staff cuts were now pushed through, although after 1966 the trend in BOAC's numbers began to rise again. The airline's finances, which had been in a mess when Guthrie took over, were gradually sorted out, due partly to the efforts of the accountant Charles Hardie who had joined BOAC's management with Guthrie and was to replace him as chairman in 1969. Guthrie also seized the nettle which Slattery had declined to grasp and negotiated a reduction of BOAC's VC10 order with Vickers. In 1966 Guthrie was able to announce a record operating surplus, predicting that BOAC was 'in the right shape for the future'.[55]

Beyond Guthrie's reforms, BOAC's recovery was attributable to rising passenger traffic combined with improved marketing during an industry-wide boom in the latter half of the 1960s (Table 3.4). However the reforms did not go deep into BOAC's underlying management culture. Although the corporation made good profits for the remainder of the decade, the financial improvement came largely from the retirement of the Britannia and the Comet 4, and the government's write-off in 1965 of BOAC's crippling £80 million debt. Underlying productivity deficiencies were not seriously tackled. A strike by BOAC pilots in the summer of 1968 drew attention to the fact that BOAC used them less, and paid them less, than other airlines. Pan American pilots received three times the salary of their BOAC colleagues for flying identical aircraft, but they flew 960 hours per year compared to the BOAC average of 550 hours. The British pilots were actually willing to work longer hours, but management was not ready for them to do so – it was perhaps one of the few strikes on record in support of a claim for longer working hours.[56]

4.4   The fruitless merger: 1974–80

The 1950s and 1960s were a golden age of growth and low inflation in the major western economies. Then came the 1973 oil shock, a major recession and much stronger inflation. This was also the moment when, after nearly 30 years of separate existence, it was decided to merge BOAC and BEA to form a single flag-carrier. The merits of a merger had been raised many times over the years, usually on the grounds that a pooling of resources and know-how would benefit both carriers. But in the 1960s there was a strong general trend towards industrial concentration and the merger should be seen against this background. In 1972 a British Airways Group was established to bring together the top management of both airlines and the following year the British Airways Board recommended a full merger of BOAC and BEA under the name British Airways (BA).

The new airline had the most comprehensive route network in the world and one of the biggest fleets. It also had a huge staff and a recent history of labour disputes. The indifferent performance of BA between 1974 and 1980 was the result not only of sharply rising costs caused by higher fuel prices, but also an unwillingness to tackle problems in the labour force. Indeed one of the first actions of the merged BA management was to 'give an undertaking to our 48,000 staff in the UK that they would not suffer lack of continuity or security of employment as a direct result of the merger'.[57] It was hoped that BOAC and BEA's respective problems would be solved without redundancies through the mere fact of fusion; the labour-saving rationalization which should have flowed from it seems to have been disregarded.

Table 3.5   BA, operations, 1975–80

Year ended March	1975	1976	1977	1978	1979	1980
			millions			
Passengers	13.34	13.79	14.51	13.37	15.77	17.32
ATK	5388	5856	6233	6408	7164	7797
LTK	2997	3249	3607	3711	4416	5035
			%			
Overall L/F %	56	55	58	58	62	65
			£millions			
Revenue	663.5	801.5	1073.9	1156.1	1403.3	1654.4
Surplus/		2.7	95.8	56.8	76.0	16.0
Deficit	*1.8*					
			£			
Cost/ATK	0.11	0.13	0.15	0.16	0.17	0.20
			thousands			
Staff	54.8	53.9	54.3	55.4	55.9	56.1
ATK/staff	106.3	115.7	120.6	122.5	134.9	145.2

KEY:

ATK	:	Total Available Tonne Kilometres
LTK	:	Total Load Ton Kilometers
Revenue	:	Total Traffic Revenue
Operating Surplus/*Deficit*	:	Before taxation and interest on capital

Source: BA Annual Reports

BA's productivity growth remained low, while passenger numbers rose at a pedestrian pace, actually falling in 1978 (see Table 3.5). One of the causes of this poor performance was the wide variety of aircraft in its fleet and the high operating costs of its remaining British types, that is, the Trident, Super VC10 and BAC One-Eleven. The effects of the 'buy-British' policy in the 1950s still impacted on the flag-carrier's accounts and BA's management admitted that the American aircraft which it had acquired were more economical on fuel, more productive and more reliable.[58] When one compares BA's performance with that of the best US airlines one sees the ground that BA would have to cover in order to turn itself into an efficient airline. In 1977 Delta Airlines carried over twice as many passengers as BA (30.7 million against 13.3 million) with little more than half the number of staff (31 000 against 56 000) and the same number of aircraft. More critically, Delta's maintenance productivity was four to five times greater than BA's, proof that the engineering problems that had plagued BOAC and BEA had still not been resolved.[59]

In their defence BA claimed in a 1977 study for the CAA that their high costs derived from the nature of European air travel: higher fuel costs and landing fees, night jet bans in Europe and the demands of multi-market advertising in Europe.[60] Nonetheless it performed poorly even in comparison with other European airlines. In 1978, from a sample of ten American and European carriers, BA had the lowest output (see Table 1.1 in Introduction)

BA's performance in the 1970s reflects the nature of the BOAC-BEA merger as a political rather than economic move. The airline acquired a new livery, featuring a stylized version of the national flag, but it did not get the more commercial workplace culture that might have accompanied the merger process. By 1980 BA had a reputation among international airline passengers for poor service and the jibe that the initials BA stood for 'bloody awful' was an indication of the airline's badly tarnished image. When BA eliminated the word 'airways' altogether from its name on aircraft and tickets, it seemed to symbolize its failure in the air transport business. 'Stunned travellers', suggested *The Economist*, 'to whom it had not occured that BA's big problem was its name, are now well primed to look for more change from the airline: lower prices, punctual flights, cleaner planes, polite and helpful service...'[61]

4.5   Into private ownership: 1981–90

In 1979 a new Conservative government under Margaret Thatcher was returned to power. An important part of its programme was the privatization of nationalized industries and BA was among the targets. But could this overweight airline be made efficient enough to attract private investors?

Some steps had been taken to improve BA's results; for example a financing deal with the government had relieved BA of the depreciation costs associated with the Concorde and attempts to force BA into accepting more major British aircraft had been resisted. The big challenge now was to raise the productivity level among its strike-prone labour force.[62]

In July 1979 plans were announced to privatize part of BA, based on a model similiar to the oil company BP. But within a year the airline industry went into a nosedive and BA's profits crashed. With sharply rising fuel costs and falling traffic growth, the airline was knocked off course. Plans to replace half its ageing aircraft fleet were delayed, the cuts in the labour force necessary to raise productivity were abandoned and privatization put on hold by the chairman, Sir Ross Stainton. By the spring of 1981 BA was heading for a monumental loss (Table 3.6) and seeking extra government funds.

This was the background to the search by the government in the summer of 1980 for a new BA chairman. The successful candidate needed to be an outsider and someone who could take the hard decisions necessary to get BA into shape for privatization. By the end of the year the choice had fallen on

Table 3.6   BA Group operations, including Caledonian, TAT and Deutsche BA, 1981–95

Year ended March	1981	1983	1985	1987	1989	1991	1993	1995
					millions			
Passengers	16.98	16.34	18.40	20.04	24.60	25.59	28.10	35.64
ATK	7930	7208	7837	8751	11868	13351	15424	18224
RTK	4812	4461	5267	5784	8002	8979	10313	12380
					%			
Overall l/f %	61	63	66	65	67	67	66	68
					£ millions			
Revenue	1760	2051	2905	3245	4257	4937	5566	7177
Surplus/*Deficit*	102	169	292	173	336	47	310	618
					£			
Cost/ATK	0.23	0.26	0.30	0.30	0.30	0.30	0.31	0.32
					thousands			
Staff	53.6	45.9	38.1	40.7	50.2	54.4	48.9	53.1
ATK/staff	153.7	181.6	212.6	221.6	236.4	245.3	315.0	343.5

ATK	:	Available Ton Kilometres
RTK	:	Total Revenue Ton Kilometres (incl. commercial freight)
Revenue	:	Total Traffic Revenue
Operating Surplus/*Deficit*	:	Before interest on capital and taxation

Source: BA Annual Reports.

Sir John King, a former ball-bearing manufacturer and a self-made million-aire. Having been made a peer by Mrs Thatcher, Lord King launched a revolution at BA. Attitudes which had typified its management since the Second World War were purged. Massive job cuts were announced, made palatable by an equally massive compensation scheme for those hit by redundancy. Applying the simple criteria of whether or not they made profits, King closed routes, maintenance stations and staff training colleges, as well as BA's all-cargo services. The unions, conscious of the airline's losses and tempted by the generous compensation package, acquiesced. Moreover BA's senior management was not spared the axe and 50 out of 240 managers took early retirement.[63]

In 1983 Colin Marshall, a former executive of the American car rental firm Avis, became BA's chief executive. It was Marshall who set about repairing BA's neglected customer image by shifting the central focus of the airline to consumer service. Particular effort was made to recapture the high-revenue business travellers who had deserted BA by the plane-load in the 1970s. To assure passengers that a different spirit ruled at BA, the airline was relaunched with the help of a new livery and the slogans of the high-flying advertising agency Saatchi & Saatchi.

BA was soon back in profit and productivity improved. Close to 20 000 staff were shed and nearly £200 million spent on redundancy payments. Loss-making domestic routes which had plagued the airline for so long were now gladly handed over to independent competitors such as British Midland. At the end of 1983 the decision was taken not to take BA on the BP route to privatization, but to sell one hundred per cent of the airline and use the money (£800 million) to pay off that part of BA's huge debt which was owed to private banks. In 1984 and 1985 BA had record profits, while productivity continued to rise (Table 3.6). In February 1987 BA's shares were finally floated on the London Stock Exchange and the airline passed success-fully into private ownership. It proved to be a good investment and in 1988 it was the world's most profitable airline with pre-tax profits of £228 million. '*The world's favourite airline*' was truly airborne.[64]

5   STRATEGY

British policy on international civil aviation in the years between 1945 and 1980 stressed *orderly* growth and development. The state-owned airlines BOAC, BEA and BA may have paid lip-service to the idea of competition, but in practice the only rivals they tolerated were designated foreign flag-carriers with whom they served common routes under a regime of strict capacity and price controls. They were innovative in the field of creative fares, particularly when faced with the challenge from non-scheduled carriers, but remained stalwart supporters of the IATA cartel and of the postwar regulatory system worked out at Chicago (1944) and Bermuda (1946).[65] As state-owned enterprises, in common with most other European airlines, they relied on government agreement rather than market forces. In comparison with the privately-owned airlines in the United States, they were high-cost businesses which sought to maintain their financial viability through tight control of the market and bilateral agreements with foreign competitors.

5.1 Collaboration

BOAC, BEA and BA collaborated rather than competed with foreign flag-carriers and this applied as much to European airlines such as Air France, KLM or Lufthansa, as it did to their fellow operators on the North Atlantic, Pan American and TWA. Through bilateral agreements between governments they constructed a regulatory regime under which the number of carriers on a route, the capacity they offered and the fares they charged, were all controlled. The guiding principle was the avoidance of overcapacity, whether through too many airlines, schedules that were too frequent, or aircraft that were too large. The British airlines sought to maintain their load factors at a high level, hopefully high enough to cover their unit costs. The alternative approach, which the Americans generally favoured and the British generally rejected, was to remove restrictions on capacity, accept lower loads while cutting costs, and stimulate demand with reduced fares.[66]

This is not to say that the British were against cheap scheduled fares, they were not. However their approach to fare reduction through the IATA conference mechanism centred on promotional inducement rather than across-the-board cuts. They were concerned to reduce the enormous seasonal variation in passenger traffic and they sought to flatten the summer peaks by luring passengers onto their winter and 'shoulder' timetables with special reduced rates. In the late 1960s and 1970s BOAC championed the advance purchase concept (APEX), against the contrary positions of many of its IATA allies, but here again the aim was not a general fare-cut as much as the chance to raise load factors and meet the challenge of the affinity and inclusive tour charter operators.

Collaboration permeated European air transport, and BEA and BOAC were an integral part of the system. They both had links to other airlines, particularly in the 1950s, which ranged from wholly-owned subsidiaries like Gibraltar Airways (BEA) to more limited stakes in Aer Lingus and Alitalia (BEA) and to technical agreements and maintenance pools with Olympic and TAP (BEA). The object was either to enhance the sales prospects of British aircraft with the target airline or fulfil political objectives, for example the control of airlines in Commonwealth territories (BOAC).

In Europe one answer to the continent's air transport needs was a single European airline. The only serious attempt to create a unified European carrier was the *Air Union* experiment undertaken in 1959 by Air France, Lufthansa, Sabena and Alitalia. BEA's membership would have been logical, but the British showed little interest in joining such a formal arrangement and Lord Douglas was sceptical that his airline would benefit from fusion with other European flag-carriers, believing that BEA was big enough to stay outside.[67] BEA never wanted a close association with other European airlines because they were all long-haul as well as short-haul carriers, and giving them priority at Heathrow airport would have harmed BEA's interline traffic with BOAC. Instead BEA championed pooling. After initial scepticism in the late 1940s, BEA became an enthusiastic advocate of the pool as a means by which two flag-carriers could enjoy orderly growth and profitability without unnecessary competition. In 1953 BEA had only two pool agreements, by 1960 it earned over half its revenue from such arrangements and Lord Douglas acknowledged that pools were 'in large part responsible for BEA now being in the dominant position in Europe'.[68] One of the reasons why BEA was so vehemently opposed to the growth of independent airlines in Britain was that their competition with BEA for scheduled traffic in Europe meant exceeding the pool quota agreed with the receiving nation and the necessity for BEA to surrender some of its traffic entitlement to the independent. According to BEA's calculations, its strong negotiating position in European traffic pools was eroded by the rights given to British independents.[69]

5.2 Competition

The main competition that BOAC and BEA faced came from these independent British airlines. They had grown up in the 1950s, often with the help of finance from shipping companies which were anxious to diversify. Beginning with simple charter work and trooping contracts for the Ministry of Defence, these airlines eventually became a major threat to the state-owned corporations.

By 1959 BOAC was sufficiently concerned about competition from British charter operators to complain about their low fare services to colonial destinations.[70] These 'colonial coach' services aimed at a new low-fare market that was exploited more fully with inclusive tours operations in the 1960s, but BOAC had already taken a confrontational position on the grounds that independents were creaming off cheap-fare passengers, when the market should have been developed 'in an orderly manner, as and when justified by the economics of the air transport industry as a whole'.[71] BEA took an identical line and protested frequently about the diversion of its traffic to independents. At the time of his retirement in 1964 Lord Douglas said of

British independents' aspirations to fly scheduled services in competition with BEA that they 'have got their sphere and their job in charters and inclusive tours, as well as trooping and car ferries; there is plenty for them to do. One thing I resent them doing is trying to muscle in on BEA's business, on scheduled services, that we have built up over a good many years now...'[72]

Douglas's argument was largely a political one and hinged on the fact that BEA met its losses on domestic 'social' services out of profits from its international network. Competition from other British airlines reduced those profits and thus BEA's ability to cross-subsidize. Douglas believed in nationalized airlines and accepted their duty to fulfil certain non-commercial obligations, even though it made more sense to let the independents try and make a profit where BEA had made a loss. In November 1963 British Eagle did start a service from London to Glasgow and Belfast, using Britannia aircraft acquired from BOAC. It was BEA's first domestic competition and it found itself having to put jet aircraft on the routes, where it had planned to use Vanguards. Having tried to stop British Eagle and lost, BEA declared war on the independent, and 'sandwiched' its flights.[73]

The biggest challenge for BEA in the 1960s came from British charter airlines specializing in inclusive tours to the Mediterranean. Britannia Airways is a good example. It began modestly in 1962 at Luton airport and within five years had become a major competitor on routes to Spain. BEA fought to restrict the expansion of Britannia and other charter carriers, and appealed in vain against the granting of their operating licences by the ATLB. If one looks at the impact of inclusive tours on scheduled European air fares, one gets some idea of the pressure on revenue experienced by BEA on routes to holiday destinations. In the summer of 1968 routes to Amsterdam, Brussels, Copenhagen, Frankfurt and Milan, where inclusive tour operations were minimal, had scheduled fares averaging 5.34 pence per mile. By contrast routes to holiday destinations such as Las Palmas, Malaga, Palma, Lisbon and Malta, where inclusive tour competition was intense, had average scheduled fares at only 4.00 pence per mile.[74]

In the 1970s BOAC and BEA felt competition from a charter airline that had become a scheduled rival and the government's chosen *second force* airline. Caledonian Airways was founded by Adam Thomson in 1961 and began flying affinity charters on the North Atlantic and inclusive tours in Europe. After it acquired Boeing 707s in 1968 it was in direct competition with BOAC, its £189 round-trip to Los Angeles being cheaper than a one-way scheduled fare to New York. In November 1970 Caledonian bought the larger British United Airways (BUA) to form British Caledonian (BCal) and the new airline started scheduled services on BOAC's West African and Libyan routes, which were transferred to it, despite protests from BOAC, as part of the *second force* programme.[75] In 1973 BCal opened scheduled services to New York and Los Angeles, but within a year it was in serious trouble and after 1974 it abandoned the route, cut its staff and grounded many of its aircraft. BCal recovered from this inauspicious start, building up its aircraft fleet and its Gatwick airport base until it had about 15 per cent of the British air travel market in 1980. At this time BCal sought to acquire more routes from the loss-making BA and was supported in this by the CAA. However, despite energetic lobbying in Westminster by Sir Adam Thomson in 1984, BCal was unsuccessful. The government accepted Lord King's argument that any further route transfer from BA to BCal would jeopardize the privatization launch. Unable to grow and in serious financial difficulties, BCal eventually fell prey to a take-over bid by BA in 1988.[76] The BA-BCal merger was investigated by the Monopolies Commission, but BA convinced the government that swallowing the independent was necessary to build the strength necessary to stand up to the newly-emerging American 'megacarriers' American Airlines, United and Delta.[77] The alternative for BCal had been a take-over by SAS, and that option was defeated by BA's jingoistic assertion that the airline could not be allowed to fall into foreign hands. Thus by the end of the 1980s a significant concentration in the British air transport market had taken place.[78] Only the appearance of Virgin Atlantic prevented BA from achieving a total long-haul monopoly.

The most memorable, if shortlived, of BA's British competitors was Sir Freddie Laker. He began his career in the aviation business after the Second World War and was managing director of BUA in the early 1960s at the time when Adam Thomson was establishing Caledonian. After leaving BUA Laker set up his own airline, specializing in affinity charters on the North Atlantic. In the early 1970s when BOAC and IATA were agonizing over how to compete with charter airlines like his, Laker came up with the inspirational idea of a walk-on, no-frills scheduled airline – *Skytrain*. Initially BOAC was not opposed to the idea; Laker was in any case blocked by the authorities on both sides of the Atlantic and BOAC was busy trying to launch the APEX fare concept. However by 1975 BA was fighting Laker both in the courts and (unsuccessfully) through the CAA. In 1977, when it was clear that Skytrain was going to be licensed for the London–New York route, BA met with a number of other carriers at IATA's headquarters to concoct a fare regime with which to attack both Laker and other charter operators, who by this stage were carrying nearly 30 per cent of all passengers from Europe to North America.[79]

Skytrain was launched in 1978 and proved to be an immediate success. Load factors were very high and during its first financial year total ticketing, sales and promotion costs came to less than 5 per cent of its operating expenditure. By contrast, the percentage for BA that year was 19 per cent. Clearly the overstaffed BA had cause to worry. In the next three years, however, Sir Freddie pushed his under-capitalized enterprise too far and too fast and did so at a time when the airline industry was experiencing one of its periodic recessions. By 1981 he was drastically over-committed to new aircraft purchases and was chasing the passengers he needed to fill Skytrain's burgeoning capacity. His load factors fell, his costs rose and he was flying more or less on the goodwill of his bankers.[80] Laker's final demise has been exhaustively analysed. Certainly he was largely reponsible for his bankruptcy in 1982. On the other hand Skytrain's end was hastened in the autumn of 1981 by what is generally agreed to have been collusion between BA, Pan American and a number of other airlines, including Lufthansa, TWA and BCal, the object being to clip the wings of the tiresome airline with 'predatory' pricing. 'They are running their flights on my routes at a loss to put me out of business', said Laker bitterly.[81] The fact that BA's transatlantic fares rose quickly after Skytrain's crash, suggests that he may have been right. The final episode in the Laker story is his court case against BA, which delayed the airline's privatization, probably by up to two years. BA initially scorned Laker and his lawyers when they took action under American antitrust laws, accusing BA and others of conspiring to put Skytrain out of business. But the case dragged on for three years, with BA facing both civil and criminal investigations, and enduring a considerable amount of adverse publicity as a result. By 1985 BA had had enough and was ready for an out-of-court settlement with Laker, aided by a high-level intervention on its behalf by Mrs Thatcher with US President Ronald Reagan.

If British Caledonian was the second force intended to compete with BA on a playing field laid out by the government, and *Skytrain* the loose cannon unleashed upon a hidebound industry by a irrepressible entrepreneur, then Richard Branson's Virgin Atlantic was a determined attempt to start an independent scheduled airline from scratch. From the financial perspective, Branson is more usefully compared with Howard Hughes, the millionaire who controlled TWA in the 1950s, than with Freddie Laker, although it is understandable that Branson is seen by many as the inheritor of Laker's mantle in so far as they both challenged the hegemony of British Airways.

Virgin established itself in the early 1990s as the only British long-haul competitor to BA. Starting with the North Atlantic and concentrating on the most lucrative intercontinental services, Virgin had built up a significant stake in the British air travel market by 1995. For some time BA refused to take Virgin seriously, perhaps because of its owner's flamboyant lifestyle. Virgin's financial collapse was regularly forecast until Branson sold his highly successful record firm for over £500 million; thereafter it was clear that, unlike Laker, Virgin's owner was not going to run out of funds. At this stage competition between BA and Virgin was clouded by what became known as the 'dirty tricks' affair. BA, anxious to neutralize Virgin's challenge appears to have used methods reminiscent of the Watergate Affair and certainly inappropriate for the nation's flag-carrier, even one in private ownership.

In what proved to be an unsuccessful campaign, Virgin's computers were tapped and its customers lured to BA with gifts of First Class flights. Branson retaliated with a legal onslaught the equal of Laker's a decade before. The affair ended badly for BA when the courts affirmed Branson's charges and forced Lord King to make a public apology. Subsequently Branson's costs and compensation of £3 million were paid by BA.[82]

Besides competition from other airlines, BA also faced a rival transport mode in the British and Continental surface transport systems, although competition from rail transport was considerably less than that experienced by other European airlines. The English Channel was a gift to the British airline industry. When Dutch or German tourists travelled to the Mediterranean they could take the train and stay in the same compartment until they reached their destination, when the British wanted to get to Spain they had to fly or else face the inconvenience of a cross-Channel ferry. Only the start of Eurostar rail services through the Channel Tunnel in 1994 diminished this bonus, but even so, the flying habit was so deeply entrenched that BA (and Air France) held their own against High Speed Train (HST) competition between London and Paris. Train travel is enjoying a renaissance and in the future HST services could eventually link every major European city, much as air networks do. However, airlines will respond with improved reservation and check-in systems, with better surface links between airport and city centre, and with more inner city airports for use by short-range feeder aircraft.

Part of the problem of judging the degree of competition between air and rail in Europe is that both have been state owned, and unfettered competition between the rival modes has rarely been permitted. Domestic air services in Britain functioned for a long time within a political straitjacket and air transport fare regimes have often been arranged to minimize their impact on the rail sector with its large capacity for providing employment. As early as 1960 concern was voiced in Britain that BEA's plans for expanding domestic trunk services and introducing lower fares would conflict with British Rail's modernization programme. Lord Douglas denied that there was any serious clash of interest, but BEA's trunk services were certainly the routes on which British Rail (BR) stood most chance of making a profit. When BEA applied in September 1961 to run a £3.3 winter night service from London to Glasgow, Belfast and Edinburgh, it was strenuously opposed by BR. It was the first time that an air ticket was below the price of a second-class rail fare and BR's chairman, Dr Beeching, took a personal interest in the appeal against BEA before the British Transport Commission.[83]

5.3 Deregulation and globalization

The 1980s represent a period of great upheaval in the history of international civil aviation. 'Deregulation' became the buzz-word of the decade and was offered as a cure for all manner of economic ills, especially in the transport sector. In Britain the combined effects of American deregulation in 1978 and the coming to power of Margaret Thatcher's Conservative government in 1979, had a transformatory impact comparable to the creation of BOAC in 1940, or even Imperial Airways in 1924.

Deregulation and privatization are intrinsically linked, although this is not to say that the one is dependent on the other. It is an obvious fact however that the liberalization of European airline regulation, which was implied by US deregulation, made little sense, and could make little headway, as long as Europe's flag-carriers were state owned and protected from the full force of the market. The irony for BA was that in their case the two ideas championed by the Thatcher government – the privatization of state-owned enterprises and the encouragement of open competition in the economy – were difficult to reconcile. Lord King recognized that only the promise of monopoly power and profits would make BA attractive to private investors and thus ensure the success of the airline's privatization. And having received the support of the government in strengthening BA's competitive position against foreign airlines, he fought remorselessly to eliminate competition from British rivals such as BCal and Virgin. His tactics may have varied, but his strategy and objective remained the same: BA had to evolve from a state-owned flag-carrier into a publicly-quoted national champion.

Having privatized BA, King and Marshall set about making the airline a global player. It was already the largest international airline in terms of routes and had a strong position on the North Atlantic, but now it needed to expand in those markets which offered the greatest potential for the future, namely Europe, North America and the Asian Pacific rim. The object was to make BA strong enough to withstand competition from the largest of American megacarriers; size would ensure the airline's survival in an age of mergers and concentration. The path to achieving this was through strategic alliances with other airlines. Initially it sought to establish a European hub in Brussels with a three-cornered share deal with KLM and the ailing Belgian carrier SABENA. The British Monopolies Commission blessed the move in 1990, but it fell apart because of SABENA's financial difficulties and a lack of agreement on how much the deal was worth. This was perhaps a missed opportunity since Brussels may have been one of the few sites which BA could have developed as a competitive Continental hub, owing to the relative weakness of the flag-carrier (SABENA) and the city's importance as the administrative capital of the European Union.[84] Undeterred BA sought and found other partners in its three target areas. In Europe it took major stakes in both French and German airlines: in TAT, a French regional carrier, and in what became Deutsche BA. On the Pacific rim it took a share of the Australian flag-carrier Qantas, which had been a traditional partner of BOAC and even of Imperial Airways in the pre-war era. In North America it secured a link with a major domestic carrier when it made a major investment in USAir in July 1992, exchanging cash, which USAir badly needed, for access to the American carrier's dense eastern seaboard network.[85]

To match its global alliance strategy, BA set about 'branding' its services in a much more rigorous manner. Indeed BA was among the first European airlines to embark on the process of converting itself from a national flag-carrier into a branded service industry. By the end of the 1980s it was already offering its products as distinctive brands, or levels of air travel service, with price differentiation to match the increasingly deregulated market. Its Club brands – Club World and Club Europe – represented a comprehensive appeal towards the business traveller, with priority reservation and check-in facilities, while its World Traveller, Euro-Traveller and Super Shuttle brands encouraged loyalty among economy class passengers. After fifty years of trial and error with different styles of management and government control, BA at last seemed to have the right aircraft and the right routes; the challenge for the future, which the branding concept aims to meet, is to get the right passengers.

NOTES

1. See Eric Birkhead, 'The financial failure of British Air Transport Companies 1919–1924', *Journal of Transport History*, vol. 4, no.3, 1960; 133–45. The creation of Imperial Airways was recommended by the 1923 Hambling Report, Civil Air Transport Subsidies Committee. *Report on Government Financial Assistance to Civil Air Transport Companies*, Cmd 1811, 1923; para. 14–16.
2. For a history of Imperial Airways, Robin Higham, *Britain's Imperial Air Routes, 1918 to 1939*, London, 1960, pp. 109–70.
3. *Report of the Committee of Inquiry into Civil Aviation* (the Cadman Report), Air Ministry, 1938, Cmd 5685; 13, para. 34.
4. See Peter J. Lyth, 'The changing role of government in British civil air transport, 1919–1940', in Robert Millward and John Singleton (eds), *The Political Economy of Nationalisation in Britain, 1920–1950*, Cambridge UP, 1995, pp. 74–80. For a history of British Airways, see Robin Higham, 'British Airways Ltd., 1935–1940', *Journal of Transport History*, vol.4, no.2, 1959, pp. 113–23.
5. The route to Portugal was operated to a large extent by KLM on BOAC's behalf. Winston Bray, *The History of BOAC, 1939–1974*, BOAC, 1975, pp. 19–22.
6. *British Air Transport*, Cmd 6605, HMSO, March 1945; para. 23. See also Lord Swinton's Memorandum to the Cabinet, Lord President's Committee, LP[45]101, 10.5.1945, PRO AVIA.2/2760.
7. See Peter J. Lyth, 'A multiplicity of Instruments: the 1946 decision to create a separate British European Airline, *Journal of Transport History*, XII, 2, Sept. 1990.
8. Very little has been published on BSAA. For a useful short history, see Don L. Brown, 'British South American Airways: Some Recollections', *Air Pictorial Magazine*, November 1974.

9.   The BOAC figures include grants paid before the 1946 Act and subsidies to BSAA, *BOAC & BEA Annual Reports & Accounts*.
10.   *Civil Aviation (Licensing) Act*, HMSO, 1960, s.2 (2).
11.   *British Air Transport in the Seventies*, Report of the Committee of Inquiry into Civil Air Transport, (the Edwards Report), HMSO, May 1969; pp. 257–67.
12.   *The Economist*, 2.8.1975, 'Airlines: no to enterprise', 81–2; 14.2.1976, 'Airlines: Shoring up BA, 100. See also the White Paper, Cmnd 6400, HMSO, February 1976.
13.   Civil Aviation Authority, *Airline Competition Policy*, CAP 500, London, 1984.
14.   *Flight Magazine*, 24.10.1952, 'Pan Am Order Comet 3s', p. 525.
15.   BOAC, *Annual Report and Accounts*, Year ended March 1955, p. 7.
16.   *Flight Magazine*, 23.11.1956, 'The hungry airlines', p. 832.
17.   Ronald Miller and David Sawers, *The Technical Development of Modern Aviation*, London, 1968, p. 43.
18.   Anthony H. Milward, 'Wasted seats in air transport', *Journal of the Institute of Transport*, May 1966, p. 362.
19.   *The Economist*, 25.4.1964, 'Last of the Subsonics', p 399–400.
20.   A.H. Milward, 'CX[58]19, BEA Vanguards', 22.8.1958, BEA Board Paper, No.132; 'DH.121', 18.6.1959, Note by Chief Engineer, BEA Board Paper, No. 143.
21.   Keith Hartley, *A Market for Aircraft*, IEA, 1974, p. 16, Keith Hayward, *Government and British Civil Aerospace*, Manchester, 1983, pp. 33–4. *The Economist*, 11.7.1959, 'BEA Jet: de Havilland dilemma'.
22.   Christopher Harlow, *Innovation and Productivity under Nationalisation: the First 30 Years*, 1977, p. 43.
23.   A.W.J. Thompson and L.C. Hunter, *The nationalised transport industries*, 1973, pp. 61–6. *The Aeroplane*, 11.8.1966, 'BEA: what happens next?'. *Flight International*, 3.7.1969, 'BEA's Compensation'.
24.   *The Economist*, 18.1.1964, 'The Wastelands of Aviation', p. 229.
25.   A number of accounts of the Concorde story have been published, see for example John Costello and Terry Hughes, *The Battle for Concorde*, Salisbury, 1971. Also Geoffrey Knight, *Concorde: the Inside Story*, London, 1976.
26.   Department of Trade, *British Airways Financial Structure*, Report of Review Group, February 1979, p. 4; British Airways, *Annual Report*, Year ended March 1974, p. 12; *Annual Report and Accounts*, Year ended March 1976, p. 6.
27.   BEA Submission to the Board of Trade Committee of Inquiry into Civil Aviation, 25.8.1967. Box 341, RAF Museum, Hendon.
28.   *Flight International*, 3.7.1969, 'Sensor, BEA's compensation'. 4.9.1969, BEA: Prospects and Preoccupations, pp. 344–5.
29.   BEA, *Group Report and Accounts*, 1970–71, p. 39; 1971–72, p. 17.
30.   There is no doubt that BA, as well as the engine-maker Rolls-Royce, preferred American aircraft to the Airbus. It was the airframe builders who were concerned to stay in the European enterprise; Hawker Siddeley remained a private contractor to the Airbus consortium and in 1979 British Aerospace officially entered the consortium with a 20 per cent share.
31.   BOAC, *Annual Report and Accounts*, Year ending March 1953; BA, *Annual Report and Accounts*, Year ending March 1973.
32.   Bray, *BOAC*; 244, 279, 316. The west coast route through the Caribbean to Lima was maintained.
33.   *The Aeroplane*, 'An airline filibuster', 10.4.1959.
34.   See Alan Dobson, *Peaceful air warfare*, Oxford, 1991, pp. 225–35.
35.   E.M. Eltis, 'The interaction between the aero engine industry and the growth of air transport', *ITA*, Paris, 1966, pp. 1–2.

36. *Flight Magazine*, 8.4.1960, 'The domestic scene: air traffic in the United Kingdom', pp. 475–478.
37. A critical inquiry was the Civil Aviation Authority report, *Air Transport in the Scottish Highlands and Islands*, March 1974.
38. In 1994 BA offered 16 destinations in Africa, nine in the Middle East and six on the Indian subcontinent; it was the only European airline serving Bangladesh. *The Times*, 25.8.1994, 'BA – 75 years'; Arthur Reed, 'BA has the whole world on its routes, *Times* Supplement *BA – 75 years*, p. XVIII.
39. Peter W. Brooks, 'A short history of London's airports', *Journal of Transport History*, vol. 3, no. 1, May 1957.
40. Rigas Doganis, *The Airport Business*, London, Routledge, 1992, p. 12. The BAA was was created by the Labour government in 1965, see *The Aeroplane*, 31.3.1966, 'Britain's airports – a new era', pp. 4–8.
41. CAA, *Passengers at London Area Airports and Manchester Airport in 1987*, CAP 560, London December 1989, p. 41.
42. Report on visit to USA and Canada, 9.3.1949 to 11.4.1949, P & D Report No.M/21, R.C. Morgan, RAF Museum, Hendon, Box.240.
43. 'Internal services – policy', Memo by General Manager Commercial, C[48]17, 15.3.1948, BEA Board Paper, No.21.
44. In November 1950 BEA did a consumer survey on 'Knowledge of BEA British Services', BEA Board Paper No.47.
45. Measured as US cents per Capacity Ton Mile, ICAO airline members' average taken as industry average, see *The Aeroplane*, 30.3.1961, p. 341.
46. *The Economist*, 19.4.1958, 'Airline manpower: root pruning at BOAC', p. 239; 24.5.1958, 'How to run an airline', p. 725; 2.8.1958, 'BOAC's loss: help or self-help?' *BOAC Annual Report and Accounts*, Year ended March 1959, pp. 4, 7.
47. *The Economist*, 4.7.1959, 'The airlines and the Minister'. Also *Economist Intelligence Unit Report*, January 1959, RAF Museum Hendon, Box 360, p. viii.
48. *BOAC Annual Report and Accounts,* Year ended March 1963, p. 14.
49. Keith Granville, 'The UK's part in the development of air transport of other nations', *Journal of Institute of Transport*, May 1962, p. 303.
50. Memorandum by Michael Custance, Ministry of Aviation, 26.4.1961, BOAC Association with British West Indian Airways, 1953–1963, Public Record Office (PRO)B T.245/329.
51. Douglas to Harold Watkinson, Minister of Transport and Civil Aviation, CH/123, 31.1.1956, BEA Board Paper, No.102.
52. Memorandum by Commercial and Sales Director, DC(58)5, Traffic and Revenue Situation, First Quarter 1958/59. BEA Board Paper, No.131.
53. BEA Engineering/Project & Development Technical Note P/250, R.J. Kiddle, September 1959, *Why are BEA Costs Higher than Airline Costs in the USA?* RAF Museum, Hendon, Box 242.
54. *Flight International*, Review of the BOAC White Paper, 28.11.1963, pp. 854–5.
55. *The Aeroplane*, 21.4.1966, 'BOAC Reshaped', p. 10.
56. *The Economist*, 25.6.1968, '550 Hours is not enough'.
57. D.L. Nicolson, 'British Airways in the Eighties', *Journal of the Institute of Transport*, May 1974, p. 84. Nicolson was the first chairman of the British Airways Board.
58. See Richard Pryke, *The Nationalised Industries: Policies and Performance since 1968*, Oxford, 1981, pp. 132–3.
59. Air Transport Users Committee, *European Air Fares*, CAA, 1979, Table B, p. 9.
60. British Airways, *Civil Air Transport in Europe*, Document prepared for the CAA, January 1977.

61.   *The Economist*, 28.6.1980, 'Delete "airways" delete', p. 76.
62.   Duncan Campbell-Smith, *The British Airways Story: Struggle for Take Off*, Coronet, 1986, p. 10.
63.   *The Economist*, 16.7.1983, 'Chopping heads off'.
64.   *The Economist*, 28.5.1988, 'Bottom Lines: British Airways', p. 77. For a detailed account see Kyohei Shibata, *Privatisation of British Airways: Its Management and Politics 1982–1987*, EUI Working Paper EPU No. 93/9, European University Institute, Florence, 1994.
65.   International Civil Aviation Conference, Chicago, 7 December 1944, Final Act & Appendices, I-N, London, HMSO, Cmd 6614.
66.   See Peter J. Lyth, 'Experiencing turbulence', in James McConville (ed.), *Regulation and Deregulation in National and International Transport*, London, Cassell, 1997.
67.   *The Economist*, 24.10.1964, 'Air Union: dis-union', pp. 415–6.
68.   Douglas to Thorneycroft, 29.8.1960, BEA Board Paper, No. 155.
69.   BEA Paper for Minister of Aviation, Cooperation with other airlines, 29.8.1960, BEA Board Paper, No.155.
70.   *BOAC Annual Report and Accounts*, Year ended March 1959, p. 11.
71.   *BOAC Annual Report and Accounts*, Year ended March 1961, p. 35.
72.   *The Economist*, 4.7.1964, Interview with Lord Douglas of Kirtleside, pp. 48–51.
73.   *Flight International*, 31.10.1963, 'It starts on Sunday'; *The Economist*, 9.11.1963, 'Room for Independents', p. 584.
74.   See A.P. Ellison and E.M. Stafford, *The Dynamics of the Civil Aviation Industry*, Farnborough, 1974, pp. 45–7.
75.   *BOAC Annual Report and Accounts*, Year ended March 1971, p. 8, Year ended March 1972, p. 3.
76.   See M. Cronshaw and D. Thompson, 'Competitive advantage in European aviation and or whatever happened to BCal?' *Fiscal Studies*, 12, 1991, pp. 44–66.
77.   Monopolies and Mergers Commission, *Report on the proposed merger: British Airways plc and British Caledonian Group plc*, November 1987, HMSO, Cmd 247.
78.   For an inside account, see the autobiography of Sir Adam Thomson, *High Risk: The Politics of the Air*, London, 1990; pp. 508–72.
79.   For the details see Howard Banks, *The Rise and Fall of Freddie Laker*, London, 1982, pp. 44–5, 136–8.
80.   Banks, *Rise and Fall*, pp. 75–88.
81.   Martyn Gregory, *Dirty Tricks*, London, 1994, p. 35.
82.   *Der Spiegel*, 5/1993, Huebsch verpackte Erpressung, pp. 101–2.
83.   Memoranda by Secretary, 21.9.1961, *BEA Board Paper, No.165*, 18.1.1962, *BEA Board Paper, No. 169*.
84.   Monopolies and Mergers Commission, *British Airways PLC and SABENA SA: a report on the merger situation*, London, 1990.
85.   British Airways plc, *Strategic Alliances Fact Book*, 1994, pp. 41–72.

4 Lufthansa: Two German Airlines

Hans-Liudger Dienel

Four years after the end of the Second World War two German states were set up in ideologically hostile political blocs, but it took six more years before either of them was allowed to have an airline. When they did, both flag-carriers reappropriated the pre-war name Lufthansa, simply because both states claimed to be the true successor of the pre-war state of Germany. By 1963 however East Germany, or the German Democratic Republic (GDR) had renamed its carrier Interflug and in a small way foreshadowed events 27 years later when the GDR was forced give up its national sovereignty altogether.

The following chapter examines the history of these two airlines. The West German Lufthansa (hereafter Lufthansa) became the larger airline and for this reason, the chapter will focus mainly on its story. However, competition between the two states influenced the development of both airlines and Lufthansa will be compared to its socialist sister (Lufthansa [East] – later Interflug).[1]Comparing the two airlines illustrates the different roles aviation played in the transport policies of the German states, one an essentially free market, the other a centrally planned economy.[2] Besides having two national flag carriers, another peculiarity of German aviation after 1945 was its late appearance. Between 1945 and 1955, the Allied authorities prohibited civil aviation in Germany. However coming late into the market did have its advantages since the Germans could wait and see what mistakes were made by the other European carriers.

The history of Lufthansa falls into four periods. Between 1945 and 1955 it did not exist as an operational entity. In the second period which lasted until 1970, Lufthansa caught up with other European flag-carriers and in the third, between 1970 and 1991, it established a stable position as one of the leading European airlines, although smaller than Air France and British Airways (BA). In the fourth period since 1991 Lufthansa's staff numbers decreased for the first time in postwar history, but in general the airline adapted successfully to new market conditions and now holds a middle position between BA and Air France. By contrast the history of Lufthansa [East] and Interflug has other milestones. As in the West, civil aviation did not start before 1955. Thereafter the two most important turning points were the end of East German aircraft production in 1961 and the end of the GDR itself in 1989 – because some months after the fall of the Berlin Wall, Interflug ceased operations.

1   ORIGINS AND POLICY

1.1   Formation of the two Lufthansas

In 1925 the government of the Weimar Republic put strong pressure on the two leading German airlines Aero Lloyd and Junkers Luftverkehr to unite. Both were already holding companies for a number of smaller airlines. While Junkers was economically the more efficient, Aero Lloyd had better contacts with the government. The merger took place in January 1926, in a similiar process to other government-induced airline amalgamations in Europe. By 1939, thanks to a policy of active market development by the former Junkers staff, Lufthansa was the biggest airline in Europe. During World War II, Lufthansa was one of the few airlines in Europe to continue services and it came close to being the first genuine *European* airline, enjoying as it did a virtual monopoly in Nazi-occupied Europe. Many managers of the postwar Lufthansas were trained during this period.[3]

The Treaty of Potsdam in August 1945 ended German sovereignty and German civil aviation for a decade. Nevertheless, right from the start, the Germans tried to rebuild civil aviation. In West Germany the restoration of civil aviation was set in motion fairly quickly because the interests of the Allies and the Germans coincided. The Allies wanted fast and reliable communications to their troops and a new market for their airlines. The Germans at both the local and national level wanted good international connections.

A turning point was the Airlift to beat the Berlin Blockade in 1948–9. Thereafter civil aviation between Berlin and West Germany remained a high priority; indeed it was the earliest mass air transport to and from Germany. The Berlin Airlift was the model for other air transport activity: students came to their Berlin universities by airlift from Hannover, children went to summer resorts in special airlifts, and so on. The success of the Airlift was the psychological foundation for the rebirth of West German aviation.

After 1949, the new government of West Germany, or the Federal Republic of Germany (FRG), established a Ministry of Transport which set up a small department of aviation in 1950. A year later Martin Bongers, a former director of the pre-war Lufthansa, was asked by the Minister of Transport to advise him on aviation, and Bongers thus became responsible for creating a new German flag-carrier. In 1953, the *Büro Bongers*, as it was known, was renamed *Aktiengesellschaft für Luftverkehrsbedarf* (Luftag) and in 1955 it finally reclaimed the old name Lufthansa. Bongers had been busy since as far back as 1945, bringing together former Lufthansa executives and dealing with American, British and French officials in the Western zones of occupation, all with the objective of re-establishing a German airline. He did not have much success at the beginning, although he secured his position within the informal Lufthansa network and was ultimately rewarded with government approval to re-establish the company.[4] Besides Bongers' informal network, there was also the old Lufthansa – a rich company in liquidation. Though it could not go into airline business again, it was able to help the new Lufthansa with land and money. Kurt Weigelt, a former director of the Deutsche Bank, was head of the advisory boards of both the old Lufthansa in liquidation and the new Lufthansa.

In the GDR civil aviation hardly existed before 1955. There was less opportunity to establish an informal network, like the one Bongers had created, since the Russians had closed all civil airports to East Germans and there was no civil airline flying to the GDR.

1.2 Government objectives

In both German states, the government played a crucial role in founding and organizing the airlines, but what aims did they pursue? Both were convinced of the need to set up national airlines as a matter of prestige and of national sovereignty.[5] The creation of a flag-carrier is a recognized means to demonstrate national sovereignty and tends to be one of the first acts of the governments of new states. In the FRG a German airline was also expected to reduce the drain of foreign currency caused by German passengers buying tickets on foreign airlines. The GDR hoped that its airline would serve as a outlet for a future national aircraft industry; its other functions however were less clearly defined. Economic considerations were of less importance to the East Germans and in its reports to the government, Lufthansa [East] always discussed the *political* dimension. Meanwhile its bigger sister in the West simply concentrated on getting out of the red.

In the FRG, executive power was divided between the national, state (Länder) and city levels. Besides the Federal Ministry of Transport, there were transport ministries in all the Länder. North Rhine-Westphalia for example, with almost a third of West Germany's population, was eager to set up an aircraft industry and a regional air transport system.[6] Even before Lufthansa was refounded, its transport minister supported a regional helicopter service of the Belgian airline SABENA between several cities in the state. Bongers also placed his office in the North Rhine-Westphalian city of Cologne in 1951, near to the new capital in Bonn. Other Länder were active as well. Karl Schiller, Senator of Commerce in Hamburg, persuaded Lufthansa to build its central maintenance base in the city-state.[7] The Bavarian state government meanwhile managed to concentrate the German aircraft industry around Munich. The Länder were competing with each other and Lufthansa could only profit from this rivalry.

The West German cities had less influence than they did over the old Lufthansa, but decisions about routes and particularly airports had to be discussed with them. The German Airport Association built its headquarters at Stuttgart airport and the mayor of Stuttgart became chairman of its board. Eventually all regional airports in the FRG had some degree of national function.

How was Lufthansa to be organized? As a government department or as a free enterprise? Although it was government pressure which had created the original Lufthansa in 1926, the postwar airline was organized as a private company, unlike the German railways, with both the Federal and *Länder* governments as major stockholders. However the influence of small towns and regions forced it to include too many local airports into its domestic network and this led to slow and cumbersome domestic 'hopper' routes.

In the 1950s, the government wanted to increase the economic independence of Lufthansa, but recognized that government supervision would be necessary and should be secured by the ownership of stock. The decision to appoint Bongers, a former Lufthansa manager and not an engineer, as chief executive, was a sign that profitability would guide the new airline. While Bongers had been optimistic about Lufthansa's financial prospects in 1951, it took him almost ten years before he reached profitability in 1964.[8] It was understandable that this 'late returnee of the German economy', as Bongers called the company, needed public funds to have a good start. Yet, as long as Lufthansa failed to make profits, Bongers had nothing to point to but the foreign currency the airline had saved the country.[9] Bongers tried hard to find private investors but was not successful; potential stockholders were not convinced that Lufthansa would make money. Only when the company made a profit in 1964 did the number of shareholders rise dramatically from 200 to over 36 000.

In 1955, the federal government held 74 per cent of Lufthansa's stock. In the following years this share rose to 87 per cent, decreasing after 1987 to end at zero in 1997.[10] Both American deregulation and BA's privatization provided a strong impetus for the privatization of Lufthansa. As so often, Lufthansa took a middle position between BA and Air France. After 1969, the German aircraft industry was a partner in the European Airbus programme and the German government felt an obligation to make the Boeing customer Lufthansa buy Airbuses. The easiest way to do this was to keep its share of the airline. After German reunification, however, when the federal Government needed money, it decided to follow the world-wide trend towards privatization and sell its share in the airline.

Bongers felt that the traditional German qualities of punctuality and reliability should be embodied in Lufthansa's character but it was also agreed that it should serve as a symbol of the new democratic Germany. Western equipment bought from the Allies emphasized this message. In some respects this meant a break with the tradition of the old Lufthansa, where punctuality had been of the utmost importance and rated even higher than safety for personnel. This was to change.[11] In both Lufthansa and the government there was a considerable continuity of personnel from the pre-war period to the late 1960s. Though Bongers stepped down in 1965, his colleague Gerhard Höltje remained on the executive committee until 1972. This continuity in senior management resembled the pattern in other Western European countries and the United States, where the first wave of retirement around 1970 saw the departure of Juan Trippe (Pan American) and Eddie Rickenbacker (Eastern); only BOAC seems to have replaced its executives more frequently. In all, Lufthansa has had only four chief executives in its entire postwar history. The economist Bongers was replaced by the lawyer Herbert Culmann in 1965 who handed over to the politician Heinz Ruhnau in 1982. Only with Jürgen Weber's appointment in 1992 did an engineer become Lufthansa's chief executive.

In the more centralized GDR civil aviation played a different role. The first secretary of the Socialist Unity Party (SED), Walter Ulbricht, was in favour of air transport and a national aircraft industry for four reasons. Firstly civil aviation was a public transport technology and thus more appropriate for a socialist state than individual car ownership. Secondly it was a modern technology and therefore proved that socialism was progressive. Euphoric enthusiasm for technology was a characteristic of the GDR and engineers could influence political decision-makers more easily than in the West. Thirdly aviation was seen as a Russian technology and the Soviet Union served as a model for the GDR. Ironically the model of American air transport entered East Germany via the Soviet Union.[12] And fourthly an air transport industry gave work to German engineers returning from Russia, who might otherwise have gone to the FRG; the aircraft industry thus became the biggest industrial investment in the GDR during the 1950s.

In West Germany on the other hand, the government was more cautious about subsidizing the aircraft industry. Only after the end of the East Germans' enthusiasm for air transport in 1961 did the West Germans subsidize aircraft development on a large scale and great plans were made to improve transport networks. Political antagonism goes some of the way in explaining this U-turn; the two Germanies were condemned to watch each other across the Iron Curtain and compare notes. It was, however, the Americans who really caught the attention of the West Germans. Gerhard Höltje had worked for the Americans at Berlin's Tempelhof airport before he joined Lufthansa. The first West German pilots were mostly instructed by the American airline TWA, although, for political reasons, some were trained by BEA.[13] America also served as an example for the GDR and popular books with American planes on the cover were widely published in East Germany.[14]

Lufthansa [East] was given the same status as the GDR railways, that is, a socialist enterprise under the control of the Ministry of Transport, although that control was relatively weak. The reason for this weakness was a personal one. In their 35 years of existence, Lufthansa [East] and its successor Inter-flug had only four directors, of whom the first was Artur Pieck – the son of Wilhelm Pieck, President of the GDR – and he had better contacts to the East German Politburo than the Minister of Transport. The last director of Interflug, Klaus Henkes, also headed the aviation department of the Ministry of Transport, managed the institute for flight control and was a general in the East German air force, developing the airline into a military unit in the mid-1970s. Interflug's contribution to national defence, rather than its commercial performance in peacetime, became a major source of legitimation for the airline. Running all of the three important institutions of East German civil aviation, Henkes was beyond the reach of the Polit-buro.[15] After Pieck's directorship, his three successors ran the airline after the decline of the East German aircraft industry and as a result Interflug lost its importance and moved into a niche of GDR society. From then on, it operated on a low budget and received little attention, as the government lost interest in aviation and moved it to the periphery of its planning arrangements. This shift in government interest is evidenced by the 1963 decision to give up the futile quarrel with the FRG over the name 'Lufthansa'. As a result Lufthansa [East] merged with Interflug, which had been founded as a charter airline in 1958, and took over its name. During the 1970s, Interflug not only ran an airline but all of the GDR's airports, general aviation, and agricultural and business flights. The airline had found itself a niche but had little potential for growth, since in gaining independence from political control it had denied itself any further financial support from the government.

1.3 Attitudes to regulation

When Lufthansa started services in 1955, more than thirty foreign airlines were already flying to Germany. The question arose, should there be only one national flag-carrier in West Germany, or should the British example be followed and two state-owned airlines created? The British solution was discussed but ultimately rejected as the division between intercontinental and European networks seemed costlier at the outset and less efficient in the long run.[16] The West Germans were convinced that Lufthansa only had a chance if it was backed by the government and that its support should not be given to more than one airline. A second, purely domestic, airline such as the French or Italians had, was another matter. But Lufthansa wanted to operate domestic as well as international services and thus a second airline would have had to compete with Lufthansa. Some West German Länder did sup-port the creation of additional regional airlines but the dominance of Luft-hansa turned out to be too strong for a successful alternative.

The German government accepted and supported a regulated market. Indeed there was no supporter of intra-European air transport deregulation except for the European Commission which opposed bilateral agreements between states on principle. The Commission saw an opportunity to get involved in European air transport in the 1970s in the wake of American deregulation.[17] Its memoranda of 1979 and 1982 proposed establishing a common air transport market which would mark the end of bilateral agreements and the subsidies. These were to be made effective in three steps in 1988, 1991 and 1993. Now, European airlines are more or less free to decide about routes and fares, although there are laws against take-overs which protect smaller flag-carriers.[18] The reunified German government has reacted to the on-going process of European deregulation by maintaining a position between that of France and Britain. In 1993, the German air traffic control authority was privatized. Lufthansa, although it had been quite content with the regulated market, did have a strong identity as a commercial organization and welcomed market freedom in the air. Unfortunately many of its employees had difficulty adapting to deregulation and as late as 1994 the former head of Lufthansa's transport policy department, Ulrich Meyer, was still arguing against it.[19]

Deregulation was not the only issue where Lufthansa took a middle position between the protectionism of the French and the free market competition of the British. In 1981 both BA and Lufthansa had a bad year. While the British airline began a major restructuring programme, the Social-Democrat Heinz Ruhnau, a former junior minister and trade union lobbyist, took over as head of Lufthansa. Under his leadership Lufthansa also underwent reforms and established a marketing directorate in 1987 and a new corporate identity for the airline. However when in 1991 Lufthansa went into deficit for the first time in 18 years, there was further change at the top and in September 1992 Jürgen Weber became chairman. Since then staff has been cut by 15 per cent, although compared to BA the cuts were introduced more gradually.

2   AIRCRAFT PROCUREMENT

Aircraft procurement is crucial for an airline. When analysing the aircraft procurement patterns of a particular flag-carrier, the close relationship between the airline and the local aircraft industry has to be taken into account. As Peter Lyth and Nicolas Neiertz show for Britain and France, the success of aircraft production can be more important for government than the profitability of the national airline. In Germany however the situation was different.

At the end of the Second World War, the aircraft industry was the largest industrial sector in Germany. Sixty per cent of the German factories were located in the Soviet Zone, which later became the GDR. The Junkers company, for instance, employed about 400 000 workers all over Europe and was based in the East German town of Dessau. In the GDR aircraft development actually continued after May 1945. In order to keep German aircraft engineers in their zone of occupation, the Russians set up special construction offices for aircraft development at former German industrial sites.[20] However in September 1946, both the machinery and the German technicians were transferred to the Soviet Union. They remained there until the early 1950s when the East German government decided to rebuild its own aircraft industry.[21] The availability of skilled labour in the GDR and the Russian promise of technical support convinced the returning German engineers that they could build aircraft again; they had built good aircraft for the Nazis and the Russians, now they would build them for East Germany. After the workers uprising of 17 June 1953 however, the government decided to limit the new industry to civil aircraft.[22] Thus the East Germans built a licensed version of the Russian Ilyushin Il-14, itself a copy of the Douglas DC-3. More than 80 planes were built before the production was discontinued in 1959. By then the GDR had developed its own aircraft, such as the four-engined passenger jet, P-152, which had its inaugural flight in 1958.

Not surprisingly Lufthansa [East] had to buy East German aircraft. It operated the German Il-14 and in 1960 it was the only airline in the world to order the jet P-152. This decision was not merely aimed at supporting the indigenous aircraft industry but was a question of national prestige. The P-152 was featured on GDR stamps and in the press, and proclaimed to be the true *Wirtschaftswunder* as opposed to the economic miracle in West Germany.[23] By 1961 more than 1.6 billion marks had been spent on the P-152, the biggest industrial project in the GDR in the late 1950s.[24] Ultimately the P-152 failed because of poor marketing and a misguided policy of self-sufficiency – unlike western aircraft companies, the East Germans even attempted to develop their own engines.

When the GDR cancelled production and transferred resources to other industrial sectors in February 1961, it signalled the end of the GDR's political and financial support for civil aviation. The unnecessary harshness of the decision (even the production of gliders was stopped) illustrates the East German preference for *big* solutions, it had, after all, been this tendency which had led to the establishment of the industry in the first place.[25] The demise of the East German aircraft industry had a crippling effect on Lufthansa [East] and it was only in 1969 that Interflug was able to buy its first jet, the Russian Tu-134.[26] After 1961 Interflug was obliged *de facto* to buy Soviet aircraft, mainly Ilyushins. It had absolutely no influence on their design and was not even able to get the instructions in the toilets written in German! Although Interflug made comparisons between Soviet aircraft and the Sud-Aviation Caravelle, no attempt was made to buy the French aircraft.

Interflug was more dependent on the Soviet Union than was LOT, the Polish national flag-carrier, which purchased British-built Viscounts at one stage. Only in 1987, in the last years of the GDR, was Interflug able to buy three Airbus A300s with credit from the FRG. These aircraft had the advantage that they could fly non-stop to Havana; hitherto Interflug had used Russian Il-62s for its Cuba service and had been forced to make a stopover in Canada where East German passengers often made an unauthorized departure from the plane.

By contrast with the GDR, the West German aircraft industry experienced more continuity. After 1945, some former aircraft manufacturers kept their factories busy by producing cars, ships, and even houses.[27] Others left the country to found companies abroad or work for American aircraft companies. Many waited for a chance to return to Germany, but the West German government was not interested in supporting the ambitions of aircraft entrepreneurs. The chancellor, Konrad Adenauer, did not want to compromise the FRG's integration into the West by creating a new German aircraft industry and he did not want to spend money on technological development.[28] Instead the West German government exerted pressure on Lufthansa to buy American and British aircraft. Adenauer recognized the American and British need to develop their aircraft industry and used aircraft procurement as a sign of Germany's orientation towards the West. The leading aircraft engineers in both East and West Germany had a common background in the Nazi aircraft industry and they were convinced that a postwar German industry, if subsidized by government, could become a major player.

The purchase of a long-range aircraft was the most important procurement decision for Lufthansa. In 1953 Luftag signed options for the Douglas DC-6 but then, surprisingly, switched to the Lockheed Super Constellation.[29] Donald Douglas was astonished, as he had arranged an informal agreement for license production of the DC-6 in Germany and appeared to have Lufthansa's executive board on his side. However Douglas had overestimated the importance of the German aircraft industry to the FRG. Lockheed did not bother negotiating with industry but simply offered an attractive finance package to the Ministry of Transport and the Deutsche Bank, Lufthansa's house bank. Although Lufthansa's executives preferred Douglas, Lockheed won the contract because it had the important advisory council of Luftag and the Ministry of Transport on its side.[30]

Given the timing of Lufthansa's relaunch, one could ask why it did not buy second-hand aircraft and wait a few years until it could operate on international routes with jets. Lufthansa was in an ideal position in relation to its competitors to choose a new fleet, as it did not have to dispose of an old one. So why did it buy expensive new Super Constellations? Its reasoning seems to have been threefold. Firstly, Lufthansa considered itself an international airline and regarded the Atlantic as its most important route. Nobody at Lufthansa could imagine waiting for five more years before commencing intercontinental operations. Operating at only the European and domestic level, and with old aircraft, did not suit Lufthansa's self image as a techno- logical leader in aviation. Secondly Lufthansa suffered from a *'latecomer syndrome'*. Hans Bongers, though he did not want to spend too much money, was nonetheless determined to catch up with competitors as quickly as possible.[31] And thirdly, Lufthansa, in common with several other Euro- pean flag-carriers, did not expect big jets to become operational as early as they did. For shorter range work within Europe, Lufthansa ordered in 1956 twelve Vickers Viscount 814 turboprops. Again, there was government pres- sure on the executive board, this time to buy British (after the long-range aircraft order had gone to the US).[32]

Despite these two cases at the beginning of its history, Lufthansa was fairly free to choose the aircraft it wanted. In the late 1950s, Gerhard Höltje, the technical director of Lufthansa, moved towards a homogeneous fleet and became an advocate of Boeing.[33] In 1960 the company took delivery of its first Boeing 707s and a year later of the smaller 720. Short and middle-haul jets were the biggest investment in the mid-1960s. Both the Ministry of Transport and the German parliament asked Lufthansa to look at the French Caravelle, the British Trident and also to talk to German aircraft builders. But Höltje resisted external pressure and sought homogeneity with Boeing. In 1964 the decision was still open, but Höltje had already talked to Boeing about developing a new short-range jet to replace the Viscounts. In 1965 Lufthansa became the launching customer of the twin-engined Boeing 737 which would become the single most important commercial jet in the world. Still in production in the 1990s, more than 2000 have been delivered. Within Lufthansa, the 737 was regarded as Höltje's work and while this might overstate the German influence, the American manufacturer certainly appreciated technical assistance from the Germans. When Höltje left Luft- hansa in July 1972, it had a one hundred per cent Boeing fleet.

Since the early 1970s the fleet has become more heterogeneous. Höltje's successor Reinhard Abraham (technical director from 1972 to 1989) was more open to Boeing's competitors and Höltje had in any case agreed to order five Douglas DC-10s, which were delivered in 1973 and earmarked for the Far East route.[34] In 1976 the first Airbus A300 was delivered. Thereafter the number of Airbus aircraft grew steadily and in 1996 made up 50 per cent of the Lufthansa fleet. The single most important aircraft however remained the Boeing 737.

The following fleet statistics shows Lufthansa's movement from Lockheed, Convair and Viscount towards Boeing and Airbus. They do not include the aircraft of charter airlines owned by Lufthansa – had they been included, the fleet would not appear so homogeneous. In the 1980s, the number of leased aircraft grew and in 1995 25 per cent of Lufthansa's aircraft were leased.[35]

Table 4.1 Fleet statistics of Lufthansa, 1955–93 (without Condor, Lufthansa Cargo or Lufthansa Cityline)

Aircraft type	1955	1957	1961	1965	1969	1973	1977	1981	1985	1989	1993
Lockheed Super Constellation L-1049G	4	8	7	7							
Lockheed Super Star Constellation L-1649A		2	2	1							
Convair 340		4									
Convair 440	4	5	11	11	11						
Douglas DC-3	3	3									
Vickers Viscount 814				9	7	10					
Fokker F-27					2						
Boeing 707				5	12	21	19	12	5		
Boeing 720				7	2						
Boeing 727					13	21	27	30	26	26	22
Boeing 737					21	28	28	41	44	60	105
Boeing 747						6	7	13	18	25	26
Douglas DC-8-73									5	5	
McDonnell Douglas DC-10						1	11	14	14	14	11
Airbus A300							5	11	5	7	11
Airbus A310									11	17	10
Airbus A320										5	33
Airbus A340											9
TOTAL	11	22	41	55	84	81	93	110	123	155	205

Source: Deutsche Lufthansa Aktiengesellschaft (ed.): *Jahresberichte 1955–1993*; Lufthansa AG (ed.), *Zahlen, Daten, Fakten. Ausgabe 1994*, Köln 1994.

One can distinguish four distinct periods in German civil aircraft manufacturing. In the first period up to 1955, the West German government was reluctant to support any attempts to re-establish an aircraft industry. After 1955 a few politicians, such as the Defence Minister, Franz-Josef Strauss, were ready to support the aircraft industry, and in the third period after 1961 there was increased government support, particularly for vertical take-off (VTO) aircraft, both military and commercial, and for commuter jets.[36] The fourth period after 1970 is marked by the Airbus programme.

In the 1960s, manufacturing big jets seemed impossible and German aircraft industry concentrated on developing short-haul commuter jets to occupy a promising niche market. Heinkel worked on the He 211, the Hamburg Aircraft Works built the Hansa-Jet and the Vereinigte Flugtechnische Werke in Bremen built the VFW 614.[37] The government supported all these aircraft although Lufthansa remained free not buy them. In fact Lufthansa was not interested in regional aviation at this time and in 1970, for example, the Boeing 737 was the smallest aircraft in its fleet. Although not the only reason, Lufthansa's rejection of these German aircraft was a major factor in their ultimate market failure. Lufthansa showed little interest in either VTO or supersonic aircraft, despite the fact that many experts predicted a bright future for the latter in the 1960s.[38] Lufthansa executives took options on both the Boeing SST and the Concorde, but in the end it never came to an order.[39] The collapse of the West German projects has much in common with the end of the East German aircraft industry in 1961, the main difference being that the outcome in the West was more positive. The FRG government stopped subsidizing the niche programmes like VTOL in the 1970s when it became clear that the European Airbus would be a success.

The Airbus project began in 1965, at a time when national governments in Europe were becoming aware of the technology gap between Europe and the USA. They became convinced that European high-tech industries had to unite in order to withstand American competition; small companies had no chance of survival. Initially the British were a member of the group, but withdrew at an early stage and in May 1969 the French and German governments signed the Airbus contract alone; a year later the French company Aerospatiale and the German consortium Deutsche Airbus GmbH founded Airbus Industrie.[40] The original intention to cooperate with American companies was dropped in 1967. The Airbus programme changed the attitude of the FRG government towards Lufthansa's aircraft procurement and it sought to persuade the airline to buy Airbus aircraft. For Lufthansa this meant an end to its exclusive loyalty to Boeing, and Höltje actually was against the Airbus. Only after he retired did his successor Abraham sign the first order in 1973. In the same year Lufthansa also ordered five McDonnell Douglas DC-10s.

Up to 1978, the first Airbus – the A300 – was not a great success and only 43 had been sold.[41] Nevertheless in that year Airbus Industrie began the construction of a new smaller version – the A320 – and secured orders from Swissair and KLM, as well as Air France and Lufthansa. Lufthansa was initally unenthusiastic about the aircraft, but the A320 actually went on to become the most successful Airbus jet. In the early 1980s Lufthansa became interested in bigger aircraft, in the 260–300 passenger range, and in 1987 it ordered fifteen A340s – a crucial decision in the development of this Airbus type. Then in the 1990s, Lufthansa was a launching customer of the smaller A319. Thus, Lufthansa bought Airbus aircraft in large numbers although it remained an important Boeing customer and continued to buy the 737 and 747 types.

3   NETWORK ANALYSIS

3.1   Domestic routes

In its first 25 years Lufthansa was chiefly interested in international services, although domestic routes did play an important role for Lufthansa [East] and Interflug. Thus although the East German airline remained smaller, its domestic network in 1960 was actually stronger than Lufthansa's and it was even possible to fly the mere 75 kilometres from Karl-Marx-Stadt (Chemnitz) to Leipzig or Dresden. According to the GDR's long-term plans, new airports were to help in the development of an 'air-taxi' service,[42] and in 1957 thirty-six such local airports were envisaged.[43] The idea of modern mobility without individual vehicle ownership was attractive to the East Germans. However the idea failed totally. The 12 Czech Super-Avros which were bought as *air taxis* in 1958, were used only for sightseeing tours. It soon became clear that flying by air-taxi did not save time in the GDR; it often took longer than travelling by car or train. Because of the very short distances and the fact that some routes required detours around the Allied air corridors into West Berlin, domestic aviation in the GDR proved to be a totally impractical proposition. Delays were another factor. In 1965, for example, more than ten per cent of domestic flights were cancelled or delayed and even when the flights arrived on time no airports except East Berlin had connecting rapid transit facilities. Instead of businessmen, domestic aviation in the GDR attracted the few tourists in the country and the passenger load factors remained under 50 per cent.[44] By 1964, the GDR government had accepted that there were no economic grounds to continue domestic services, although it retained them to avoid compromising itself before the East German people.[45] The routes were gradually cut back until the last domestic flights were abandoned in 1980.[46]

Meanwhile in the FRG domestic flight had become a means of mass transport. In 1949 the Ministry of Transport estimated that revenue from domestic services in Germany were 40 per cent of the long-distance earnings of the German railways.[47] For a long time Lufthansa regarded domestic routes as a means to feed passengers into its international services; it was an important part of its business but its official position remained that profits were made with intercontinental routes, while domestic flights were loss-makers. In fact Lufthansa's costing procedures were partly to blame since connecting flights were priced so low that even fully booked aircraft could not make money. Additionally, domestic routes were habitually counted as part of international routes, Hamburg-Düsseldorf, for instance, was part of Hamburg–London or Hamburg–South America, and thus services seldom had a load factor above 50 per cent.[48] The only exception was Berlin. Because of the restricted surface connections to the Western half of the city there was always strong demand for air services to Berlin, although this did not benefit Lufthansa since the route remained a monopoly of the Allies' flag-carriers (Pan American, BA and Air France) until German reunification in 1990. Only in 1988 did Lufthansa start a joint venture with Air France, called EuroBerlin; after reunification Air France sold its share to Lufthansa, while BA founded a new airline for German domestic services, aptly named Deutsche BA.

So far as regional short-range air services were concerned, Lufthansa was even more sceptical. In the 1950s the idea of intercity helicopter services was popular throughout western Europe.[49] In 1953, for example, the state of North Rhine–Westphalia began to subsidize the Belgian flag-carrier SABENA to provide helicopter services between eight cities including Bonn and Brussels.[50] Lufthansa remained cautious and shunned the helicopter business, and SABENA finally gave up its German services in 1966.[51] By then both the FRG government and the German aircraft industry were encouraging Lufthansa to enlarge regional and domestic services with VTOL aircraft,[52] However Lufthansa rejected the idea[53] and suggested instead the foundation of a subsidiary to be funded by the government.[54] In the following years Lufthansa began to cooperate with small feeder airlines to provide regional services; the routes became part of Lufthansa's network although the airlines were partly funded by local authorities. Above all Lufthansa wanted to prevent the emergence of a new German domestic airline that might compete with it.

In the years up to 1980 Lufthansa seemed to have good economic reasons for its opposition to domestic and regional routes. However in the 1980s it was challenged by the appearance of new independent regional carriers offering non-stop flights between provincial airports with fast turboprop aircraft. Lufthansa's traditional view that domestic services fed traffic into its international routes was now reviewed, while at the same time the deregulation of regional air transport in Europe boosted the number of intra-European routes.[55] Without Lufthansa's deliberate planning, German domestic air transport was transformed from an unloved but necessary activity into a profitable and growing business, with further impetus coming from German reunification and the fact that East German surface transport – its motorways and railways – were in such bad shape.

3.2 International routes

The differences between the two German flag-carriers were even bigger on international routes, since they basically had different functions. Lufthansa was designed primarily as an international carrier, and in the first years 80 per cent of its investment capital was spent on long-distance aircraft for the Atlantic routes. This was regarded as absolutely crucial for the development of the West German economy[56] and the FRG's political integration into the West.[57] Lufthansa executives modelled their airline on Pan American Airways and TWA. Hans Bongers argued that Lufthansa's share of the international air transport market should equal the German share of international trade and favoured a commercial route policy, which was not typical for all European flag-carriers, some of which developed their routes according to political imperatives. In 1953, two years before starting operations, Luftag estimated that the North Atlantic would bring revenues of DM30.5 million while DM6.7 million could be expected on domestic and European routes.[58] The North Atlantic was not only the most commercially important region for Lufthansa, but also the central political focus of West German foreign policy. Moreover the United States, for political reasons, gave more encouragement to the new German airline than, for example, to the long-established Dutch flag-carrier KLM.[59]

Lufthansa's North Atlantic service opened in July 1955, followed by routes to South America and the Middle East in 1956. In April 1956, Lufthansa opened an additional service to the US, via Manchester and Montreal to Chicago. This innovative route, which was the first direct service from the North of England to North America, was in part a solution to Lufthansa's lack of slots at London Airport. In 1961 the Middle East route was extended to Hong Kong and Tokyo and in 1965 to Sydney. In 1962, Lufthansa began services to South Africa. While other European flag-carriers focused on long-distance routes to reach colonies and former colonies, Lufthansa flew them because they were the most profitable to operate and because it was most anxious not to limit itself to purely European operations at the beginning of its new existence.

Like other airlines, Lufthansa depended on the government to secure it new routes through bilateral agreement. Being a latecomer brought both advantages and disadvantages for the FRG. Certainly each foreign airline that already flew to West Germany could be used as a bargaining chip to acquire reciprocal rights for Lufthansa, but in most cases it had to accept an inferior status in bilateral agreements. For instance, Lufthansa did not get as many routes to the United States as American airlines enjoyed to West Germany but did receive more landing rights in the USA than it could actually serve. In 1955 Lufthansa already had the right to fly via New York to South America, while KLM had to wait ten more years to gain this right. Likewise it had full rights to Hong Kong and Singapore, because of the British rights in Germany, years before it actually flew to those destinations. In the case of the Dutch and Scandinavians, Lufthansa put pressure on the FRG government to cancel the rights of KLM and SAS to serve the West German domestic market, which would have meant that those countries would have had little to offer in return, however the government refused to take that step and even threatened to dismiss Bongers if he continued his agitation against the other airlines' privileges in Germany.[60]

After bilateral agreements between governments, the airlines usually entered into pooling agreements between themselves.[61] Pools were the kind of market instrument that European flag-carriers loved because they secured their monopoly and made life easier.[62] On the other hand they also underscored the flag-carriers' dependence on their respective government. Bilateral agreements often include hidden understandings about joint ventures, royalty payments or contracts for training and maintenance. Early on Lufthansa signed somewhat disadvantageous pooling agreements. For example Lufthansa did not manage to get landing rights in Caracas until 1971 and blamed KLM for this, since the Dutch flag-carrier supported the Venezuelan airline VIASA and wanted to keep the Germans out of that particular market.[63] Lufthansa frequently had to accept less than optimum pool arrangements because other European pool partners had more attractive airports to offer, the chief exception being the French – Air France accepted equal pooling agreements with Lufthansa from the beginning.

Founded at the height of the Cold War, Lufthansa not surprisingly had little opportunity at the outset to expand in Eastern Europe. The FRG government sought recognition by East European states as *the* German representative, but it was only in the 1960s, as a sign of *détente*, that Lufthansa got landing rights in Prague and Warsaw, with Yugoslavia, Romania and Hungary following suit in 1967. As the West German state became more prosperous and economically successful, the Comecon states grew more interested in economic contacts and after the first routes were established, others followed quickly. Understandably, the GDR tried to prevent this development, but had little success. In 1973, the first Lufthansa aircraft landed in the GDR itself, bringing visitiors to the international fair at Leipzig. However it took until 1988 before scheduled services between the two Germanies were operational.

A political aim of West German aviation policy was to persuade foreign countries not to grant the GDR landing rights and thus isolate the East Germans internationally. This was a main reason why western states refused to sign bilateral treaties with East Germany with the consequence that international connections to the West never played an important role for Lufthansa [East] and Interflug. Routes within the Comecon states were also not highly developed although that was for economic rather than political reasons. After 1962, air fares within the Comecon states were fixed at such a low level that the GDR government soon realized that it was cheaper to let its citizens use the airlines of the other socialist countries than develop an international network of its own. Where international routes did have a function for the GDR, they tended to be defined in political terms, usually as a step towards recognition of the GDR as an independent state. Intially it failed to receive this recognition but after 1970 was more successful. In addition, the opening of international routes was a sign of solidarity with other socialist states and for this reason Interflug flew regularly to Cuba, Vietnam, and Angola but never to London, Paris or New York. Not surprisingly, these services lost money continually.

The expansion of tourism in the 1960s changed the international market structure for both German airlines. At first, the East Germans did not like the idea of tourism as a stimulus for its airline: the flag-carrier was supposed to serve businessmen and promote economic development. But in the 1970s, private consumption became a more prominent goal within the GDR and recreational travelling with Interflug was encouraged. By the early 1980s, Interflug's official assignment had become tourist transport.[64]

3.3   German airports and the Lufthansas

In contrast to most other European states which had capital hubs like London, Paris or Amsterdam, Germany had a polycentric network of airports similiar to the United States. In West Germany the airports were organized as private companies, although their stock often belonged to local authorities. Up to 1945, Berlin-Tempelhof had had an outstanding position as a central hub for European air transport. It was by far the largest German airport, although it did not have a total monopoly on international connections. After the war, however, Berlin could no longer fulfil this role. Air transport between East and West Europe, the source of Berlin's strength, was now in decline and for the more important Atlantic routes Frankfurt was better positioned, further to the west. Moreover German airlines were not allowed to fly to Berlin and West Berlin itself was a political island without a hinterland.

Table 4.2   The importance of German airports
Per cent of total air traffic counted in travel units (TU)

	1950	1960	1970	1980	1989
Berlin	30.7	21.0	16.4	8.0	7.0
Düsseldorf	7.7	11.2	10.8	13.1	12.8
Frankfurt	30.0	29.6	34.0	42.4	43.8
Hamburg	17.7	12.5	9.4	8.7	7.7
Munich	7.5	9.3	10.2	11.2	12.6
Stuttgart	2.8	3.7	5.2	5.2	4.7
Other	3.6	11.7	14.0	11.4	11.4

1 TU = 1 passenger = 100 kg freight = 100 kg post
Source: Werner Treibel, *Geschichte der deutschen Verkehrsflughäfen. Eine Dokumentation von 1909 bis 1989*, Bonn, Bernard und Graefe, 1992, p. 35.

In some countries, the establishment and enlargement of airports demanded close cooperation between airport authorities and airlines. In Germany, airports were the first institutions in commercial aviation to be returned to German control in the late 1940s, at a time when there was no German airline with which to cooperate. This is one reason why German airports have aquired more influence in German air transport than else-where.[65] The American Treasury Secretary Henry Morgenthau demanded in

1944, in his *Plan to Solve the German Problem* that 'no German shall be permitted to operate or to help operate any aircraft'.[66] But the Potsdam Treaty laid down no conditions regarding airports and in 1945 the Allied authorities were already hiring German experts to help build their airports in Germany. In late 1946, the first airports were handed over to local German administration and in 1947, two years before the foundation of the Federal Republic, the Association of German Airports (ADV) was founded – the first official German civil aviation organization and its most powerful lobby and research institution for some time to come.[67] The ADV had around ten first class members which were international airports.[68] Three of these lay in the American Zone of Occupation (Munich, Frankfurt and Bremen), three in the British Zone (Hamburg, Düsseldorf and Cologne), one in the French Zone (Stuttgart) and three in West Berlin (Tempelhof, Tegel, and Gatow).[69] Later in the 1970s new members (Nürnberg and Münster-Osnabrück) joined the first class round and after German reunification in 1990 the former East German international airports joined.

With the foundation of the FRG all the international airports in West Germany sought financial support from the federal government, but only Frankfurt, Munich, Berlin, Hamburg and Cologne received help and that was through part ownership. With the establishment of Lufthansa in 1955, the airports competed with each other to become the airline's home base. For geographical reasons, Hamburg and Munich seemed to have the best chances because Lufthansa was trying to combine domestic and international routes in its early years in order to save aircraft. Lufthansa's international routes therefore started from the periphery – Hamburg or Munich – and not from Germany's centre. The director of the ADV recommended Munich as Lufthansa's central maintenance base, but Hamburg was actually chosen because of the better financial terms it offered. In the following years Düsseldorf and Frankfurt also received direct connections to the US and Lufthansa's administrative headquarters was moved to Cologne, while the airline's pilot school was situated in Bremen. To complete the distribution of the prizes, Stuttgart became the home of ADV.

Many aviation experts predicted that Germany's decentralized airport structure would change with the arrival of jets. Intercontinental jets required longer runways, and in 1958 the federal government decided to support the construction of a single airport for jets in the centre of the country.[70] Frankfurt was the only airport which met this requirement, although the new airport at Cologne-Bonn was discussed as a second possibility on account of its close proximity to the West German capital in Bonn.[71] The other airports naturally opposed this decision[72] but in the long run, it turned out that the technical obstacles had been overestimated and within a decade all of West Germany's international airports had extended their runways to cater for jets. Nonetheless Frankfurt profited from the delay in the construction of longer runways at other airports and has managed to maintain its predominant position in passenger traffic ever since – in 1995 forty percent of all Lufthansa flights began or terminated at Frankfurt.[73]

In the Soviet Zone, the only civil airport that existed was Leipzig, to which international aircraft were allowed to fly for the International Fair in spring and autumn.[74] Therefore in the GDR airports were not able to revitalize air transport as in West Germany and Lufthansa [East] had no partners with whom to cooperate. The lack of independent airports remained an important obstacle to the expansion of East German civil aviation right up to 1990. Karl-Marx-Stadt (Chemnitz), Erfurt, Barth and Heringsdorf did not even have concrete runways and this was one of the reasons why East German domestic air transport was abandoned in 1980; the old Russian Antonov aircraft were obsolete and their successors required concrete runways. Leipzig and Dresden were somewhat better provided for, but only Berlin-Schönefeld came close to meeting the standards of a western airport. Schönefeld had been a military airport until 1955, when the Soviet Union returned part of it to German control. It was not until 1970 however, when the East German airports became part of Interflug, that plans to enlarge Schönefeld were carried out.

It is a mystery why the GDR's Politbüro, despite of its ideological preference for public transport, did not invest in the necessary infrastructure for airports, not to mention the railway system. In comparison with industrial investment, the GDR's transport infrastructure was totally neglected. The only big project that was carried out was the new passenger terminal at Schönefeld and that was not a success. The air conditioning system, for example, was so poor that the director of Interflug threatened to break the windows in order to get fresh air into the building, and no baggage transport system was ever installed. Interflug's staff had to handle baggage manually in a rotating system. From the early 1970s, Interflug tried to attract Western customers to West Berlin to Schönefeld airport in order to earn currency, but the poor facilities hindered the attempt.

In West Germany the construction of new airports was discussed as the older ones became outdated. In the first years of the FRG it was easier to move an airport and modernize it and in Cologne the old city airport was simply transferred to the other side of the Rhine, while in Hannover the airport was transferred from Vahrenwald to Langenhagen. In 1970 Cologne got a modern drive-in terminal, and others followed in Hannover (1973) and Berlin-Tegel (1975). As this was the height of the motor age the airports did not receive a connecting urban transport system.[75] Gradually it became more difficult to alter the location of an airport in Germany and only Munich succeeded in doing so when, after 30 years of planning, the city's airport was moved from Riem to Freising in 1993. Plans to build a big southern airport between Munich and Stuttgart and another in the north outside Hamburg, were given up after years of fruitless quarrels between the competing *Länder*. Therefore, a North German alternative to Frankfurt has never emerged. After 1990 Berlin and the new *Land* of Brandenburg planned to build a new airport for the two states.[76] Brandenburg favoured a site 50 miles south of Berlin, but finally agreed to the existing site at Schönefeld after the plan to merge the two *Länder* failed in May 1996.

Perhaps not surprisingly, airport siting in Germany has a lot to do with state politics. Germany has been lucky to have had so many *Länder*, for it now has no less than sixteen international airports and they prosper because of the rapidly growing air transport market. In the pre-war period, the great number of airports brought losses to their owners, but now Germany's federal structure has become an important aid in the expansion of civil aviation.

4 PERFORMANCE

4.1 Operating and financial performance

Profitability has played a vital role in Lufthansa's management but not with its East German counterpart. However, performance is more than profitability: reliability has always been a part of Germany's national image and Lufthansa's public relations has traditionally valued it more than any other selling point.

In its early years Lufthansa needed more government subsidies than Bongers had anticipated and the reason was simple. The running costs of the 12 Lockheed Constellations and Superstars were substantially higher than expected and not competitive with the costs of competing Douglas aircraft. Moreover the Constellations quickly lost their value after the first Boeing jets arrived on the market. 1961 was an especially bad year because the oversupply of seats on the North Atlantic, caused by the new jets, led to a fall in Lufthansa's load factor from 71 per cent to 56 per cent. Although the German flag-carrier rose in 1963 from 21st to 13th in the list of world airlines (measured in passenger-miles), it had a bad press.[77] In the long run, however, jets made air transport much more profitable as efficiency rose and maintenance costs fell. With the write-off of losses on the Constellations backed by the government, Lufthansa made a profit for the first time in 1964.

In its first year Lufthansa had only 2 per cent of first class passengers. By 1960 the figure had risen to 8 per cent but then decreased again to 3 per cent in 1964.[78] In the early 1970s when all the major IATA airlines had to reduce fares to fill empty seats, Lufthansa managed to keep a bigger share of the First Class market than others with a successful marketing campaign aimed at businessmen. Its high quality image seems to have been so convincing that

Lufthansa had to counterbalance its marketing strategy in the 1980s to convince customers that it could also offer competitive low fares.

For a long time Lufthansa was neither able nor willing to offer the low budget fares which would have been necessary to achieve a greater share of the tourist market. It therefore decided to found its own charter airline, and after some disappointments with German partners in the tourist trade, established a charter airline which became known as Condor after 1961.[79] Gradually Lufthansa was transformed from a business airline to a business *and* tourist carrier. For tourist destinations it used charter airlines like Condor while on its scheduled services it pursued a policy of offering the lowest fares possible.

Table 4.3   Lufthansa's operating and financial performance, 1955–81

	Passengers (millions)	Employees	Revenue DM million	Deficit/Surplus DM million
1955	74.1	2 040	23.6	−20.5
1956	229.7	3 443	80.6	−20.1
1957	429.9	4 940	128.8	−27.1
1958	622.5	6 040	178.2	−35.2
1959	786.6	7 441	230.7	−42.1
1960	1 237.6	9 564	351.6	−4.3
1961	1 553.4	11 981	453.2	−109.1
1962	1 859.0	12 434	529.1	−46.4
1963	2 152.2	11 224	622.6	1.8
1964	2 567.4	11 963	767.1	35.7
1965	3 218.4	14 990	940.3	43.0
1966	3 688.4	16 483	1 165.8	47.9
1967	4 267.4	17 970	1 381.3	39.4
1968	4 970.7	18 261	1 525.2	0.0
1969	5 874.2	19 745	1 772.0	12.3
1970	6 958.3	21 948	2 033.0	12.9
1971	7 539.0	22 841	2 391.4	−34.0
1972	8 523.2	22 888	2 537.6	56.8
1973	7 984.4	23 761	2 775.8	−45.7
1974	9 602.1	24 441	3 443.6	64.5
1975	10 147.9	25 340	3 760.2	33.5
1976	11 223.4	26 451	4 300.4	112.3
1977	11 704.9	27 677	4 561.6	39.7
1978	12 576.2	29 400	5 015.0	42.5
1979	13 736.9	29 838	5 644.8	68.8
1980	13 943.9	30 664	6 404.0	5.6
1981	13 894.4	30 696	7 738.9	5.6

Table 4.3 continued

	Passengers (million)	Employees	Revenue DM million	Surplus/Deficit DM million	CTM	LTM
1982	13 848.9	30 712	8 113.8	45.0	6 358.8	3 768.4
1983	14 307.9	31 575	8 821.0	63.0	6 753.0	4 226.1
1984	15 334.9	32 316	10 373.5	162.0	7 186.2	4 747.1
1985	15 921.2	34 905	10 689.6	63.0	7 767.9	4 937.7
1986	16 618.1	37 920	10 191.7	64.0	9 800.9	6 458.8
1987	18 392.0	39 139	9 831.0	86.0	10 961.4	7 497.1
1988	19 385.7	40 684	10 606.0	99.0	10 417.3	6 984.1
1989	20 447.5	43 565	11 812.0	124.0	11 169.9	7 550.3
1990	22 367.0	47 619	12 806.0	9.0	12 514.2	8 209.1
1991	25 070.6	49 641	14 318.0	−444.0	13 002.0	8 418.7
1992	27 760.2	48 351	14 955.0	−373.0	14 064.7	9 169.9
1993	28 438.5	44 149	14 967.0	−111.0	14 429.6	9 986.2

Source: Lufthansa AG (ed.): Zahlen, Daten, Fakten. Köln, 1966–94.

In the 1980s Lufthansa also became one of the biggest freight carriers in the world. Initially it had merely added cargo to ordinary passenger flights, but then as early as 1957 it began its first all-freight service between Frankfurt and New York. Twenty years later it founded German Cargo as a freight charter airline.[80] And in 1987 it purchased 25 per cent of Cargolux, a non-scheduled cargo carrier located in Luxembourg. Lufthansa's freight business has actually grown more quickly than the passenger business and since 1986 the airline has been the biggest international air cargo carrier in the world.[81] The same picture of German strength in cargo can be seen at the airport level. In 1988 Frankfurt airport was tenth largest in terms of passenger numbers in the world, but second in freight.[82] This achievement was partly a reflection of West Germany's strong export economy and partly a sign of worldwide confidence in German *reliability*, certainly Lufthansa's marketing of its cargo services plays on this perception of German thoroughness.

A comparison with other European flag-carriers shows the relative strength of Lufthansa in freight in 1993 (see Table 4.4).

In East Germany, civil aviation's performance was quite different and its flag-carriers made continual losses. Until the end of the GDR, Interflug could rely on a total subsidy on tickets as well as further support from the government for investment in aircraft and buildings as well as fuel supply. In its reports the airline was presented as a profitable business, because these subsidies were hidden.[83] In the late 1970s, when funds were short in the GDR, Interflug came under pressure to make profits and it gave up domestic aviation for this reason.[84] At the same time, it extended its charter business

Table 4.4 Lufthansa's strength in freight transport

1993	Passenger-kilometers (thousands)	Cargo tonne-kilometers (thousands)
Lufthansa	52 657	4 635
Air France	43 534	3 581
British Airways	80 085	2 733
KLM	36 806	3 070
Alitalia	24 520	1 326

Source: AEA (ed.): *AEA Yearbook 1994*, Bruxelles, 1994, pp. 74–8.

for Western tourists from Schönefeld airport.[85] In the eyes of the Politbüro, Interflug was an expensive political necessity, but growth of the airline was not necessary since East Germans could travel with other cheap Comecon airlines, and be thus subsidized by other Comecon governments. It was this attitude which kept Interflug small.[86]

Compared to American airlines, all the European flag-carriers look inefficient when it comes to staff productivity. However Lufthansa's overmanning problem was never as serious as that of BA in the 1970s or Air France in the 1990s. The airline had started modestly and it kept its lean structure for quite a time. During the airline industry crisis years of 1982–3 and 1991–2, Lufthansa's top management was changed promptly and staff productivity rose; both times the executive board proved able to transform the situation. In the years from 1978 to 1995 staff productivity at Lufthansa rose from 103 000 tons per kilometre per year to 347 000.[87] The changes after 1991 were more fundamental. The new chairman Jürgen Weber quickly reduced the number of aircraft by 26, the number of staff by a thousand, the number of managers by 12 per cent and training costs by 95 per cent.[88] The programme that Lufthansa had been funding to help passengers with their fear of flying, the *Seminar für entspanntes Fliegen*, was also reduced.[89] In the following years Weber went further and announced staff reductions of 7000 to be carried out by the end of 1994. He also cut old privileges which had given employees of Lufthansa a similar status to civil servants.[90] Weber believed that Lufthansa was in a position between the losers (Air France) and the winners (BA) of deregulation and that it had to make sure it ended up among the latter. By mid-1993 he had reduced Lufthansa's staff by 4200 and had additionally managed to cut the fees the airline paid to German airports.[91] A further step towards cost reduction was the structural reform of the company in 1994/95 which organized Lufthansa as a group, with Lufthansa Technology, Lufthansa Cargo and Lufthansa Systems becoming independent companies with their own wage systems.

By 1997 40 per cent of the group's revenue was made outside the core Lufthansa company.

4.2 Marketing and service

The Germans are seen as precise and reliable, if a little dull, and Lufthansa has always tried to use these characteristics positively for its corporate identity. In air transport reliability can be measured in terms of punctuality and in this respect Lufthansa quickly became one of the leading airlines in the world. Another element in its marketing strategy has been safety and although this can be a two-edged sword for an airline, Lufthansa has managed to successfully exploit the fact that its aircraft fleet has a low average age (5.3 years in 1994),[92] and that the aircraft are well maintained. In its modern history, Lufthansa has suffered only three major accidents.[93]

Although not famous for its German cuisine, Lufthansa tries to give good service. From the Americans it learnt the value of Club Programmes and it revived the Lufthansa *Senator Klasse* as early as 1958. Under this scheme every passenger received a rose on board and could use the *Senator Lounge* in airports.[94] In 1992 it started the *Miles&More* frequent-flyer programme which it now shares with a number of other allied airlines. Lufthansa's corporate identity has been renewed more than once in its postwar history. However it was not before 1987 that a director of marketing was appointed.[95] He was responsible for the new look presented in 1988 which presented Lufthansa in white, silver and grey instead of the traditional colours of blue and yellow. All aircraft, airports and offices received a new look.

By the 1980s, Lufthansa seemed to have been too successful in convincing passengers of its high standards and survey results showed that its fares were thought to be much higher than they actually were. Budget-conscious tourists were convinced that they could not afford Lufthansa and were staying away. The airline had to reorientate its marketing to make it clear that Lufthansa was an airline for everybody and yet do so without losing its precious first class passengers.[96]

In contrast to the *lean* concepts of its American competitors, Lufthansa has always provided support services like pilot training, catering, and reservation systems for its own use and for other airlines. An example is the pilot school Lufthansa founded in Bremen in 1956, which has since trained more than 4000 pilots. The school not only served the needs of Lufthansa, but also trained pilots for Swissair, Austrian Airlines, Iberia and All Nippon Airways, among others.[97] Since 1993 the school has been open to external students who train at their own expense. Lufthansa also has a long history of cooperation with travel agencies and in the 1980s it offered independent travel agencies the opportunity to become Lufthansa 'City Centres'. It was equally active in computer reservation systems (CRS), launching the European *Amadeus* system with Air France and Iberia in 1992.[98] Like many airlines Lufthansa acquired substantial interests in hotels and in the early 1980s owned around fifty. However since then it has divested itself of most of them. In catering Lufthansa has had its own company, Lufthansa Service Ltd (LSG), since 1966 and is now one of the largest airline caterers in the world with more than 10 000 employees (1990). Unlike the independent American airline caterers such as Caterair and Dobbs, European caterers tend to belong to airlines.[99] LSG however is something in between. Since the 1980s, it has worked for customers outside Lufthansa and by 1990 was already doing 60 per cent of its business with other airlines. Because Lufthansa now leases many of its aircraft, it founded its own leasing company in 1989. In short, Lufthansa has not followed the American or British pattern since 1980 and limited itself to its core business. Instead it has maintained a wide field of associated operations, although it has placed services outside its core business in 'profit centres,' which have to compete for Lufthansa's orders.

5 STRATEGY

5.1 Competition and collaboration between the carriers

When Lufthansa was relaunched in 1955 the German domestic market was divided between airlines of the former Allies and other European countries. Because these airlines already had landing rights in Germany, the West German government could hardly offer them in bilateral exchange and Lufthansa usually got less than a 50 per cent share in agreements with other airlines. Moreover German airports, travel agencies and private customers already had good contacts to foreign airlines, so that in 1959 for example Swissair, SAS and KLM still accounted for nearly 40 per cent of the international air transport in Germany.[100] Lufthansa recognized that, as a German airline, it would have to work hard to achieve a cooperative atmosphere with its European rivals, and in the 1950s it was not easy for the Germans to join the established community of airlines, or international organisations like IATA and ICAO. From the start Lufthansa executives told their colleagues in other western airlines that they needed technical and economic advice, and Lufthansa's search for American and British pilots in the 1950s should really be seen as a step towards international integration. American airlines in particular, served as models for Lufthansa although eventually the German flag-carrier became quintessentially European in management and outlook.

German desire to be part of Europe was so strong that it was initially suggested that not a German, but a European airline should be founded.[101]

As early as 1951, Max Hyman of Air France had spoken of the need for a similar undertaking and when Hans Bongers visited Hyman's successor, Lesieux, the two managers agreed on initial collaboration between Air France, Lufthansa, SABENA, Alitalia, and Swissair.[102] KLM was also interested for a while but was never fully involved in the negotiations. Before the end of 1958, a joint venture was set up under the name Air Union, as well as two committees to represent airlines and governments. When Bongers was injured in a car accident in 1959, Kurt Weigelt, the Lufthansa board chairman, took over and he made Air Union his personal project. The German Ministry of Transport, as well as the German Treasury, were less enthusiastic. At the time they were covering Lufthansa's losses and they saw no reason for spending more money. Indeed the same scepticism of Air Union existed at the government level among all four partners (France, Germany, Belgium and Italy). They were more cautious and nationally minded than the airlines and this ultimately became an insuperable obstacle.[103] In addition, Air France and SABENA were more or less government departments, while Lufthansa and Alitalia, although partly owned by the state, were nonetheless more commercially oriented companies.

The biggest obstacle to Air Union was disagreement over the market shares of the four partners, which was never solved and finally destroyed the project.[104] As a new airline, Lufthansa was reluctant to accept long-term limits to its growth and when it was suggested that the 1958 distribution of the market be frozen, that is, Air France 46%, SABENA 20%, Lufthansa 17%, and Alitalia 16%, Bongers responded with a very different set of figures (Lufthansa 36,3%, Air France 34%, Alitalia 19.4% and SABENA 9.8%) which was almost bound to upset the French.[105] Eventually the airlines agreed on a compromise and the discussions became increasingly a matter of negotiations between governments.[106] Weigelt, the most powerful protagonist of Air Union at Lufthansa, had stepped down as chairman in 1961 and the Germans gradually lost interest in the project. Air Union was finally dropped in 1967, although the four partners did build up a maintenance support network for the Boeing 747, called ATLAS, which lasted until the early 1990s.[107] Air Union never took off, but the idea was an important indication that Lufthansa was ready to cooperate with its European competitors and this cooperation continued in the form of pooling agreements, technical cooperation, and joint ventures of various sorts.[108] In Europe the 'cooperative blocks' from the 1950s survived deregulation and continue as three groups: Swissair, SAS and Finnair; BA, KLM and SABENA; and Lufthansa, Air France and Iberia.[109]

Charter airlines have been particularly successful in Europe and have been one of the strongest sources of competition for the flag-carriers. However while in Britain they emerged as rivals to BEA and BOAC, in Germany Lufthansa sought to control them.[110] Indeed Lufthansa founded the most successful German charter airline, Condor Flugdienst GmbH in 1955, then called Deutsche Flugdienst. The idea was that it would exploit the booming tourist market, and use old Lufthansa aircraft. Lufthansa asked the two biggest German shipping lines, the Hamburg-Amerika Line [Hapag] and Norddeutscher Lloyd, to participate but after initial losses they quit the enterprise. In 1961 Lufthansa bought a small charter airline called Condor in order to use the name for Deutsche Flugdienst; the great eagle of the Andes had been adopted as the name of Lufthansa's famous pre-war subsidiary in South America.[111] In the following years, Lufthansa continued to buy other charter companies and add them to the Condor enterprise, e.g. Südflug in 1968. The reason was to *reduce* competition. As the national flag-carrier, Lufthansa wanted no other German airline to compete with it either in scheduled or charter services.

By 1975 West Germany was producing more foreign tourists than any other country in Europe. Tourism had become the biggest section of the air transport market and Lufthansa controlled the holiday charter business in its home country through Condor, while remaining a high-fare scheduled carrier itself. This was in contrast to other north European countries like Britain where the charter carriers were independent and in competition with the flag-carrier. However Lufthansa did face some local rivals and the largest was LTU (Luft Transport Unternehmen) which had been founded in 1955 with private capital. Until 1969 LTU flew charters for big German tour operators but in that year the biggest operator, TUI (Touristik Union International), pressured LTU not to sell tickets to small travel agents. LTU felt that it was big enough to ignore TUI and in the years that followed it began buying tour operators, taking over 40 per cent of Tjaereborg in 1986, for example, and more recently 90 per cent of Thomas Cook in 1992.[112] A third important German charter carrier was founded in 1972 by the shipping line Hapag-Lloyd. Being a co-owner of TUI and other tour operators, Hapag-Lloyd took advantage of the split between TUI and LTU to found its own airline. In this case Lufthansa accepted the new competitor because of the traditionally good contacts between the shipping company and itself, and in 1992 Lufthansa actually bought 10 per cent of Hapag-Lloyd.[113]

In general Lufthansa's policy has been to stifle competition from West German charter carriers by buying stakes in them. Even Hans Bongers felt the long arm of Lufthansa, when he was involved in a charter airline after his retirement from Lufthansa. He had underestimated the opposition of Lufthansa and its allies to any new airlines and the charter company was quickly forced into liquidation. For very different reasons, Lufthansa [East] also founded a charter airline. It was named Interflug and was intended to fly to destinations where the name Lufthansa was not appreciated. Interflug was needed for political rather than economic reasons. In 1963 Lufthansa [East] merged with Interflug, retaining the latter name, and thereafter it was the only airline in the GDR.

In most European countries the domestic air transport market has been more or less reserved for the flag-carrier. From the geographical point of view West Germany was large enough for a significant domestic system, but for a long time Lufthansa showed little interest.[114] On the other hand it was interested in protecting its domestic air transport monopoly.[115] After 1980 the situation changed and new regional airlines emerged in the FRG which attracted more and more passengers.[116] New small jets and turboprop aircraft became available, combining the merits of big aircraft with a carrying capacity suitable to low-traffic routes. Lufthansa was caught unprepared for this development and was surprised that it was possible to make money in this sector. However it reacted in typical fashion by buying into its new competitors. In 1978 it acquired 25 per cent of Deutsche Lufttransport GmbH (DLT), increasing its stake to 100 per cent by 1992 and renaming it in the process Lufthansa Cityline – later Lufthansa *Express*.

In the 1990s Lufthansa's most determined domestic competitor has been Deutsche BA. Originally known as Delta Air, Deutsche BA got its new name when Lufthansa's arch-rival British Airways acquired a 49 per cent stake in the airline in 1992. British Airways had been involved in West German domestic air transport since the end of the war, but its other partners on the Berlin route either went out of business (Pan American) or collaborated with Lufthansa (Air France), BA decided to compete with Lufthansa head-on in its home market; in 1995 Deutsche BA had 21 aircraft and was growing rapidly.[117]

The number and importance of new regional airlines is growing in Germany. German Air offers a Berlin–Bonn shuttle for politicians and government officials while Eurowings offers more than 200 regional international scheduled routes. These new airlines got their chance partly through Lufthansa's negligence of market opportunities in Germany but also because German federalism has helped the smaller airlines thanks to the support from independent airports and *Länder* which want their own carriers, for example Nürnberg has supported NFD (Nürnberger Flugdienst), Augsburg helps Interrot, and Kiel supports Cimber Air. For the smaller airports in Germany, regional air transport has offered the chance of strong and sustained growth.[118]

5.2 Intermodal competition and collaboration

Lufthansa initially earned its money on the North Atlantic, so how did the well-established German shipping companies react to this competition in the air? Before the war the two main German shipping companies Hapag and Norddeutscher Lloyd had a close relationship with Lufthansa.[119] After the war Germany was allowed to have a merchant shipping fleet only in 1953. Hapag and Lloyd concentrated on freight and had neither time nor money for risky investments in passenger services. Moreover they doubted that passenger sea transport had a future.[120] Lloyd saw the future in cruise shipping and in collaborative ventures with air transport, in which Lufthansa would be a partner, not a competitor.[121] Lloyd's director, Emil Kipfmüller, was an old friend of Hans Bongers and he not only joined the board of the old Lufthansa in liquidation, but also helped the new Lufthansa with staff and ticketing. He was so loyal to Lufthansa that he refused to join a charter airline except as a Lufthansa subsidary.[122] In the long run this policy was successful for when Hapag-Lloyd founded its own airline, after the merger of the two shipping companies in 1970, it could count on Lufthansa for support. and, not surprisingly, eventual co-ownership.

Unlike the shipping companies the German Railways (DB) quickly identified Lufthansa as a major competitor. Being state-owned, the DB used their connection to the German Ministry of Transport to limit the growth of domestic air transport. The first Minister, Hans-Christoph Seebohm (1949–66), tried to make Lufthansa and DB cooperate with each other.[123] Since both enterprises were dependent on the federal government, the Ministry wanted them to cooperate, but they remained competitive.[124] The DB was more affected by the competition than Lufthansa because it was more interested in the domestic market and suffered a steady loss of passengers to air transport. The DB was in fact a founding owner of Lufthansa with one per cent of the stock and successfully demanded a seat on Lufthansa's advisory board in order to control the new competitor. The DB's Trans-Europa-Express (TEE) which began in 1957 was explicitly aimed at air transport customers, yet Lufthansa responded in 1963 with cheap Air-Bus fares between West German cities. Without check-in formalities or costly reservation systems, these services promised to be cheaper than a first class ticket on the TEE. The DB was alarmed and in a defensive move managed to get the Ministry to disallow the Air-Bus's 'dumping fares'. Without them they were no longer viable and the scheme was cancelled in 1966.

Until the 1970s Lufthansa was open to collaboration with the German railways because it did not consider domestic aviation as especially profitable and it recognized the railways' strength in the Ministry.[125] Only after domestic aviation became more profitable in the 1980s did competition intensify.[126] One of the few successful joint ventures between the two transport modes was the express train between Düsseldorf and Frankfurt airports, founded in 1982 and run jointly by DB and Lufthansa. However it was cancelled in 1993 and replaced by one of the DB's own high-speed trains.[127] The long-awaited baggage check-in facility on railway stations for international flights did become a reality in 1991 but only on a small scale.[128] And it remains something of a mystery why there was not greater cooperation between Lufthansa and the DB in the face of the common danger from the motor car, why for instance did the new West German airports of the 1960s (Cologne, Berlin-Tegel and Hannover) get better road access than railway connections? The likely answer would seem to be that there were no joint ventures in place between the railways and Lufthansa that would have profited from collaboration.[129]

One of the results of Lufthansa's domestic services, combined with increasing car ownership in the FRG, has been the development of high-speed trains. The TEE was a first step in 1957 and was followed by the Intercity system in 1979. The DB learnt from the airline how to serve businessmen and in 1981 it hired Lufthansa's marketing expert, Hemjö Klein, to make its passenger services more competitive. It was Klein who invented the discount Bahn-Card and pushed for the introduction of the latest German high-speed train, the ICE, which started operations in 1991. The ICE Sprinter, which, for example, takes only 3 hours 16 minutes from Munich to Frankfurt, brought about a genuine revival of business travel on the railways. Not surprisingly Lufthansa re-hired Klein as its new sales manager in 1993. With increased speed the DB has been experiencing growth in its long-distance business for the first time since the war and some experts even predict that the railways will reclaim a large share of the German domestic market from air transport in the years to come.[130]

However the main competitor of domestic air transport is the privately owned motor car. The number of cars per thousand citizens in West Germany rose from 12 in 1950, to 412 in 1984, with a comparable trend in the GDR. The process has been encouraged by the West German government which has continously expanded the Autobahn network since 1957.[131] Indeed in the late 1960s the Minister of Transport, Georg Leber, promised that no German would live more than 25 kilometres from an Autobahn ramp.[132] Lufthansa reacted to competition from the car with competitive pricing and tried to set fares that allowed car owners, driving alone, to save money by flying. Lufthansa also collaborated *with* the car *against* the railways by improving airport connections to the Autobahnen. For the middle classes living in the suburbs, it is often quicker to reach the local airport in their car than reach the city centre railway terminus by public transport. Lufthansa's ties with car rental companies are a further manifestation of this link between wings and wheels.

5.3 Does the European flag-carrier have a future?

Lufthansa is now one of the biggest airlines in the world, but it has changed fundamentally since the 1950s and in the 1990s it has become a *global player*. Two developments have made this change possible: privatization and strategic alliances. In 1960 the West German federal government owned 85 per cent of Lufthansa, thirty years later the government's stake was below 40 per cent and is falling further. This privatization process has taken place more smoothly than in Britain because Lufthansa was always, notionally at least, a private company, but nonetheless it has taken place. Lufthansa has also been active in the other field of strategic managment that has become critical in the present era of international airline deregulation and that is the formation of global alliances. With such alliances former national flag-carriers can become global carriers, not by changing their route structures, but by enhancing them through calculated marriages to other airlines with complementary systems – in other words, strengths in different parts of the world. Lufthansa's most important international partner in the 1990s has been the American megacarrier United Airlines, the alliance to which has opened the US domestic market to the German flag-carrier. In addition the Lufthansa-United alliance now includes other complementary airlines such as SAS, Air Canada and Thai International.[133]

If we look at the years from 1955 to 1995, we can see the national percularities of the German flag-carriers and the differences between the two Lufthansas. Initially the GDR supported civil aviation as the realization of a modern socialist transport system, meanwhile in the West the government pursued a more reserved policy towards the aircraft industry and air transport. In the 1960s, however, the GDR abandoned its plans for aircraft manufacturing and air transport, while in the West the aircraft industry received funds and Lufthansa became a profitable business. This development was the result of political antagonism. The two Germanies were obliged to face one another and differ. However there were some similarities between the two national airlines and government air transport policies. Both airlines wanted to stand in the tradition of Lufthansa and this showed in their rival claims to the famous old name. And both German governments also saw a fundamental responsibility in the 1950s to re-establish and support a national flag-carrier. Similar technological conditions shaped developments on both sides of the Iron Curtain, as American airlines and American aircraft served as a model for both Germanies.

Lufthansa's current boss Jürgen Weber has stated that the airline has no future as national flag-carrier, only as a global player.[134] On the other hand, he has stressed that the traditional characteristics of Lufthansa's service must continue. It is an open question how well the shift from 'made in Germany' to 'designed in Germany' will work. It is likely however that the national characteristics that have made Lufthansa successful in the past will remain important in the future.

NOTES

1. The only books on Lufthansa (East) and Interflug are Hansjochen Ehmer, *Der zivile Luftverkehr der DDR*, Berlin, 1983; Karl-Dieter Seifert, *Weg und Absturz der Interflug: der Luftverkehr der DDR*, Berlin, 1994.

2. There is no academic history of Lufthansa or Interflug which covers the whole period from 1945 to 1995. For Lufthansa see: Joachim Wachtel, *Die Geschichte der Deutschen Lufthansa*, Köln, 1980. Deutsche Lufthansa (ed.), *Zehn Jahre Deutsche Lufthansa*, Köln, 1965. Hans Bongers wrote two important books, an autobiography: Hans M. Bongers, *Es lag in der Luft. Erinnerungen aus fünf Jahrzehnten Luftverkehr*, Düsseldorf, Econ, 1971, and Hans M. Bongers, *Deutscher Luftverkehr. Entwicklung, Politik, Wirtschaft, Organisation. Versuch einer Analyse der Lufthansa*, Bad Godesberg, 1967. See also Rudolf Braunburg: *Die Geschichte der Lufthansa. Vom Doppeldecker zum Airbus*, Hamburg, 1991; Georg Reul, *Planung und Gründung der Deutschen Lufthansa AG, 1949–1955*, Köln, 1992.

3. Hans M. Bongers, *Deutschlands Anteil am Weltluftverkehr*, Leipzig, 1938.

4. Correspondence between Bongers and Treibel 1946–1955. Personal achives of Werner Treibel, Leinfelden-Echterdingen.

5. Luftverkehr. Bundesarchiv Abteilungen Koblenz, B 108, 1961.

6. Leo Brandt, *Verkehrstechnik und Verkehrspolitik. Ausgewählte Kapitel aus einer Vorlesung an der Technischen Hochschule in Aachen im Wintersemester 1949/50*, Düsseldorf, 1950. Heinz Röhm, *Die Entwicklungstendenzen im deutschen Luftverkehr*, Düsseldorf, 1952.

7. The base employed many more personnel than the airport. 1961: 1717 (airport) 3550 (base); 1990: 4371 (airport), 8739 (base). Patrick Rodeck, *Flughafen und Luftwerft in Hamburg als Determinanten der Stadtentwicklung*, Diplomarbeit, Hamburg, 1991, pp. 45–8 and 131.

8. Already in 1925, Bongers was convinced that air transport should and could be profitable. Together with Kurt Weil he wrote a memorandum: Hans Bongers and Kurt Weil, *Betrachtungen über die Möglichkeit eines eigenwirtschaflichen Luftverkehrs*, unpublished, November 1925. Lufthansa AG Archives.

9. This was an important argument, for foreign currency was scarce and the German Mark was not freely convertible before December 1958 (Declaration of the International Monetary Fund). Not before the London Debt Treaty of February 1953 could German institutions get international loans. Manfred Pohl, *Wiederaufbau. Kunst und Technik der Finanzierung 1947–1953*, pp. 137–40.

10. '*Die Entwicklung des Grundkapitals der Deutschen Lufthansa AG*'. Lufthansa AG, *Presse und Öffentlichkeitsarbeit*, 21.5.1996.

11. 'Die gefährdete Deutsche Lufthansa', *Die neue Zeit*, 10.5.1952.

12. See: *Travel reports to the United States* of Werner Treibel of 1952, 1954, 1959 and 1964. *Travel report of Kurt Knipfer*, Abteilungsleiter L of German Ministry of Transport, October 1954, Arbeitsgemeinschaft deutscher Verkehrsflughäfen. Archiv, folder, Reiseberichte USA. Erik Grasser, *Die staatliche Regulierung des Wettbewerbs im internationalen amerikanischen Luftverkehr. Ein Beispiel gelenkter Konkurrenz in der Wirtschaft*, Bern, 1953.

13. Rudolf Braunburg, *Die Geschichte der Lufthansa. Vom Doppeldecker zum Airbus*, Hamburg, 1991, pp. 181.

14. Karl-Dieter Seifert, *Flugzeuge überall*, Berlin, 1960. Wolf-Dieter Picht, *Strassen der Zukunft*, Berlin, 1957.

15. Karl-Dieter Seifert, *Weg und Absturz der Interflug: der Luftverkehr der DDR*, Berlin, 1994, pp. 97.

16. *Abschlußbericht des Vorbereitungsausschußes Luftverkehr*, unpublished, Bonn, October 1952, Lufthansa AG Archives.

17. Bernd J. Höfer, *Strukturwandel im europäischen Luftverkehr. Marktstrukturelle Konsequenzen der Deregulierung*, Frankfurt, 1993, pp. 135–48.

18. Bernhard Wiezorek, *Luftverkehrsbeziehungen zwischen der Europäischen Union und den Vereinigten Staaten*, Bonn, 1994.

19. Ulrich Meyer, 'Die Rolle der Luftfahrtunternhemen bei der Luftverkehrsentwicklung in Europa nach 1945', in *Beiträge zur Geschichte der Binnenschiffahrt, des Luft- und Kraftfahrzeugverkehrs*, Bergisch-Gladbach, 1994, pp. 248.

20. Head of the most important Sonderkonstruktionsbüro became Brunolf Baade. Helmut Bukowski and Manfred Griehl, *Junkersflugzeuge 1933–1945. Bewaffnung, Erprobung, Prototypen. Der illustrierte Original Bericht des Professor Brunolf Baade an die Sowjetische Militäradministration, Dessau 1946*, Friedberg, 1991.

21. Jürgen Michels and Jochen Werner (eds): *Luftfahrt Ost 1945–90. Geschichte der deutschen Lulftfahrt in der sowjetischen Besatzungszone (SBZ), der Sowjetunion und der Deutschen Demokratischen Republik (DDR)*, Bonn, 1994. Hans-Liudger Dienel, 'Das wahre Wirtschaftswunder. Flugzeugproduktion, Fluggesellschaften und innerdeutscher Flugverkehr im West-Ost-Vergleich, 1955–80'. in Johannes Bähr and Dietmar Petzina (eds), *Innovationsverhalten und Entscheidungsstrukturen. Vergleichende Studien zur wirtschaftliche Entwicklung im geteilten Deutschland 1945–1990*, Berlin, 1996, pp. 341–72. Burghard Ciesla, 'Von der Luftkriegsrüstung zur zivilen Flugzeugproduktion. Über die Entwicklng der Luftfahrtforschung und Flugzeugproduktion in der SBZ/DDR und UdSSR 1945–54', in Hans-Jürgen Teuteberg (ed.): *Beiträge zur Geschichte der Binnenschiffahrt, des Luft- und Kraftfahrzeugverkehrs*. Bergisch-Gladbach, 1994, pp. 179–203. Gerhard Barkleit and Heinz Hartlepp, *Zur Geschcihte der Luftfahrtindustrie in der DDR 1952–1961*, Dresden, 1995 (Berichte und Studien aus dem Hanna-Arendt-Institut für Totalitarismusforschung 1).

22. Jürgen Michels and Jochen Werner (eds), *Luftfahrt Ost 1945–90. Geschichte der deutschen Lulftfahrt in der sowjetischen Besatzungszone (SBZ), der Sowjetunion und der Deutschen Demokratischen Republik (DDR)*, Bonn, 1994, pp. 73.

23. Kindscher, Rolf, 'Entwicklungstendenzen im Luftverkehr. Eine politisch-ökonomische Betrachtung'. Berlin 1960, pp. 114–17; 'Der Luftverkehr in Westdeutschland'; in: *Fliegerjahrbuch 1958*. 'Eine internationale Umschau des Luftverkehrs'. Berlin, 1959, pp. 43–6; 'Die westdeutsche Luftfahrtindustrie im Dienste der Wiederaufrüstung', in: *Fliegerjahrbuch 1959*. 'Eine internationale Umschau des Luftverkehrs'. Berlin, 1960, pp. 55–63.

24. Bundesarchiv, SAPMO, ZPA, DY 30/J IV, 2/2A/805.

25. Critique on the ambitious programme was refused by the planning commission. See the comments of the planning commission to the ideas of the aircraft works, Bundesarchiv Potsdam, DE 1/8050, DE 1/14433, DE 1/8049.

26. Joachim Wölfer, *Deutsche Passagierluftfahrt von 1955 bis heute*, Hamburg, 1995, pp. 103.

27. In 1951 a *Verband zur Förderung der Luftfahrtindustrie* was founded by the old manufacturers. *Gründungsversammlung am 13.11.1951*, DASA Archives Bremen, Feilcke-Papers.

28. The western 'Wirtschaftswunder' relied more or less on pre-war technology. Joachim Radkau, *Technik in Deutschland. Vom 18. Jahrhundert bis zu Gegenwart*, Frankfurt, 1989, 339–48.

29. Luftag Advisory Board meeting 14.5.1953, and *Schluβbericht der Arbeitsgruppe 'Lockheed Dokumente'* of 22.12.1977. Lufthansa Archives Köln.

30. Hans M. Bongers, *Es lag in der Luft. Erinnerungen an fünf Jahrzehnte Luftverkehr*, Düsseldorf, 1971, p. 238. It was never clarified whether bribery might have played a role too, as in the Starfighter case. Arbeitsgruppe 'Lockheed Dokumente,' of 22.12.1977. Lufthansa Archives Köln, without signature. Rudolf

Braunburg, *Die Geschichte der Lufthansa. Vom Doppeldecker zum Airbus*, Hamburg, 1991, pp. 188.

31. Hans Bongers, *Luftverkehr mit Turbinenflug*, Hof, 1960, pp. 36.
32. Hans M. Bongers, *Es lag in der Luft. Erinnerungen aus fünf Jahrzehnten Luftverkehr*, Düsseldorf, 1971, pp. 284.
33. Gerhard Höltje, *Aktuelle technische Probleme der Luftfahrt*, Köln, 1963.
34. DC *Zehn. Der 100. Flugzeugtyp in der Geschichte der Lufthansa*, Steinbach, 1977.
35. Lufthansa AG (ed.) *Finanzdaten 1991–1995*, Frankfurt, 1996.
36. Rolf Stüssel, *Die Senkrechtstarttechnik im zivilen Luftverkehr unter besonderer Berücksichtigung der Verhältnisse in Europa*, PhD Diss., Berlin, 1965.
37. *He 211 folder*, Heinkel Archives Stuttgart. *HFB 320 folder and VFW 614 marketing studies*, DASA Archives Bremen.
38. Edger Rößger, *Der Überschall Luftverkehr. Grundlagen der Technik und des Betriebs von Überschall-Verkehrsflugzeugen für den gewerblichen Luftverkehr*, Köln, 1966.
39. Andreas Canal, *Das Überschallflugzeug im kommerziellen Luftverkehr*, PhD Diss., St Gallen, 1974.
40. Ulrich Kirchner, 'Das Airbus-Projekt (1965–1990). Genese, Eigendynamik und Etablierung am Markt; in Johannes Weyer et al. (eds), *Technik, die Gesellschaft schafft. Soziale Netzwerke und Technikgenese am Beispiel von Airbus, Astra-Satellit, Personal Computer und Transrapid*. (In press), p. 96.
41. Ibid., p. 111.
42. *Prognosis of the transport department of Interflug of 1967*, Interflug Archives A 160, p. 4. Ministerrat der DDR, MfV (ed.): Verkehrszweigprognose der zivilen Luftfahrt für den Zeitraum 1970–1980, Bundesarchiv SAPMO DY 30/J IV/A2/605/119, Anlage 8. *Analyse der volkswirtschaftliche Effektivität des Inlandsfluggastverkehrs der Interflug*, Interflug Archives III/27.
43. Archiv BMV-Ost M 1/3631. On each of the 32 airports there should wait two air-taxis. 1963, Walter Ulbricht was personally involved in the air-taxi project. See *Liquidierung der Deutschen Lufthansa*, Interflug Archives A 143.
44. In 1963, the average load factor of Interflug's domestic routes were 33.2 per cent. Archives BMV-Ost M2/5072. Only the tourist route Leipzig–Barth reached load factors of 85 per cent. See *Analyse der volkswirtschaftlichen Effektivität des Inlandsfluggastverkehrs der Interflug und dessen mögliche Weiterentwicklung bis 1980*, p. 7. Interflug Archives III/27.
45. Joachim Grenzdörfer, *Analyse zur Stellung des Luftverkehrs im einheitlichen sozialistischen Transportwerden der Deutschen Demokratischen Republik (1961/62)*, PhD Diss., Dresden, 1964; 164.
46. Archiv BMV-Ost, M 2/8998. *Zukunft Inlandsflugverkehr nach 1980. Studie 1979*, Interflug Archives I 523; *Vermerk an Arthur Pieck über US Flughafenprojekte vom 5.3.1956*, Interflug Archives A 143.
47. *Referat über Luftverkehr auf Verkehrsministerkonferenz vom 18.11.1949*, Bundesarchiv, Abteilungen Koblenz, B 108/1897.
48. Deutsche Lufthansa AG (ed.), *Jahresbericht 1961*. Köln, 1961, p. 10.
49. Helmuth Trischler, *Luft- und Raumfahrforschung in Deutschland 1900–1970. Politische Geschichte einer Wissenschaft*, Frankfurt, 1992, pp. 327–9.
50. Arbeitsgemeinschaft Deutscher Verkehrsflughäfen (ed.), *IATA Hubschrauber-Konferenz Puerto Rico 24–26.4.1953*, Stuttgart, 1953. Arbeitsgemeinschaft Deutscher Verkehrsflughäfen (ed.), *Der Hubschrauber und andere Senkrecht- und Steilstartflugzeuge im gewerblichen Luftverkehr. Bisherige Entwicklung und Wachstumsprognose bis 1965 und 1970*, Stuttgart, 1962, p. 26. W. Just, *Einführung in die Hubschraubertechnik*, Stuttgart, 1955, p. 178, Werner

Lambert, 'Verkehrswirtschaftliche Fragen des Nahluftverkehrs', in Paul Berkenkopf (ed.), *Der Verkehr in der wirtschaftlichen Entwicklung des Industriezeitalters. Festschrift zum 40-jährigen Jubiläum des Instituts für Verkehrswissenschaft an der Universität Köln*, Düsseldorf, 1961, p. 159. Werner Lambert, *Bedarf und Aussichten eines Nahluftverkehrs im südwestdeutschen Raum*, Stuttgart, 1963.

51. Hans M. Bongers, *Es lag in der Luft. Erinnerungen an fünf Jahrzehnte Luftverkehr*, Düsseldorf, 1971, p. 286.

52. Vereinigte Flugtechnische Werke–Fokker GmbH (ed.), *Strecken- und Kostenanalyse der VFW 614 im projektierten innerdeutschen Flugnetz. Erstellt für den Herrn Bundesminister für Verkehr*, Bremen, June 1971. DASA Archives Bremen.

53. Ibid., p. 108. Rolf Stüssel, *Möglichkeiten des wirtschaftlichen Einsatzes von Strahlverkehrsflugzeugen im Regionalluftverkehr*. VFW-Berichte 1272, Bremen, 1969, p. 42.

54. Hans Scharlach, *Ergänzungsluftverkehr 1966*, Köln, Lufthansa 1966. Hans Scharlach, *Vorschläge zum innerdeutschen Ergänzungs-Luftverkehr. Eine Studie der Lufthansa*, Lufthansa Pressedienst, 20.2.1967.

55. Since 1987, the regional airlines, organized in the European Regional Airlines Organisation (ERA), grew almost twice as fast as the scheduled airlines organized in the Association of European Airlines (AEA).

56. John L. Kneifel, *Der Wettbewerb im nordatlantischen Luftverkehr. Eine Untersuchung der Wettbewerbsverhältnisse und Wettbewerbsfaktoren*, PhD Diss., München, 1967, p. 33.

57. On his first trip to the US, the German chancellor Konrad Adenauer used TWA planes for the domestic routes and went back to Hamburg by PanAm. Max Martin Brehm, *Mit dem Bundeskanzler in USA*, Bonn, 1953, 94.

58. *Ertrags- und Aufwandsplanung 4.7.1953*, Bundesarchiv, Abteilungen Koblenz, B 108/1906.

59. *Air Transport Agreement between Federal Republic of Germany and the United States of America, signed on 7.7.1955*, National Archives, STD, 59, 611.62A94/7-655. Jochen Vogt, *Entwicklung und Zukunft des Flugverkehrs auf dem Nordatlantik. Dargestellt an einer Strategie für die Deutsche Lufthansa*, Diplomarbeit, München, 1991, pp. 31–9.

60. Hans M. Bongers, *Es lag in der Luft. Erinnerungen aus fünf Jahrzehnten Luftverkehr*, Düsseldorf, 1971, p. 296.

61. A list of all pooling agreements of Lufthansa valid in 1966 is given in, Deutsche Lufthansa AG (ed.), *Informationen. Zum persönliche Dienstgebrauch. Vertraulich. Exemplar 71*, Köln, Presseabteilung der Deutschen Lufthansa, 1966, Lufthansa AG Archives.

62. Wilhelm Pompl, *Luftverkehr. Eine ökonomische Einführung*, Berlin, 1991, p. 55; Hans M. Bongers, *Deutscher Luftverkehr*, Bad Godesberg, 1967, p. 226.

63. Bongers, *Deutscher Luftverkehr*, p. 56.

64. *Tourismuskonzeptionen*, Interflug Archives, A467.

65. Andreas Spaeth, *Flughafen und Luftverkehr. Wirtschaftliche Aspekte einer Partnerschaft*, Landsberg, 1995.

66. Article 13 of the Morgenthau Plan. See Pohl, *Wiederaufbau. Kunst und Technik der Finanzierung 1947–1953*, pp. 227–30.

67. Werner Treibel, *Praxisnahe Forschung. Ein Rückblick auf fünfzehnjährige Tätigkeit*, Stuttgart, 1967.

68. Besides, ADV had a growing number of up to a hundred second class members, the regional airports.

69. *Verkehrsministerkonferenz 18.11.1949*, Bundesarchiv, Abteilungen Koblenz, B 108/ 1897.

70. This decision was widely discussed between airport experts throughout the 1950s and early 1960s. Werner Treibel, *Anforderungen der Düsenverkehrsflugzeuge an die Flughäfen*, Bielefeld, 1956. Werner Treibel, *Entwicklungstendenzen im internationalen Luftverkehr im Hinblick auf die erfolgten Bestellungen und Optionen auf DC-8 und Boeing 707 durch Luftverkehrsgesellschaften, Stuttgart, 1956. Landesplanungsbehörde Nordrhein-Westfalen (ed.), Zur Frage des Standortes eines Verkehrsflughafens für den internationalen Luftverkehr in Nordrhein-Westfalen, Düsseldorf, 1956.*

71. Hans-Christoph Seebohm, *Die Aufgaben des Bundes auf dem Gebiet der Luftfahrt*, Bielefeld, 1956.

72. Werner Treibel, *Anforderungen des Düsenluftverkehrs an die Flughäfen,.* Stuttgart, 1956. Werner Treibel; *Technisch-planerische Voraussetzungen für die Anpassung der Flughäfen an den Düsenluftverkehr*, Stuttgart, 1958, pp. 10–12.

73. Jürgen Weber, *Bericht des Vorstandes der Lufthansa AG auf der 43. ordentliche Hauptversammlung am 3. Juli 1996*, Lufthansa AG, Public Relations Department, July 1996.

74. Consequently, the aviation department of the East German Ministry of Transport was called 'Fair Aviation' in 1951, Archives BMV-Ost, M1/1244.

75. Werner Treibel, *Geschichte der deutschen Verkehrsflughäfen. Eine Dokumentation von 1909 bis 1989*, Bonn, Bernard und Graefe, 1992, pp. 71–2.

76. Landesregierung Brandenburg, *Internationaler Flughafen Berlin-Brandenburg. Ergebnisse der Standortsuche. Überarbeiteter Auszug aus der Präsentation der Ergebnisse der Sitzung der interministeriellen Kommission Luftverkehr am 20. August 1992 in Potsdam*, Potsdam, 1992.

77. Lufthansa AG (ed.), *Weltluftverkehr. Lufthansa Konkurrenz*, Köln, Lufthansa, 1963; 19. Lufthansa AG Archives, Köln.

78. Deutsche Lufthansa AG (ed.), *Informationen. Zum persönlichen Dienstgebrauch. Vertraulich*, Köln, Presseabteilung der Deutschen Lufthansa, 1966, Part G, Lufthansa AG Archives, Köln.

79. In fact, it bought a small airline called Condor, from the German industrialist August Oetker, *Ordner Condor*, Lufthansa AG Archives Köln.

80. Gerhard Gompf, *Linien-Luftverkehrsgesellschaften in Luftfracht Logistikketten. Strategien von Luftfracht Linienverkehrsgesellschaften unter besonderer Berücksichtigung der Schnittstellenproblematik in Luftfracht-Logistikketten am Beispiel der Deutschen Lufthansa AG*, PhD Diss., Gieaen, 1994.

81. Lufthansa AG (ed.), *Statistik und Information STA 1987*, Köln, Lufthansa, 1987; 58.

82. Ibid., p. 49.

83. *Preisanalysen Interflug 1977–1991*, Interflug Archives, A 261 and A 262.

84. The cost was more than four times higher than the revenues. *Bericht Arbeitsgruppe Inlandsflugverkehr vom 17.3.1963*, Archiv BMW-Ost M 2/5915 and *Studie von 1971*, Interflug Archives III/93.

85. Joachim Wölfer, *Deutsche Passagierluftfahrt von 1955 bis heute*, Hamburg, 1995, p. 104. *Möglichkeiten der Steigerung des Westtourismus 1986*, Interflug Archives, A 467.

86. *Konzept Verkehrsflug 1982, Konzept Liniennetz Interflug 1985*, Interflug Archives, A 465.

87. A passenger counts for 100 kg. ICAO (ed.), *Civil aviation statistics of the world*, Montreal, ICAO, 1984. Lufthansa AG (ed.), *Finanzdaten. Lufthansa Konzern 1991–1995*, Frankfurt: Lufthansa 1996.

88. Letter of CIO Jürgen Weber to Lufthansa executives on 2 September 1992. *Ordner Jürgen Weber 1992*, Lufthansa AG Archives, Köln.
89. Rudolf Braunburg and Rainer-Joachim Pieritz, *Keine Angst vorm Fliegen*, Niedernhausen, 1979.
90. It was almost impossible to fire Lufthansa employees. Lufthansa employees were insured at the VBL (Versorgungskasse des Bundes und der Länder) to receive additional pensions like civil servants.
91. Bilanzpressekonferenz of CIO Jürgen Weber on May 13, 1993. *Ordner Jürgen Weber 1.4.–30.11. 1993*, Lufthansa AG Archives, Köln
92. Lufthansa mentions this fact in most of its brochures. Lufthansa AG (ed.), *Zahlen, Daten, Fakten. Ausgabe 1994*, Köln, 1994, p. 2.
93. In 1959 a Super Constellation crashed near Rio de Janeiro (36 dead, 3 survivors), in 1966 a Convair crashed in Bremen (46 dead) and in 1974 a Boeing 747 crashed near Nairobi (97 dead, 156 survivors). To the public, the airline has always counted as very safe and reliable. Günter Femmers, *Das Image der Lufthansa in den gedruckten Medien. Ein komparativ-statistischer Vergleich mit ersten Ansätzen zur Ableitung gezielter Marketing-Maßnahmen*, Diplomarbeit, Duisburg, 1980, p. 68. Lufthansa AG Archives Köln.
94. John L. Kneifel, *Der Wettbewerb im nordatlantischen Luftverkehr. Eine Untersuchung der Wettbewerbsverhältnisse und Wettbewerbsfaktoren*, PhD Diss., München, 1967.
95. Werner Stapp, *Die Absatzpolitik der Deutschen Lufthansa AG*, Diplomarbeit, Göttingen, 1963, pp. 111–14. Lufthansa AG Archives, Köln.
96. Since the late 1980s, Lufthansa advertised special offers, for example, and at the same time invited master chefs from famous European hotels to come on board and celebrate their art before the customers. Adrian von Dörnberg, 'Schlagkräftige Verkaufs- und Werbestrategie', in *Lufthansa Jahrbuch 92*, Köln, 1992, pp. 157–65.
97. Tilman T. Reuss (ed.), *German Aerospace Annual* 43(1994), p. 138.
98. Werner Claasen, 'Die weltweitern Computer-Vertriebssysteme', in *Lufthansa Jahrbuch 90*, Köln, 1990, pp. 81–9.
99. Helmut Wolki, 'Wachstumsmarkt Flugcatering am Beispiel der LSG', in *Lufthansa Jahrbuch 91*, Köln, 1991, pp. 115–21.
100. Hans M. Bongers, *Es lag in der Luft. Erinnerungen aus fünf Jahrzehnten Luftverkehr*, Düsseldorf, 1971, p. 334.
101. *Auszug aus dem Protokoll des 276. Sitzung des Deutschen Bundestages am 25. Juni 1953. Rede Dr. Vogel (CDU)*, Bundesarchiv, Abteilungen Koblenz., B 108, 40803.
102. Swissair dropped out fairly quickly.
103. *Korrespondenz Weigelt-Seebohm*, Bundesarchiv, Abteilungen Koblenz. B 108, 43126.
104. Hans M. Bongers, *Es lag in der Luft. Erinnerungen aus fünf Jahrzehnten Luftverkehr*, Düsseldorf, 1971, p. 393. Kurt Weigelt, *Von der alten zur neuen Lufthansa*, Bad Homburg, Privatdruck, 1966.
105. Hans M. Bongers, *Es lag in der Luft. Erinnerungen aus fünf Jahrzehnten Luftverkehr*, Düsseldorf, 1971, p, 391.
106. Air Union. Bericht, *Sitzung Aufsichtsrat der Deutschen Lufthansa AG vom 20. September 1960;*. 22. Lufthansa AG Archives.
107. On 7.3.1966 the German Minister of Transport suggested to Lufthansa that it should transform the Air Union into a maintenance group, *Letter of Kreipe to Lufthansa*, Lufthansa AG Archives, Bestand Air Union 1958–67. 20 Jahre Atlas, *Luftansa Artikeldienst 1.2.1989/1*, Lufthansa AG Archives, Bestand Atlas.

108. Michael Hammes, *Wettbewerbsregeln für den Luftverkehr in den Europäischen Gemeinschaften*, PhD Diss., Frankfurt, 1982. Rolf J. Haupt, *Liberalisierungsbestrebungen im europäischen Luftverkehr*, Köln, 1985. Alexander Ditze, *Start ins Chaos. Liberalisierung im Europäischen Luftverkehr*, Frankfurt, 1986.

109. Wolf-Rüdiger Uster, 'Kooperationen und Fusionen im Luftverkehr', *Lufthansa Jahrbuch 1990*, Köln, Lufthansa 1990, pp. 38–48.

110. Lufthansa AG (ed.), *Statistik und Informationen*, Köln, Lufthansa 1989, p. 40.

111. Hans M. Bongers, *Es lag in der Luft. Erinnerungan aus fünf Jahrzehnten Luftverkehr*, Düsseldorf, 1971, pp. 415–16.

112. Additionally, the company began to buy hotel companies like the Spanish Prinotel group. Joachim Wölfer, *Deutsche Passagierluftfahrt von 1955 bis heute*, Hamburg, 1995, pp. 97, 112.

113. Hapag-Lloyd AG (ed.), *Hapag-Lloyd Jahresberichte 1970–1995*, Hamburg/Bremen.

114. Hans M. Bongers, *Deutscher Luftverkehr. Entwicklung, Politik, Wirtschaft, Organisation. Versuch einer Analyse der Lufthansa*, Bad Godesberg, 1967, p. 83.

115. Susanne Doms, *Der Wettbewerb um den innerdeutschen Linienluftverkehr*, Diplomarbeit, Köln, 1990, pp. 86–93.

116. German Wings (since 1989), Aero Lloyd (since 1980), Deutsche Luftverkehrsgesellschaft DLT, Nürnberger Flugdienst NFD, Arcus Air YY, Cimber Air (GW), Südavia FV, Delt Air DI, Interrot IQ, Regionalflug GmbH VG, Contact Air VJ, Rheinland Air Service ROA.

117. Joachim Wölfer, *Deutsche Passagierluftfahrt von 1955 bis heute*, Hamburg, 1995, pp. 56.

118. Werner Treibel, *Geschichte der deutschen Verkehrsflughäfen. Eine Dokumentation von 1909 bis 1989*, Bonn, 1992, pp. 107.

119. Hapag director Kipfmüller closely cooperated with Hans Bongers of Lufthansa before 1945. In a letter to Bongers on 23.3.1942 he wrote 'after the end of the war we shall work in the same partnership to rebuild German transport'. Hapag-Lloyd Archives, Folder Lufthansa-Verträge.

120. Ernst Hansen, *Das Flugzeug im Wettbewerb mit Eisenbahn und Überseeschiffahrt*, Hannover, 1951.

121. In April 1964 Lufthansa, Norddeutscher Lloyd and Hapag signed an agreement on combination tours (one way by plane, one way by ship). Hapag-Lloyd Archives, Folder Lufthansa Verträge.

122. *Deutsche Bedarfsluftverkehrsgesellschaft*. The airline had not then been founded. Hapag-Lloyd Archives, Folder Lufthansa 1953. The Lufthansa subidiary was named Deutsche Flugdienst GmbH. Hapag and Norddeutscher Lloyd had 30 per cent each, Lufthansa and the German Railways 20 per cent each. Folder Deutsche Flugdienst GmbH.

123. Hans-Christoph Seebohm, *Die Aufgaben des Bundes auf dem Gebiet der Luftfahrt*, Bielefeld, 1956, p. 9. Hans-Christoph Seebohm, *Stand und besondere Probleme der deutschen Zivilluftfahrt*, Bonn, 1964, p. 75.

124. Rigas Donanis, 'Air versus Rail in Western Europe', *Flight*, 1964, pp. 614–16.

125. Hans Scharlach: *Eisenbahn und Inlandsluftverkehr. Folgerungen aus Entwiclung und Planung der Deutschen Bundesbahn für den Luftpersoenverkehr in der Bundesrepublik Deutschland*, Köln, Lufthansa, 1969, p. 27. Lufthansa AG Archives. G. Müller, *Wettbewerb zwischen der Deutschen Bundesbahn und den Luftfahrtunternehmen im innerdeutschen Eisenbahn- und Flugreiseverkehr*, PhD Diss., Mainz, 1966, pp. 10–11.

126. Dieter Giese: *Regionalluftverkehr und Eisenbahn als Wettberwerber im deutschen und grenzüberschreitenden Personenverkehr*, Berlin, 1992.

127. R.J. Haupt, *Linienluftverkehr*, Deutsche Lufthansa AG, Firmenarchiv, Folder Airport Express.
128. 'Airport Express auf dem Abstellgleis', *Frankfurter Allgemeine Zeitung*, 18.2.1993.
129. In the last years, this changes. See for example Prognos AG (ed.), *Bedeutung von Umweltwirkungen von Schienen und Luftverkehr in Deutschland*, February 1995. The study was financed by the Deutsche Bahn, and Lufthansa AG. This underlines the common interests against motor transport.
130. Enquete-Kommission 'Schutz der Erdatmosphäre' des Deutschen Bundestages (ed.), *Mobilität und Klima. Wege zu einer klimaverträglichen Verkehrspolitik*, Bonn, 1994, pp. 313; Prognos AG (ed), *Bedeutung und Umweltwirkungen von Schienen- und Luftverkehr in Deutschland*, Februar 1995; A 32–36. Lufthansa Archives Köln.
131. Dietmar Klenke, *Bundesdeutsche Verkehrspolitik und Motorisierung. Konfliktträchtige Weichenstellungen in den Jahren des Wiederaufstiegs*, Wiesbaden, 1993.
132. Dietmar Klenke, *Freier Stau für freie Bürger. Die Geschichte der bundesdeutschen Verkehrspolitik 1949–1994*, Darmstadt, 1995.
133. Lufthansa AG (ed.), *Konzernbericht zum 1. Halbjahr 1996*, p. 11. Lufthansa Investor Relations 1996. Lufthansa AG (ed.): *Geschäftsbericht 1995*, Köln, Lufthansa, 1996, p. 24.
134. Jürgen Weber in a speech on 19 October 1995, Lufthansa AG Archives, Ordner Jürgen Weber 1.5.1995–1996.

5 KLM: an Airline Outgrowing its Flag

Marc Dierikx

In May 1945 only a few European airlines were still flying. In Britain BOAC was run as a civilian branch of the RAF. In Germany the last remaining civil operations of Deutsche Luft Hansa had ceased amid the final battles of the war. Air France had gradually disappeared from view and would not re-emerge before 1946. Swissair had ceased flying in 1943. Apart from BOAC, this left only two of the pre-war European airlines still operating: the Swedish carrier ABA, and the Royal Dutch Airline, KLM. That the latter should still be around in the European skies was extraordinary. Shortly after Holland's capitulation in May 1940, KLM's representatives in London had offered the services of KLM's staff and what equipment the airline still possessed outside Holland to the British government. As a result, KLM was commissioned in August 1940 to operate a wartime service between Bristol and the Portuguese capital Lisbon under a charter agreement with BOAC. When the service was suspended in January 1946, KLM had carried 18 108 passengers on its last remaining European route.[1]

1 POLICY

The Netherlands' air transport policy has been remarkably constant in its focus on advocating freedom in the air. Initially, this policy was necessary to ensure continued air communications with overseas colonial territories, but after 1945 it increasingly served to ensure optimal market access abroad for the commercial interests of KLM.

1.1 Origins, 1919–45

KLM can justly claim to be the oldest airline in continuous operation. The company was founded in The Hague on 7 October 1919 as *Koninklijke Luchtvaart Maatschappij voor Nederland en Koloniën* (Royal Aviation Company for the Netherlands and Colonies), but its roots go back some months further. In the aftermath of the First World War, Dutch shipping lines, trading banks, and entrepreneurs had joined forces in an effort to participate in the glorious future that was prophesied for air transport. They put up just over a million Dutch guilders, one-fifth of the agreed total possible share

126

issue, to start an airline company. It was envisaged that after an initial period of losses, the airline would become solvent within a few years.

Surprisingly, the founders entrusted the day-to-day running of the fledgling company to a 30-year-old commissioned officer in the Netherlands Air Corps, Albert Plesman. Bursting with a forceful energy that in time would make him the personification of KLM, Plesman was no random choice. He had already attracted nationwide attention as a skilled organizer of a large international aeronautical exhibition in Amsterdam during the summer of 1919.

KLM's earliest operations were limited to chartering aircraft from the British carrier Air Transport & Travel Co. for a service between London and Amsterdam that started in May 1920, but further European expansion was intended. However such growth was handicapped by two factors. To begin with, Holland was not a party to the International Convention on Aerial Navigation of 13 October 1919 that was signed by the Allied and Associated powers in Paris. This meant that KLM ran into serious problems in its efforts to obtain the necessary traffic rights for international scheduled services. Secondly, like all early airlines, it was handicapped by the available aircraft technology, which did not allow profitable operations. Before a single KLM airplane left the ground, the founders of the airline admitted that their initial calculations had been wrong, and turned to the Dutch government for financial assistance. Stressing the role that a national airline could play in establishing a Dutch air link to the colonial territories in the Dutch East Indies (Indonesia), they found a receptive ear in the government. In the spring of 1920 the Cabinet granted aid to KLM, ostensibly on a temporary basis. Apart from the colonial motive, which was important in the context of the medium-size power that Holland claimed to be, the fear that without government funds international aviation would bypass Holland also played a role in the government's decision to support KLM.

In the following years KLM's losses rose at roughly the same pace as the expansion of its services and by the spring of 1927 Plesman was fighting for the airline's survival. It was clear that the long-term future of the airline depended on additional government funds. In return for these, the state demanded that KLM should attract new private capital as a sign of continued confidence in air transport. Only then was the government prepared to come to the rescue of the national airline, take a position of major shareholder, and ensure its continued existence. The new balance of interests henceforth showed in government representatives occupying key positions on the Board. Over the course of the next two years KLM was transformed from a private company into a semi-state enterprise, dependent for most of its decisions on ministerial approval. In March 1929 the government acquired the majority of the shares.[2] Despite continued losses, things looked up for KLM from then on. Governmental subsidies and other forms of state assistance were henceforth secure. During the 1930s KLM was allowed to expand and become one of the world's leading airlines. With support from the government, it was even able to set up a modest *West Indisch Bedrijf* (West Indies Branch) on the Dutch Caribbean island of Curaçao in 1935. However, the airline's pride was a scheduled service between Amsterdam and the capital of the Dutch East Indies, Batavia, which began in 1931, after series of test flights at regular fortnightly intervals in 1929 and 1930. Thanks to generous payments for airmail by the Post Office, the Indies route became the lifebelt that lifted the airline out of its long immersion in heavy losses. In the late 1930s KLM was unique among its European counterparts in rapidly approaching a financial break-even point. In 1938, the last regular year of operations before the war, KLM recovered just over 91 per cent of its costs on its European services, and 96 per cent of its overall costs, including its Amsterdam–Batavia service and West Indies Branch. Taking into account the direct government subsidy, KLM's accounts even showed a small profit of 73 609 guilders ($29 400 in 1938 values) on a total revenue of just under 11.3 million guilders, or 0.65 per cent.[3]

In May 1940 the German attack on Holland spelled the end of KLM's regular operations in Europe. In the confused days of the military struggle the official seat of the company was transferred to Batavia. There the affairs of KLM became the responsibility of KLM's sister company in the East Indies, KNILM (*Koninklijke Nederlandsch-Indische Luchtvaart Maatschappij*). From Batavia, KNILM also managed the remnant of KLM's pre-war operations: a scheduled service between Batavia and Lydda in Palestine, where passengers could connect with BOAC's 'horseshoe route' across Africa to Britain. Until early February 1942 this service was kept going, despite increasing wartime difficulties. With the collapse of the Dutch Asian empire before the advancing Japanese forces later that month, KLM's seat was moved again, this time to Willemstad in Curaçao, although the management came to reside in London. Here a new board of directors was appointed by the Netherlands' government in exile in 1943. Former KNILM manager Hendrik Nieuwenhuis became chief executive of KLM's European operations.

1.2 Government aims since 1945

The last phases of the war caused severe damage to the infrastructure of Holland, with vital bridges, roads and rail connections severed. As in other newly liberated states, aircraft provided the only means of transport between the various parts of the country. Air connections were important in the first stages of reconstruction and the return to normal government. The Cabinet recognized the value of the airline in the allocation of funds to buy transport aircraft and maintain key domestic routes. In September 1945 KLM started an embryonic domestic network, although it was ill-prepared for this role, as it found itself in the midst of a managerial crisis. In Holland the pre-war board of directors had never been dissolved, yet the seat of the company was now in Curaçao, where the West Indies Branch not only saw to its own management, but also handled some aspects of KLM's general administration. In addition, the government-appointed board in London also claimed control, as did the original board in The Hague. The dispute extended to the executive level. Plesman, who had always considered himself to be the heart of the company, reclaimed his position as managing director when he returned from internal exile at the cessation of hostilities. He had difficulty accepting that in his enforced absence the airline's affairs in Holland had been supervised by his deputy-director, Hans Martin. Moreover, personal relations between Plesman and Nieuwenhuis, who claimed a similar position of authority as KLM's chief executive in London, had a long history of conflict. Straightening out these disputes took until 1946, but eventually KLM's headquarters was moved back to The Hague and the authority of the old board of directors, and of the pre-war managers Plesman and Martin, was restored.

If KLM was of special significance to the government in the process of re-establishing its peacetime authority within the Netherlands, its weight was even greater in the wider context of Dutch colonial rule. With strong aid from the government, which again provided aircraft and finance, KLM restarted its service to Batavia in November 1945 in the guise of the Netherlands Government Air Transport organization. This service soon became of vital importance because of the crisis that was developing in the Dutch East Indies, where independence had been declared by the Indonesian nationalists on 17 August 1945. The Dutch rejected this declaration and in the following months the contours of an emerging colonial war became visible. The conflict emphasized the importance of KLM to the government, which responded with large-scale financial assistance for KLM's reconstruction and subsequent expansion as an international airline.[4]

As a result of the fighting in Indonesia, KLM was amalgamated with KNILM on 1 August 1947, a reversal of wartime practice. KLM, which had held a stake in KNILM since its foundation in 1928, took over the remainder of KNILM's stock at face value, compensating the shareholders with KLM shares. The operation of a network of air services in the Indonesian archipelago henceforth became KLM's responsibility, although the colonial government in Batavia footed the bill.[5] In the event, KLM's Indonesian activities were short-lived. Two years later they were turned over to the newly-founded national air carrier *Garuda Indonesia* (Indonesian Bird of Paradise) following Dutch recognition of Indonesian independence. Despite the deep rifts caused by the four-year conflict between the Dutch and the Indonesians, the two governments agreed that KLM participate in Garuda's capital. In spite of the sorry state of the nation's finances, the Netherlands' government provided the 15 million guilders needed.[6] KLM also provided personnel, technical, and managerial assistance to the new airline – to such an extent indeed that Garuda was dubbed an acronym, in aeronautical circles, for 'Going All Right Under Dutch Administration'. KLM's direct financial involvement in Garuda did not last long. In 1954 the Indonesian government bought out the Dutch, replacing co-ownership with a six-year remunerated contract under which KLM provided technical assistance only. Until the Indonesian government withdrew KLM's landing rights in Jakarta in December 1957, some 350 KLM employees worked for Garuda.[7]

KLM continued to operate in southeast Asia as a contract carrier for the Dutch authorities in New Guinea, the last remaining outpost of the Dutch empire in that part of the world. These services were gradually reorganized under the guise of a new subsidiary, *Kroonduif* (Crown Pigeon), which was founded in 1955. Kroonduif's operations continued until the final collapse of Dutch rule in New Guinea in 1962, after which the company's activities were taken over by Garuda.[8]

From its base on the Dutch island of Curaçao, KLM also maintained the Caribbean network of its *West Indisch Bedrijf*. Following the Kroonduif example, Caribbean operations were restructured in a semi-independent KLM-subsidiary, known as *Antilliaanse Luchtvaart Maatschappij* (ALM: Antillean Aviation Company) in August 1964. In the second half of the 1960s ALM's Caribbean network gradually developed with the expanding economy of the region. On 31 December 1968 the Antillean government, keen to use the airline as a symbol for the growing independence of the islands from Holland, took over 91 per cent of ALM's shares from KLM.[9] After that the companies gradually parted ways until the Antillean government took full control of the airline in 1974, in the wake of further rearranged relations between the Netherlands and its Caribbean possessions – a reflection of the changed position of the Netherlands in the post-colonial world. Only in 1991 did the Dutch national carrier reinvest in ALM, taking a 40 per cent share in order to expand its Caribbean interests.

1.3 Attitudes to regulation

Although KLM always enjoyed a protected position in the home market, the Dutch maintained a very liberal position internationally. Representing a relatively small country and occupying a position of only minor importance in the playing field of international politics, a liberal stance was the only realistic position the Dutch could take, given KLM's size and its dependency on foreign markets. In the *quid pro quo* world of bilateral air transport agreements, success in securing new commercial landing rights for KLM depended on promoting the attraction of Holland as a gateway to Europe for foreign air carriers, and on the level of service offered by KLM. Through the 1950s this policy was highly successful, but when the stakes in air transport rose during the 1960s, with the expansion of the market and the increased financial burden on airlines caused by the change-over to jet aircraft, KLM's growth became increasingly dependent on reaching financial agreements with other airlines concerned with access to markets.[10] To add to these pressures, KLM's international position also suffered from the increased activities of charter airlines in Europe and on the North Atlantic. When air transport slumped following the oil crisis of 1973, the Netherlands government heeded KLM's calls for some measure of protection as a scheduled carrier. Within the ECAC the Dutch representatives led the field in attempts to bring charter air traffic under the provisions of the bilateral treaty system.[11] Nevertheless, when the United States announced their deregulation measures in 1977, the Dutch were quick to change their position and embrace the American policy, because the benefits of wider access to the US market outweighed the dangers of a further increase in competition. In March 1978 the Dutch were the first to conclude a bilateral air transport agreement with the US that introduced a deregulated environment for carriers from both sides. In exchange for admitting an unspecified number of US scheduled and charter airlines to, and through, Amsterdam, KLM received additional gateways in the USA.

American-style liberalization crossed the Atlantic as a result of these and other European-American contacts on air transport. But although the Europeans adopted liberalization, it was in a gentler form, aimed at a gradual relaxation of governmental controls preceding a full liberalization of the market in the future.[12] The first real attempt at liberalizing air transport within the EC was the bilateral Anglo-Dutch treaty of June 1984 that brought complete freedom of routes and frequencies for British and Dutch carriers between the two countries. The result was noteworthy. Between 1984 and 1990 the number of airline passengers travelling between Holland and the United Kingdom increased by 50 per cent. Moreover, Dutch and British advocacy within the European Community led to the adoption in 1986 of a package of deregulating measures aimed at increasing competition between the airlines of the member states. This was confirmed in the Single European Act of July 1987 that brought aviation within the scope of the competition clauses of the Treaty of Rome.[13] KLM was one of the airlines which sought to profit from this relaxation of the rules, taking a substantial shareholding in the British regional carrier Air UK in 1988, and in the French regional airline Air Littoral in 1989. The aim was to use the increased opportunities for commercial air transport within the EC to channel British and French intercontinental traffic through KLM's hub in Amsterdam when the full scope of EC liberalization came into force after 1992.

2   AIRCRAFT PROCUREMENT

Through the postwar period KLM has almost exclusively operated American aircraft, a reflection of its focus on operating economy. This, however, did not mean that the airline always enjoyed freedom of choice and, especially in the 1950s, KLM struggled to determine which aircraft it needed.

2.1   Propeller aircraft

In May 1945 KLM was lucky to have any operational aircraft at all. Its European fleet consisted of two pre-war Douglas DC-3s, three converted Douglas C-47s, and a lone DC-2 that had survived the war. The airline needed new aircraft urgently and these were acquired surprisingly fast. In April 1945, even before fighting on Dutch soil had ceased, Plesman had managed to make his way through the front line and make contact with the Netherlands' government-in-exile in London. He then managed to obtain priority seating aboard one of the few flights from Britain to the USA, carrying in his wallet government documents requesting that the Americans make available to its newly liberated ally transport aircraft that could be put to use in Holland's reconstruction. This turned out to be a shrewd move and Plesman was able to leave Washington with delivery promises for a batch of 14 ex-military Douglas C-54J Skymasters, the military version of the DC-4. With these, KLM was able to restart its route to Batavia in November 1945 in the guise of the *Netherlands Government Air Transport*. But Plesman had done more than persuade the US government to part with a few surplus transport aircraft; he had also signed a contract with Douglas for four civil DC-4s, to be delivered after conclusion of military production. He also met with Lockheed representatives, whom he reminded of the agreement between KLM and Lockheed, made in the last months before the outbreak of the war in western Europe, for the sale of six L-049 Constellations.[14] Now Plesman was able to claim early delivery dates for these aircraft.

These purchases indicated KLM's continued preference for American aircraft. Even though the Netherlands' government had agreed to play a role in the postwar resurrection of the Fokker Aircraft Factory in Amsterdam, the new Dutch aircraft company had neither the resources nor the know-how to produce civil airliners for the time being. This meant that KLM was free to buy the best available aircraft 'off the shelf' wherever it saw fit. KLM never had any doubts that such shelves would be American, preferably Douglas. The close relationship between KLM and Douglas, and between Donald Douglas and Albert Plesman made KLM one of Douglas's most favoured foreign customers. The airline would continue to pride itself on having bought every type of Douglas airliner since Donald Douglas entered the business with the DC-2 in 1933.

To assemble a fleet of aircraft for new scheduled operations in Europe, KLM spent the latter half of 1945 acquiring no less than 35 ex-military C-47s, refurbished for civil use as DC-3Cs. European services were gradually reopened from December 1945 onwards. Demand for seats was high in the war-stricken countries of western Europe, as surface communications had been badly disrupted. Within a year the airline was making a profit and it decided to buy a number of twin-engined Convair CV-240s that could seat up to 40 passengers, compared to the DC-3's maximum of 28. Thus on its busiest routes KLM's capacity could be expanded rapidly. The new Convairs were delivered at the end of 1948. They were among the first short-range airliners to be equipped with a pressurized cabin and offered standards of passenger comfort comparable to the larger intercontinental aircraft. That way KLM hoped to attract even more passengers onto its European services, using them to feed its intercontinental routes.

The acquisition of the CV-240s was the logical follow-up to the earlier expansion of KLM's fleet of long-range aircraft that had resulted from a less obvious decision-making process. In 1947 and 1948 KLM replaced its initial C-54As, DC-4s, and L-049 Constellations with Lockheed L-749A Constellations, seating up to 61 passengers (as compared to 43 for the L-049s), and Douglas DC-6s that seated 62 passengers. But if the initial postwar order of two different types of aircraft with comparable specifications (DC-4 and L-049) had been dictated by the need to secure modern long-range aircraft to start scheduled operations as soon as possible, the later order of both Douglas *and* Lockheed aircraft was more puzzling. It showed that buying aircraft 'off the shelf' presented its own difficulties in deciding which aircraft best fitted the airline's requirements. Like other airlines in this period, KLM never managed to achieve standardization of its long-range fleet and it continued to operate the graceful Constellation and Super Constellation alongside the more bulky-looking Douglas aircraft. The L-749 Constellation beat the early DC-6 in maximum take-off weight, cruising speed and range, but the improved DC-6A and DC-6B aircraft, both of which KLM bought, had higher payloads. Indeed, the Constellation needed a special 'speedpack' attachment to its belly in order to provide cargo space comparable to that of the DC-6B. This made the DC-6B a more versatile plane in a commercial airline environment, especially as it could be used more easily for medium and short-range services as well. In April 1951 the Netherlands' government agreed to a KLM proposal to order seven Douglas DC-6Bs. At the same time KLM bought nine L-1049 Super Constellations from Lockheed.[15] KLM was never quite able to resolve the issue of choice between these competing aircraft, and even in the late 1950s ordered the penultimate L-1049H version of the Super Constellation next to Douglas's final long-range propeller airliner, the DC-7C.

2.2   Procurement muddles: piston engines, turboprops or jets?

Aircraft procurement was difficult throughout the 1950s because a range of airliners appeared on the market in rapid succession. Beyond the problem of choosing between comparable types such as the Constellation and the DC-6, airlines had to decide which propulsion technology was likely to dominate the field in the future: piston engines, jet engines or a hybrid between them, the turboprop. The course of future technological development did not become discernible until the late 1950s. Without government pressure to favour aircraft and engines developed by a domestic industry, KLM should have found it comparatively easy to translate its requirements into aircraft procurement choices. However this was not so and KLM's management found itself increasingly bewildered by the choices open to them. The problem was approached in terms of aircraft speed versus operating costs, yet KLM's management never arrived at a clear choice between these two factors, even though marketing studies indicated that the adoption of jet aircraft by KLM's competitors would have adverse effects on its international position.[16] KLM's traditional reliance on American aircraft complicated matters further. Whereas British and French aircraft constructors went for speed and the early introduction of jet propulsion to 'leap-frog' the American producers that dominated the market, the Americans themselves stressed size, fuel economy, range and passenger comfort. Buying non-American aircraft represented not only culture shock to KLM, it also meant swapping well-known American engineering standards and maintenance practices for new, and as yet unproven, British standards that KLM's engineers tended to regard as inferior. Plesman, for example, had serious doubts about the durability and power of the de Havilland Ghost and the Rolls-Royce Avon jet engines that powered the de Havilland Comet airliner. Moreover, the operating characteristics and the limited capacity of the Comet 1 were such that the aircraft fitted none of KLM's main routes.[17] Matters were made worse for KLM when, on New Year's Eve of 1953, Albert Plesman died. In the years that followed, KLM continued to struggle with aircraft procurement and Plesman's forceful opinions were clearly missed.

It took until 1957 for KLM to recognize the potential of the British turboprop concept and follow other European carriers in ordering the Vickers Viscount for its European route network. The aircraft was introduced by KLM in April 1957. In doing so, KLM became a late but fervent convert to the turboprop and it would also order the larger, Lockheed L-188C Electra for its intermediate range operations to Africa and the Middle East. For its long-range operations however, KLM continued to rely on the proven technology of the piston-engined airliner. At a time when most of its competitors were queuing up for jets in the USA, KLM bought ten DC-7Cs for its transatlantic operations. By the time KLM put the first of these into service

Table 5.1 KLM fleet statistics

	1946	1951	1956	1961	1966	1971	1976	1981	1986	1991	1996
Douglas C-47/DC-3	40	20	14	5	–	–	–	–	–	–	–
Douglas C-54/DC-4	24	10	6	–	–	–	–	–	–	–	–
Lockheed L-049	4	–	–	–	–	–	–	–	–	–	–
Lockheed L-749A	–	–	16	10	–	–	–	–	–	–	–
Douglas DC-6/6A/6B	–	7	14	6	–	–	–	–	–	–	–
Convair CV-240	–	12	7	–	–	–	–	–	–	–	–
Convair CV-340	–	–	10	10	–	–	–	–	–	–	–
Lockheed L-1049C/E/G/H	–	–	–	18	5	–	–	–	–	–	–
Douglas DC-7C/F	–	–	–	–	15	4	–	–	–	–	–
Vickers Viscount 803	–	–	–	9	–	–	–	–	–	–	–
Lockheed L.188 Electra	–	–	–	11	11	–	–	–	–	–	–
Douglas DC-8-30/50	–	–	–	14	18	15	4	–	–	–	–
Douglas DC-8-63	–	–	–	–	–	–	11	11	10	–	–
Douglas DC-9-10/30	–	–	–	–	6	19	18	20	13	–	–
Fokker F-27	–	–	–	–	2	3	6	8	7	–	–
Fokker F-28	–	–	–	–	–	–	–	4	4	4	4
Douglas DC-10-30	–	–	–	–	–	–	7	6	5	6	–
Boeing B.747-206B	–	–	–	–	–	–	6	9	6	–	–
Boeing B.747 Combi	–	–	–	–	–	–	2	7	10	6	10
Boeing B.747-306	–	–	–	–	–	–	–	–	3	3	3
Airbus A.310-203	–	–	–	–	–	–	–	–	10	10	5
Boeing B.737-300	–	–	–	–	–	–	–	–	10	21	29
Boeing B.737-400	–	–	–	–	–	–	–	–	–	10	12
Boeing B.737-200	–	–	–	–	–	–	–	–	–	8	4
Boeing B.747-400	–	–	–	–	–	–	–	–	–	2	5
Boeing B.747-400 Combi	–	–	–	–	–	–	–	–	–	8	11
Fokker 100	–	–	–	–	–	–	–	–	–	6	6
Boeing B.757	–	–	–	–	–	–	–	–	–	–	4
Boeing B.767-200	–	–	–	–	–	–	–	–	–	1	–
Boeing B.767-300ER	–	–	–	–	–	–	–	–	–	–	3
McDonnell Douglas MD-11	–	–	–	–	–	–	–	–	–	–	9
Fokker 70	–	–	–	–	–	–	–	–	–	–	2
Fokker 50	–	–	–	–	–	–	–	–	–	10	10
Saab 340B	–	–	–	–	–	–	–	–	–	6	11
TOTAL	68	49	67	83	57	41	54	65	78	101[*]	128[*]

Source: KLM Annual Reports
[*] Figures for KLM Group, including participations.

in April 1957, Douglas was already contemplating closing down production of its last propeller-driven type. Two months later KLM surprised everyone by ordering five more DC-7s. The airline needed extra aircraft to initiate operations to Houston, KLM's second US destination (after New

York), which had finally been granted by the Americans after 12 years of negotiations.[18] KLM accepted that it would have to depreciate the DC-7Cs at a much faster rate than its normal practice, but assumed it would be easy to find a buyer for the piston-engined planes when the airline acquired jets in a few years time. However, no sooner had the contract for the second batch of DC-7Cs been signed in California, than the KLM board realized it was pursuing a course quite different from all its rivals, who were entering the jet age at top speed.[19] Using its long-standing favoured relationship with Douglas, KLM now managed to become Europe's first airline to take delivery of the DC-8-30 jet airliner in 1960.

This left the airline with the awkward problem of disposing of the almost brand-new DC-7Cs that were rapidly becoming obsolete. In the airline industry crisis of 1961–3, KLM's losses were exacerbated by the burden of having too many aircraft in its inventory; in 1961 the airline possessed over a hundred. A reduction of the fleet was desperately needed and KLM put all of its 17 Super Constellations up for sale, eight of them parked in protective wrapping at Schiphol airport. However, buyers were few and five of them eventually had to be sold for scrap in 1965. To make matters worse, passenger surveys showed that KLM's customers preferred to fly in the new jets. The Lockheed Electras were actually hated by passengers because of their comparatively high noise level and vibration.[20] After much discussion, KLM resolved to cut its losses in December 1962 and bend to the demands of its customers to operate jets. It was decided to fly all intercontinental and Middle Eastern routes with KLM's thirteen DC-8s even if this meant reducing the frequency of operations. The Electras and Viscounts were put onto KLM's European network, replacing the piston-engined Convairliners, while the DC-7Cs were put up for sale, four being kept and converted to DC-7F freighters. The older Convair CV-340s and DC-6Bs were practically given away, and the Dutch charter carriers Martins' Air Charter and Transavia hurried to purchase them for their inclusive tour operations. Now KLM needed to buy four additional DC-8s in 1964 and 1965.[21] Thus, the airline finally resolved a decade-long muddle in aircraft procurement.

2.3 Jets in all sizes

The lesson had been very expensive. A lack of clear strategy, and an unfortunate separation between commercial management and aircraft procurement that had led the airline to focus exclusively on aircraft economics while ignoring the public's preference for jets, had cost KLM dearly. The new board, which was appointed at the height of the crisis in 1962, was determined that KLM would not suffer from similar mistakes in the future. KLM now became almost over-eager to regain its pre-war reputation of being the first European carrier to introduce new aircraft. Early in 1964, six months after President

Kennedy's announcement of an American government-sponsored supersonic airliner (SST) programme, KLM contacted the Federal Aviation Administration and the main contenders for the project, Boeing and Lockheed, declaring the airline's interest in acquiring three such aircraft. Once again, KLM showed a clear preference for American technology and a willingness to wait for the second generation of supersonic airliners, rather than risking its luck with the Anglo-French Concorde, which was summarily rejected for having the wrong size and operating characteristics. On 28 September 1967 KLM's supervisory board agreed to place an additional order for three supersonic Boeing 2707s, to be delivered by 1976.[22]

However KLM's aircraft difficulties were not yet over. Following the success of the French Caravelle twin-engined jet, aircraft constructors in Britain and the USA developed aircraft to similar specifications in the early 1960s. For airlines that had chosen not to buy turboprops, the arrival of medium-haul jets brought substantial gains in aircraft utilization at a time when their older piston-engined aircraft were ready to be phased out. KLM, however, faced the necessity to accelerate the depreciation of its Viscounts and Electras to make room for the jets that the public preferred. Still loyal to Douglas, KLM ordered sixteen DC-9 twin-jets to be used on its European network. The first of these aircraft, a DC-9-15 with a seating capacity of 80, was delivered in May 1966. The remainder were of the stretched DC-9-30 version, which had a capacity of 105. In keeping with KLM's large cargo operations, six of the latter aircraft were of a 'rapid change' configuration, so that they could be easily converted into freighters. After daylight operations as passenger aircraft, KLM used these machines at night for cargo. According to KLM's president at the time, Gerrit van der Wal, no other passenger airline in the world received more of its revenue from cargo than KLM. Consequently, the airline's policy remained focused on the development of cargo in the future, and would be a vital element in future aircraft choices.[23]

With these issues of long-term policy finally resolved, KLM conformed to the drive for standardization of equipment that followed the success of the big American jets. While the supersonic option was monitored closely, the more immediate challenge of the expanding market took priority, and KLM ordered the 233-seat stretched version of the Douglas DC-8 long-range jet, the DC-8-63, in 1965. The first these was delivered in 1967. KLM was among the first European operators to order the Boeing 747, introducing it into service in 1971 on its New York service. But if the 747 was an attractive aircraft in terms of operational economics, it was in its 387-seat all-passenger configuration – very large indeed for an airline without a large captive domestic market. The risks involved in ordering the 747 were considerable for an airline that could only use the 'Jumbo' on its busiest routes. Nevertheless, KLM ordered seven of the 206B version, which featured an increased take-off weight and extended range over the original 747. The aircraft were delivered in 1971. For its less voluminous long-range services, the 747s were supplemented with six 240-passenger Douglas DC-10s that were delivered between 1972 and 1975. In this way KLM continued its ties with Douglas.

The wisdom of caution showed in a second period of overcapacity that followed the introduction of the Boeing 747 in the early 1970s. It was a problem that Boeing itself had recognized and which reduced the attraction of the 747 to those airlines which had no immediate operational need to replace their 707s or DC-8s with jumbos. The issue was resolved in a brilliant engineering concept that paired the large payload capacity of the 747 to the requirements of thinner traffic routes – the 747 combi. This aircraft combined seating for around 200 passengers with the cargo volume of a Boeing 707 in full freighter configuration. For KLM it was an ideal solution and it bought seven 747-206B combis, with large side cargo doors in the rear of the fuselage. These aircraft were delivered to KLM between 1975 and 1981. To increase passenger seating in Business Class without reducing the plane's cargo capacity, all seven were fitted with a stretched upper deck in the early 1980s, identical to that of the 747-306 *combi* version, of which KLM acquired another three. Towards the end of the decade, when KLM began preparing for further expansion, an order was placed for Boeing 747-400 aircraft as replacements for the 747-200 series. KLM acquired five standard 747-400s, plus fifteen 747-400M combi aircraft to cater for continued mixed passenger and cargo operations into the twenty-first century.

KLM followed the same policy on new aircraft for medium-range routes. In May 1979 it announced it had placed an order for ten A310 Airbus aircraft, with an option on a further ten. The plan was to operate the aircraft on KLM's European, Near Eastern and North African services for a period of 10–15 years. KLM preferred the Airbus's good cargo capacities over those of its competitor, the Boeing 767. Besides, as state secretary Neelie Smit-Kroes admitted in the Dutch parliament, the French government offered an extremely flexible financial package for the purchase, on which Airbus Industrie actually stood to lose money.[24] Nevertheless, when KLM faced further fleet expansion ten years later, it let standardization on Boeing prevail over the A310's cargo features, and chose to lease ten Boeing 767-300ER aircraft, the first of which were delivered in 1995.

In 1986 and 1987, the Douglas DC-9s were replaced on the shorter, more passenger-intensive routes, with the more economical Boeing 737–300, of which KLM bought ten, followed by an order for a further five, plus twelve of the stretched (129-seat) 737-400 version. KLM preferred the Boeing, redeveloped from the 737 short-range jet of the late 1960s and specially adapted to use economical, low-noise turbofan engines, over the conceptually older McDonnell Douglas MD-80, successor to the DC-9. The choice of the 737-300 was a timely one, because it enabled KLM to keep ahead of the 1992 EC regulations on aircraft noise, under which airports began to penalize airlines operating older, more noisy aircraft like the DC-9. The fact that the Boeing could not be easily adapted to double as a freighter like the McDonnell Douglas seems to have played little role, despite KLM's continued reliance on cargo operations. Air cargo within Europe had in any case lost considerable ground to competition from trucks.

In the short-range market, KLM was eventually subjected to pressure to buy locally produced aircraft, much as the British and French airlines had been pressured in the past, in this case the twin-jet Fokker 100. Swissair had actually been the launching customer for the Dutch-built Fokker and KLM had to be given government incentives to acquire six of them in 1985, as a part-replacement for the DC-9s.[25] KLM initially had no clear plans for the Fokkers, as they broke its intended standardization on Boeing 737s, and for that reason it leased them to the French regional carrier Air Littoral in 1991, in which KLM held a 35 per cent stake at the time. When the link with Air Littoral was severed after the French government was unforthcoming with new routes to link French regional centres with Amsterdam, the Fokker 100s were taken back into KLM's inventory to operate less voluminous short-range routes, along with its older Fokker F-27s and F-28s that were operated by its subsidiary KLM Cityhopper.

3 ROUTES

In the absence of a large home market, KLM's growth potential depended on the success of its international network and the airline's ability to schedule its services so as to offer optimal connections in Amsterdam. This required close cooperation between the airline and its home base, Schiphol Airport.

3.1 Network development

In the first months of peace in 1945, KLM served a vital function in maintaining a domestic network connecting Holland's two western provinces with the northeastern and southern parts of the country across the big rivers. It also reopened the service to the Dutch East Indies for the government, with the first aircraft destined for Batavia taking off from Schiphol Airport on 28 November 1945. As the number of available aircraft grew in the early months of 1946, KLM was able to start operations across the north and south Atlantic to New York and Curaçao. A service to Rio de Janeiro was initiated in October and extended to Montevideo two months later.

In Europe the service to London was the first to be restored. When the British opened London for international flights early in 1946, KLM had

already been waiting for months and was ready to start operations immediately. By the end of the year, the service to London had been joined by other
scheduled flights from the pre-war route structure: to Brussels, Paris, Copenhagen, Stockholm, Oslo, Basle, Zurich, and Prague. In addition, new services
were initiated to Geneva, Madrid and Lisbon, and to Glasgow, where transatlantic airliners made a refuelling stop at Prestwick. In the absence of
adequate surface communications, there was a strong demand for air transport, and the expansion of KLM's European network was limited only by the
availability of aircraft. Since Allied restrictions on Germany prevented the
re-emergence of Lufthansa, KLM found it profitable to open as many routes
to Germany as possible, flying passengers and cargo through its home base at
Schiphol, where connections were available to KLM's services to North and
South America. This strategy became so successful that KLM marketed 'the
Amsterdam connection' and made it the cornerstone of KLM's international
operations. The idea of carrying passengers between two foreign countries
through Schiphol eventually became known as 'sixth freedom'. But while
very profitable sixth freedom operations posed problems for KLM and the
Netherlands government at the negotiating table, where reciprocal rights for
air services between countries were exchanged on a strictly bilateral basis.
This was particularly true in negotiations with the United States.[26] KLM's
negotiators never tired in their efforts to explain that the practice of sixth
freedom operations existed only in the minds of their misinformed partners
across the table, who had difficulty in accepting the size of the Dutch air
transport market in the *quid-pro-quo* dealing over landing rights and schedules. Pressing for a breakdown of statistics showing how much air traffic
had its true origin or final destination in the Netherlands, the Americans
restricted KLM's operations for decades. Employing legal arguments as to
what 'origin' and 'destination' really meant, the Americans observed that in
practice KLM's schedules and operational practices inflated the size of the
Dutch air transport market. Passengers would arrive in Amsterdam one day
and leave the next. Yet both airline and airport authorities registered the two
segments of such a stopover journey as individual flights, so that KLM's
statistics for 'Dutch' air transport figures showed more than Amsterdam's
true origin-and-destination traffic.

As the aircraft situation improved, KLM was able to initiate more intercontinental services that fed into its European network, which continued to
depend on services to Germany. In the late 1940s Dutch negotiators managed to procure operating rights for African destinations such as Cairo,
Tripoli, Tunis, Casablanca, Kano, Dakar and Leopoldville (Kinshasa). A
scheduled service to Johannesburg, embodying the linguistic and cultural
ties between Holland and South Africa, was initiated in 1947. In the 1950s
attention focused on expanding the KLM network to South America. After a
service had been opened to Buenos Aires in 1948, further expansion

included Santiago, Sao Paulo, Guatemala City, Guayaquil, San Salvador and Lima. The end of the Indonesian conflict in 1949 cleared the way for the opening of a KLM service to Sydney, which began in December 1951 and marked the end of 22 years of Anglo-Dutch and Dutch–Australian negotiations on the issue.[27] The number of services to the Middle East and Far East were also increased around 1950, though the biggest expansion in this region was effected in the next decade. By the mid-1960s, KLM possessed a world-wide network of some 217 000 kilometres, with its centre of gravity in Europe and on the North Atlantic.

From the mid-1960s, route expansion focused on 'thin' European routes, with predominantly business passengers, through KLM's subsidiary NLM City Hopper, Then the 1978 'open skies' treaty between the USA and the Netherlands gave KLM improved access to the North American market with routes to Los Angeles (1979) and Atlanta (1981). A second breakthrough followed with KLM's US $400 million investment into Northwest Airlines in 1989, resulting in new KLM routes to Orlando (1989), Baltimore (1990) and Minneapolis/St. Paul (1991). Further expansion in the USA followed in January 1992 with a service to Detroit. In the summer of 1993 San Francisco and Washington DC were added and, in cooperation with Northwest, Boston.

3.2 Airport environment

To rise to prominence in the field of air transport, a national airline from a small country needs a mutually advantageous relationship with the national airport which forms its home base. Such a symbiosis existed in the Netherlands between KLM and Amsterdam's Schiphol Airport. KLM's expansion depended on bilateral agreements on commercial traffic rights. The difficulty in obtaining such rights was what to offer in exchange. It was in the Dutch interest therefore that Schiphol Airport should offer optimal technical facilities, and a good selection of transfer connections to offset the small size of its hinterland. Schiphol Airport had to be promoted as one of the prime nodes of Europe's air transport system. And as the growth of KLM's network depended on the success of the sixth freedom concept, it was important to be able to present Schiphol as an attractive destination for scheduled operations. KLM's expansion was, therefore, linked to the ability to offer connections from Amsterdam and take passengers and freight to their ultimate destinations. Thus Schiphol's development was of great importance and every effort was made towards providing the highest possible standard of service there.[28]

The foundation of Schiphol itself preceded that of its main customer by three years, having been laid out as a military relief airbase in 1916. After KLM commenced operations from Schiphol in May 1920, mixed commercial and military use of the facilities caused problems. Even in those early days civil and military requirements of an airport were very different. It took the Amsterdam municipal authorities six years to acquire control of the airport in 1926 and start a programme to provide the standard of service that airlines required. By the end of the 1930s Schiphol, with its four concrete runways, terminal building, spacious hangars and radio facilities, was among the most modern airports in Europe. At the end of the war in 1945, however, very little was left; Allied bombing and German demolition prior to retreat had wrecked the airport beyond recognition.

Immediate postwar reconstruction efforts concentrated on repairing the main east–west runway, the hangars, and erecting prefabricated wooden sheds, on loan from the Swedish carrier ABA, as improvised terminal buildings. Later attention focused on expanding the airport and adapting it to the requirements of KLM's intercontinental operations. Under the guidance of the airport's managing director, Jan Dellaert, a plan for future expansion was devised between 1947 and 1949 as the basis for Schiphol's development in the next half century. The creation of optimal facilities took centre stage, yet the conception of those facilities was dominated by the technology of the 1940s which required aircraft to take off and land against the wind. Schiphol's planners felt that the airport should have eight tangential runways in the prevailing wind directions, emanating from a centrally positioned 'traffic island'. The idea was based on airport design trends developed in the United States towards the end of the war and Chicago's O'Hare airport served as the prime example. Schemes for a new Schiphol with up to eight runways were studied until, in 1956, the City of Amsterdam finally approved a scaled-down version with four tangential runways. The next year construction began with the building of a new east–west runway. Though aircraft technology had changed substantially in the decade since the first sketches had been drawn, the airport's plan was not substantially altered, and the new Schiphol that was constructed next to the existing airport between 1957 and 1967 suffered the consequences as jet aircraft tearing down its runways spread jet noise in all directions.[29]

More important in the short run were the costs of constructing the new airport. Most of these were incurred after the government and Amsterdam's municipal authorities had agreed to found a publicly owned, yet privately organized airport company, the *NV Luchthaven Schiphol*, in January 1958. Although the new company continued the policy of optimizing public service to its users, it also had to recover more of its costs. After an immediate increase in landing fees of 50 per cent in 1958, these and other charges were increased year after year. KLM was taken by surprise and, as the biggest user of the airport, had to foot the bill. Partly as a result of this, relations between KLM and Schiphol gradually changed; instead of two parties working towards a common goal, they became two independent enterprises, each with its own dynamic and each pursuing its own interest. As aviation became more commercialized in the

1970s, the relationship became somewhat ambiguous and the Ministry of Economic Affairs was required to play a mitigating role when the airport tried to use its monopoly to raise landing charges beyond KLM's willingness to pay. Nevertheless, there continued to be areas where the two sides worked together, the most important being terminal and cargo-handling, and the encouragement of tax-free sales to passengers. The new terminal building that was opened with expanded facilities in 1967 not only provided for easy passenger transfers, but also furnished ample space for a tax-free shopping centre that became one of the airport's major selling points and was used by KLM to entice passengers to travel on their aircraft.

On the operational side, the airport's management was concerned to safeguard the capacity of the runway system and terminal facilities to ensure foreign carriers scheduled their flights through Amsterdam. By the mid-1980s, Schiphol had joined the ranks of Europe's busiest airports – fourth after London, Paris and Frankfurt. To keep this position into the twenty-first century was in the interest of both Schiphol and KLM, and a masterplan for a large-scale expansion programme that more than doubled the terminal's capacity was initiated in 1989. The range of shops and services offered at the airport was increased, and Schiphol, and the area around it, was actively promoted as a prime location for international business, logistic enterprises and light industry. Plans for the construction of a fifth runway were finally approved after much public debate in 1995. At the same time, the airport also pursued an active policy in preparation for its envisaged role as connecting node for other transport systems, enlarging the existing underground rail terminal in advance of the arrival of high-speed trains. In the terminology of Dutch airport planning, Schiphol Airport would thus develop into one of five envisaged European 'Mainports' that would function as a gateway, offering interconnectivity for intercontinental and European air and surface transport, as well as being a prime international business centre.[30]

4 PERFORMANCE

After the loss of the Dutch colonies in southeast Asia, KLM was able to focus its policy on commerce and profits. Creative financing and a clear sense of the importance of marketing the airline product abroad formed the basis of KLM's development from the mid-1950s onward, even though the actual performance of the airline was not exceptional.

4.1 Financial and Operating Performance

When KLM resumed operations in September 1945, it tried to reconcile two fundamentally different approaches to air transport. With the few aircraft the airline had left, and 18 Douglas C-54s on loan from the government, it served, first of all, as an instrument of government offering transport within a war-ravaged country and restoring links to Holland's colonial possessions. But KLM was also strongly geared towards a commercial future. This showed in the airline's eagerness to boost its dollar-earnings, which it needed to buy more aircraft in the USA. KLM was the first European carrier to operate a scheduled service across the North Atlantic. It did so on the basis of a provisional understanding between the United States and the Netherlands. Trial flights between Amsterdam and New York commenced on 25 February 1946, with scheduled services following on 21 May 1946. A scheduled service to Montreal began a week later.

By the end of KLM's first full postwar year, 1946, the airline had not only already carried more passengers and cargo than it had done in 1938, but its new network surpassed that of 1938 in length. Moreover 1946 was the first year in the history of the airline in which KLM recorded a profit – 821 985 guilders on a total revenue of 66.6 million guilders, a marginal 1.2 per cent.[31]

Despite these initially favourable results, the future spelt hard times for KLM. With the need for immediate aircraft investment, the airline's finances were fragile. In the last weeks of 1948 the new Lockheed Constellations flew into heavy turbulence, brought about by the old political constellation of Dutch civil aviation policy. While for Albert Plesman serving the profitable transatlantic route to New York represented KLM's prime task, in the eyes of the government and many of the older employees, KLM's purpose was to provide services to Holland's colonial possessions. To the politicians who had grown up with the colonies, the Netherlands had a responsibility resulting from the centuries of Dutch rule in the Indonesian archipelago, and to give up this territory to a bunch of rowdies who had all too willingly put their eggs in the Japanese basket, was simply inconceivable. Despite the sorry state of the Dutch economy, and the need for immediate physical reconstruction at home, the Netherlands mounted a full-scale war against the Indonesian freedom fighters. In that struggle, KLM became vital for the maintenance of communications between The Hague and the Dutch government in Batavia. Unfortunately the Dutch refused to admit that the days of colonial wars had passed. When the Netherlands Army launched a big offensive against the Indonesian nationalists in December 1948, the action was not only condemned by the United States, but also prompted immediate action by the newly independent, and fellow-Muslim, state of Pakistan. The Pakistanis closed their air space to all Dutch aircraft on the pretext that KLM was carrying arms and equipment to the Dutch forces in Indonesia. India soon followed suit. The Dutch could not afford to have their air link severed at this crucial point, and the government and KLM worked out an alternative route to Batavia, avoiding the hostile Asian countries, and using the colonial possessions of the French to divert KLM's planes through the island of

Mauritius. From there, KLM flew its Constellations non-stop across the Indian ocean to Batavia, which was just within range if they operated with a considerably reduced payload. This made the service completely uneconomic, but the government gave KLM a special subsidy to keep the air link operational. When in 1949 the Dutch could no longer hold out against the Indonesians at the negotiating table, the prohibitively expensive Indian Ocean operation was abandoned.

Meanwhile, the pace of KLM's fleet expansion had enabled it to take a lead over its European competitors. KLM was able to offer more services than rivals like BEA. In 1948 KLM had already captured 87 per cent of the air travel market between London and Amsterdam, offering 1350 seats per week against BEA's 190.[32] In 1951 KLM's sales rose by 24 per cent, with its average load factor up from 65 per cent to 67.9 per cent.[33] With the absence of a home market large enough to warrant the rapid growth that Plesman was aiming for, KLM was heavily dependent on the success of its overseas marketing. To attract passengers who would choose the airline for its service and not because they had a business appointment in Amsterdam, KLM pioneered the new concept of 'sixth freedom': the carriage of passengers between two foreign destinations through the national home base. In this KLM was greatly helped by the post-war ban on German aviation. Lufthansa's absence as a competitor enabled KLM to develop the German market as a basis for its own expansion. After flying its passengers from various German and other European cities to Amsterdam, to spend a night in a local hotel, KLM would then welcome them back on board one of its intercontinental flights the next day. A similar practice was developed for passengers travelling in the opposite direction on intercontinental flights to ultimate destinations beyond Amsterdam. This sixth freedom traffic helped KLM grow to be one of the world's largest international airlines by the end of the 1950s.[34]

KLM's development of its air cargo business, another consequence of the small size of its home passenger market, also helped it grow. Between 1950 and 1965 the market for international air cargo increased by an annual average of 18 per cent. By the mid-1950s KLM had become Europe's biggest cargo airline, receiving a quarter of its total revenue from freight, a share that had risen to 29 per cent by 1960.[35] The rapid expansion of the international economy in the 1960s contributed still further to KLM's cargo success. For intercontinental cargo traffic, the airline used four DC-7F freighters, replacing them with the combi DC-8, plus one full-freighter, in the second half of the 1960s. For its European services KLM acquired six rapidly convertible DC-9-33RCs for overnight air freight.

Capitalizing on an industry-wide movement towards integrated distribution chains that offered worldwide door-to-door transport of goods, KLM expanded its ground facilities for cargo handling at Schiphol in cooperation with the airport's management. In the 1970s, this trend towards integrated cargo services contributed to the growth of inter-airport road haulage of air cargo in Europe, transported under regular flight numbers and known in the industry as 'trucking'. The development was furthered by the increases in fuel prices after 1973. In evidence of KLM's decade-long cooperation with road transport firms, and the general interlinking of transport modes, KLM participated on a fifty-fifty basis in the founding of XP Express Parcel Systems, a company specializing in integrated air/road courier services in 1983.[36] KLM increased its shareholding to 100 per cent in 1987, before selling XP Systems to the Australian TNT group in 1988. That same year, KLM acquired a minority shareholding in Frans Maas Beheer, the Dutch logistics, shipping and trucking company operating in 14 countries.[37]

Throughout this period the government continued to be KLM's majority shareholder. In the first postwar decade the Dutch state owned over 95 per cent of KLM's stock. The airline was also helped along by a series of low-interest loans, for which the government acted as guarantor, thus providing additional finance for KLM's short-term development. However, the airline's long-term growth necessitated the issue of new stock, for which KLM cast its eyes beyond Holland's borders. In August 1956 the airline's Articles of Association were amended, reducing the special powers and rights of the government, which would henceforth hold only ordinary shareholder's rights. KLM also bought back 12 per cent of the government stockholding. The aim of these changes was to make KLM more attractive for foreign, particularly American, investors.[38] In May 1957 KLM shares were traded for the first time on the Amsterdam stock exchange. The common share capital was increased 37 per cent by an issue of over forty million guilders worth of new stock. In an unprecedented move for a European airline, KLM was, at the same time, given permission by the government to float two-thirds of the new stock issue, representing a value of 25 million guilders (17 per cent of the total issued share capital) on the New York stock exchange.[39] KLM would remain the only non-American airline to have its shares listed on Wall Street until the 1990s. It was a move that was not so much dictated by the need to acquire dollar-investment in the company, as by the necessity to improve KLM's negotiating position *vis-à-vis* the United States government, because KLM's expansion had brought the Dutch into serious conflict with the United States.

For decades, negotiations on a bilateral air transport agreement, of over-riding importance to the Dutch because transatlantic air travel represented the largest intercontinental market, were thwarted by the American demand for what they called 'an equitable exchange of economic benefits'.[40] Inter-mittent discussions between the two countries took place from 1946 until 1957. KLM's 'sixth freedom' operations were the main stumbling block for the Americans, who wished to restrict KLM's frequencies to the United

States to a level that would reflect air traffic between the US and Holland itself. The Dutch needed more political strength than they possessed to get the Americans to turn a blind eye to the nature of KLM's operations and in the second half of the 1950s KLM was denied commercial landing rights on the American west coast, which were allocated instead to its rivals SAS, BOAC, Lufthansa and Air France.[41] Only as a consolation prize did Washington agree to let KLM serve Houston in 1957 to soothe the strained relations between the two countries that had resulted from the controversy.[42]

Table 5.2 KLM Productivity

	Passengers (1000s)	Cargo (metric tonnes)	Available ton/km (millions)	Revenue ton/km (millions)	Pass. load factor	Overall load factor	Total Staff	Revenue ton/km prod. per employee
1946	313	2181	57	42	76.4 %	73.1 %	–	–
1951	499	12870	225	153	68.7 %	67.9 %	11900	12857
1956	822	23739	407	235	61.5 %	60.5 %	16008	14680
1961	1398	44696	763	399	50.1 %	55.6 %	18065	22087
1966	1913	79006	1148	625	54.8 %	56.3 %	13987	44684
1971	3130	129128	2396	1246	50.7 %	52.0 %	16690	74655
1976	3708	167000	3282	1811	57.4 %	55.2 %	16481	109884
1981	4461	244000	4111	2615	64.3 %	63.6 %	19042	137328
1986	5655	360000	5229	3490	65.4 %	66.7 %	21235	164351
1991	8222	385000	7073	4988	72.0 %	70.5 %	25596	194874
1995	12339	598000	11194	8108	74.4 %	72.4 %	25528	317612

Source: KLM Annual Reports

The emphasis on marketing, economics, and the acquisition of new routes made the KLM management strangely oblivious to two other important aspects of airline operation: timely procurement of the right aircraft and changes in the environment at its home base. For an airline whose success depended on attracting passengers in the international market, such a lop-sided approach was potentially dangerous. When recession struck the airline industry in 1961, KLM was hit badly. The effects of the crisis were all the more severe since KLM struggled with excessive costs for aircraft deprecia-tion – the price of ordering the DC-7C in the spring of 1957, instead of focusing on jets. The Netherlands government had to step in not only to resolve the financial crisis in which KLM found itself, but also to shake up the company's management, which was at a loss as to how to resolve the difficulties it faced. KLM's president, Ernst van der Beugel, who had only joined the board 18 months before, after a career at the Ministry of Foreign Affairs, resigned in December 1962 to make way for the banker Horatius Albarda.

After writing off substantial losses, Albarda and (after his death in a plane crash) his successor Gerrit van der Wal, reorganized the company's finances and organizational structure to put the airline back on a profitable track. In 1965 an agreement on the reduction of outstanding debts was reached, converting debentures into issued common stock, which thus increased by as much as 26 per cent.[43]

In the second half of the 1960s and the early 1970s KLM benefited from the transformation the company had undergone between the introduction of its first DC-8 jets in 1960 and the onset of the 'age of the jumbo' towards the end of the decade. KLM had been successful in gaining new traffic rights, particularly in the new African countries. The combination of high economic growth in the western world, the appeal of Dutch friendliness and reliability, and the introduction of mixed passenger and cargo services, with a further fine-tuning of 'sixth freedom' operations, enabled the airline to expand both its operations and its profits. The distinctive blue-top livery, which KLM introduced on its aircraft in 1971, set the airline apart from other carriers and enhanced its recognition in the eyes of the public. None of the other European carriers – apart from British Airways in the 1980s – have broken with the practice of having predominantly white aircraft.

When the oil crisis hit the airline industry late in 1973, KLM was in good financial shape. However, because of the traditionally pro-Israeli view prevailing in the Netherlands, KLM was hit harder than most airlines by the oil embargo that followed the Yom Kippur war.

In the 1970s KLM decided to diversify its activities and combine its airline operations with other fields related to international travel. Investment in hotels at key locations in KLM's international network began in the mid-1960s and grew steadily. Although the hotel involvement proved to be a financial liability, the number of hotels in which KLM held a stake increased. When recession hit the airline industry in 1974, KLM held a financial interest in 22 hotels in the Netherlands and abroad, which had contributed nothing but losses to the airline.[44] Nevertheless, in view of the adverse conditions brought about by the rise in oil prices, KLM's management decided to further diversify the airline's interests. In 1975 KLM participated in the fast-growing Golden Tulip hotel chain, conducting marketing, sales and reservation activities for some 60 hotels in 19 countries. A year later KLM expanded its shareholding to 50 per cent and began coordinating its hotel activities with the Golden Tulip group.[45] Shares in individual hotels were eventually sold off as KLM's interest in the hotel business decreased with the recovery of the air transport market, and the airline's management reverted to concentrating on its core business.

Another development was the expansion of KLM's European network in the 1970s through its subsidiary NLM Cityhopper. Cityhopper operated smaller aircraft such as the Fokker F-27 turboprop and F-28 jet, and was able to work 'thin' routes with marginal profitability, carrying passengers who transfered to KLM's intercontinental services in Amsterdam. KLM's strategy in this field fitted the larger pattern of emerging regional air services in Europe.

Table 5.3   Financial performance of KLM, 1946–95

Year	Total operating revenue (million f)	Net profit (million f)	Net loss (million f)	Net profitability (net profit as % of revenue)
1946	67	0.8	–	1.2
1951	234	10.5	–	4.5
1956	406	23.0	–	5.7
1961	544	–	76.7	–14.1
1966	819	81.4	–	9.9
1971	1 425	–	96.3	–6.7
1976	2 500	77.5	–	3.1
1981	3 772	32.0	–	0.7
1986	5 376	301.0	–	5.6
1991	7 913	125.0	–	1.6
1995	9 536	547.0	–	5.7

Source: KLM Annual Reports

In the 1980s, the long-term effects of KLM's 1957 Wall Street share issue, and subsequent issues, levelled the path to American banking institutions for leasing arrangements for five American-registered aircraft. Thirteen years after KLM had leased its first aircraft in November 1967, with the acquisition of the first of four DC-8-63s,[46] this type of financing was taken a step further, when KLM acquired its first American-registered Boeing 747. These types of financial arrangement reflected a further increase in KLM's commercial orientation in international aviation. In subsequent years, leased aircraft, which provided the airline with more financial flexibility, became a dominant feature of KLM's operations. Indeed by 1996 the majority of KLM's planes were leased: while 61 aircraft were owned by KLM, 82 were kept under financial lease arrangements, with a further 16 as operational leases provided by other airlines.[47]

Meanwhile, the government shareholding in KLM, which had dwindled to 54.9 per cent by 1985, was reduced to a minority shareholding. KLM and the state agreed to a safety net to preserve the requirements of substantial ownership and effective control, an accepted practice in the international airline industry. Because bilateral air service agreements, in which particular airlines are designated to operate specified routes, are drawn up between governments, those governments need to be able to control their national carriers operating under such an agreement, both as regards technical stand-ards and safety, and in the political sense. In the event that a foreign country required proof of nationality through majority ownership by the state, then the state kept the right to convert warrants and preferred shares into a straightforward majority holding of common shares.[48] An agreement was reached that the Dutch state should exercise its option if and when it would be necessary and reasonable

> that KLM, under either one or more international agreements, either one or more licenses, issued by whatever country, would have limiting or aggravating conditions imposed in the operation of scheduled flights as a result of the situation that a predominant share or rather the majority of the KLM capital is not demonstrably in Dutch hands; or in the case that it is necessary to prevent one person or company or a group of persons or companies from obtaining such a share in KLM that an undesirable power is amassed in the General Meeting of Shareholders.[49]

Subsequently, the government shareholding was reduced to 38.2 per cent in 1987.

Further evidence of KLM's vigour showed in its efforts to improve its position in the global market for air transport, which led to the unprecedented step of joining in the leveraged take-over of the troubled American carrier Northwest Airlines in 1989.

4.2 Marketing and service

KLM's marketing has always been skilled and aggressive. In the immediate postwar years it began to sponsor the publication of the periodical *ABC World Airways Guide* for travel agents, and it ensured that the slogan 'Call KLM first' was printed on the back of these tools of the trade.[50] Inside the *Guide*, the travel agent would find large advertisements under the heading 'Comfort first and fast', continuing:

> Flying schedules are carefully arranged for the maximum convenience of passengers who travel in luxury aboard the great KLM airliners, with free food and drink and the finest service in the world for their every comfort.[51]

By contrast, the new Scandinavian carrier SAS contented itself with the rather meaningless message: 'The selling line to anywhere in the world'. And this was probably an improvement on Iberia's 'Fly with us to the sunshine'. Along with this information to travel agents, KLM's advertising characteristically emphasized features of its destinations along with the air-line's regularity and comfort of service.[52]

In the following years, the focus of KLM's advertising shifted away from the aviation environment and followed the example of other European carriers in using national stereotypes – in KLM's case Dutch tulips, windmills, clogs and traditional costumes. After 1960, this approach was seen as too sentimental and out of touch with the modernity expressed by the new jet aircraft. Instead a distinctive new approach toward a 'house style' was launched. The KLM logo was redesigned with a stylized crown on top. The new logo, stressing geometric modernity, suggested the characteristics of an emerging high-tech society. It was followed with advertising that shifted attention away from the aircraft, stressing 'the difference' it made to fly the *reliable* Dutch carrier, a logical step in a period in which jet aircraft had become commonplace and mass travel had replaced the old style of first class airline travel.[53] In tune with the times, a subtle humour was also added in to the advertisements. But as the Boeing 747 stressed the mass travel business that air transport had become, new marketing was needed to distinguish the Dutch carrier from its competitors. In 1972 the phrase 'fly the difference' was introduced, augmented by a reference to KLM's reliability. The latter proved to have lasting appeal in the ever more complex environment of air transport and has continued to feature as KLM's prime marketing gambit. After KLM's link-up with Northwest Airlines in 1989 – the world's first strategic alliance – the two carriers proclaimed a 'seal of partnership', which carried the words 'worldwide reliability', around the logos of the two airlines. In the 1990s the idea of the seal has been developed as a quality trademark and appears on the aircraft of airlines subcontracted to KLM and Northwest, or operating feeder services on their behalf.

5 STRATEGY

Collaboration and alliances between airlines have a history as long as commercial air transport itself. Because of the small size of its home market, and the scope of its operations, partnerships have always been important for KLM. Indeed much of the postwar history of the airline was characterized by a long quest to find the right strategic alliance for further growth.

5.1 Collaboration and competition

Over the years KLM has cooperated with a number of emerging carriers from non-European nations. This practice began in the immediate aftermath of Indonesian independence in 1949, when the new Indonesian government needed its own airline to take over the various services in the archipelago that had hitherto been operated by KLM. Under KLM's management the new airline, Garuda, operated from 1949 until political differences between the Netherlands and Indonesia brought the cooperation to an end in 1954.

Various other cooperative alliances followed in the same decade. As the first steps towards creating a worldwide network of associated airlines feeding into KLM's intercontinental network, KLM took a share in Air Ceylon and in the Colombian carrier Rutas Aéreas de Colombia (RAS). Both associations were severed in 1961, when the air transport market slumped and KLM re-evaluated its strategic interests. The commercially promising Venezuelan airline VIASA became KLM's new partner and there was also a brief association with the Panamanian carrier PIASA in 1967. In Asia, a close relationship was established with Philippine Air Lines in 1962. Expansion along similar lines was also pursued in Africa, where KLM entered into a partnership with Nigerian Airways in 1972.

Closer to home, KLM followed an erratic strategy with regard to cooperation within Europe. In 1955 KLM was involved in the creation of Air Austria, one of two rival airlines that struggled for take-off in that country. KLM chartered several of its ageing DC-4s to the new airline, hoping to use Air Austria's neutrality status as a means to expand into the Eastern Bloc countries.[54] When Air Austria merged with Austrian Airways to form Austrian Airlines in 1957, KLM pulled out because it felt the Austrian government offered insufficient financial safeguards to protect a long-term investment in the new carrier.[55] KLM's position as Austrian's supporter was taken over by SAS and the Norwegian shipping line Fred Olsen.[56]

In 1957, when Lufthansa and Air France opened discussions, soon expanded to include SABENA, about the founding of a pan-European airline called Air Union, KLM watched distrustfully from a distance. SABENA approached the Dutch in 1958 to participate in the project, but it was only after the airline crisis of 1961 that KLM became interested. With active support from the Netherlands' government, KLM negotiated to join Air Union, but when it failed to receive satisfactory assurances over its traffic share, its interest waned again. In response to government insistence KLM continued to participate in the intermitent Air Union discussion until the project finally collapsed in 1967.[57]

Cooperation in aircraft maintenance was easier to achieve. This was effected in the maintenance consortium KSSU (KLM, SAS, Swissair, UTA), a group of European Douglas customers, which grew out of the initial cooperation in aircraft and engine maintenance between KLM, Swissair and SAS in 1966. In the 1970s the alliance was extended to include maintenance of the DC-10 and Boeing 747. Cooperation ended when UTA became part of Air France in 1993.

Within Holland KLM set up a subsidiary company for the operation of helicopter services, KLM Noorzee Helikopters (KLM North Sea Helicopters), in October 1965, but this new branch remained dormant until March 1968. Its main purpose was to offer services to oil rigs in the North Sea.

When it became clear that there would be a market for domestic air services and other 'thin' routes to secondary destinations in neighbouring countries, KLM founded a new subsidiary, Nederlandse Luchtvaart Maatschappij (NLM: Netherlands Aviation Company), in September 1966. In 1964 it had already taken a 40 per cent share in the fast-growing Dutch charter airline Martin's Air Charter (now Martinair Holland), the aim being to protect itself from domestic competition to its own scheduled and charter operations. A second Dutch charter airline, Transavia, founded by the shipping concern Royal Nedlloyd Group in 1966, remained beyond KLM's reach until it ventured into the realm of scheduled services in 1986. In 1988 KLM declared its intention to the European Commission (EC) of taking a 40 per cent share in Transavia, acquiring it from Nedlloyd in 1989. In fact formal approval by the EC was withheld until 1991, when KLM was also given the go-ahead to acquire a further 40 per cent of the Transavia shares on condition it did not monopolize the Dutch air transport market any further.[58]

Financial participation in competing enterprises in Holland was KLM's favoured approach to keep rival airlines under control. In similar fashion it took over the recently formed but loss-making regional feeder airline Netherlines in 1988. After several years of substantial losses KLM merged Netherlines and NLM City Hopper into a new domestic and regional carrier, KLM Cityhopper, in April 1991.

5.2 Globalization

None of these initiatives solved KLM's basic weakness in the international market, which was the lack of a home market big enough to grow into a major player in the field. An alternative strategy was called for, and having discovered the potential of high-revenue, business class feeder services, KLM started expanding its interests abroad. In June 1988 it bought a 14.9 percent interest in the British feeder carrier Air UK. This move fitted the larger perspective of liberalization in the Anglo-Dutch air transport market, and the efforts by Schiphol Airport to advertise itself as 'London's Third Airport'. Subsequently, KLM's investment in Air UK was increased to 45 per cent, and Air UK's services were fine-tuned to offer optimum connections at Schiphol for British passengers from Stansted and nine other British airports, to continue to their ultimate destination on the KLM/Northwest network.[59]

In a further effort to improve its strategic position in the European market, KLM joined British Airways in an effort to found a new European air carrier, Sabena World Airlines late in 1989. In this reincarnation of the ailing Belgian flag-carrier, both partners were to hold a 20 per cent stake, while the remainder resided with the parent company Sabena SA. However, when it became clear the following year that the necessary restructuring of

SABENA and the creation of a 'Euro-hub' at Brussels Airport would take years to complete, and that the three-cornered hat of Sabena World Airlines was therefore not going to fly anytime soon, KLM withdrew its interest.[60] The collapse of Sabena World Airlines served as a warning that the proposed megamerger of British Airways (BA) and KLM, over which negotiations began in the autumn of 1991 after BA had become interested in a financial stake in KLM's American partner Northwest Airlines, would also be difficult. As American legislation prevented BA from taking a direct interest in Northwest since the quota for foreign investment in that airline was already taken by KLM, an Anglo-Dutch joint venture, tentatively christened Royal World Airlines in Holland, would have given BA a stake in Northwest through the back door. Six months of painstaking negotiations on what would have been a path-breaking fusion between two long-established European flag-carriers yielded no agreement, as the two sides failed to reach agreement on the value of each other's assets as well as on control of the new venture. In February 1992 negotiations were broken off, when KLM apparently concluded that the gap between their offer of a 60/40 division of the shareholding in Royal World Airlines, and BA's demand for a 70/30 split, could not be bridged. The deeper reason for the failed merger was that BA and Northwest could not agree on how to accommodate a British stake in the American airline.[61]

Even before that, KLM tried to improve its position in the European market, which had slumped to a mere 3 per cent, on its own accord, and bought a 35 per cent share in the loss-making French regional carrier Air Littoral in 1991. KLM's French excursion did not work out however. The French government refused to allow Air Littoral to expand the number of services to Amsterdam, because these would have presented unwelcome competition to the services of the Air France/Air Inter group. In November 1992 KLM sold its interest to the French carrier Euralair, a company operating services for Air France under a partnership agreement. The demise of the KLM/Air Littoral partnership illustrated that, despite the European Commission's efforts to liberalize air transport, Europe's airspace was still very much divided by national sovereignties.

Likewise, in 1993 national and corporate interests, and different airline styles, worked against the attempted pan-European Alcazar partnership, which KLM hoped to achieve with Austrian Airlines, SAS, and Swissair. Here too, strategic alliances across the Atlantic were the complicating factor: KLM was allied to Northwest, Swissair with Delta, SAS with United, while Austrian was considering a deal with All Nippon Airways. The project proved that European alliances, hinging on agreements about the division of a market already served by the participating partners, were far more difficult to negotiate than global alliances, which aimed at a mutual expansion of new markets. As a result Alcazar was abandoned in December 1993.

Globally KLM was more successful in finding a strategic ally and finally decided to participate in a leveraged buyout of the fourth-largest US carrier, Northwest Airlines, in July 1989. KLM was one of the major players in Wings Holdings Inc., which acquired all the shares of the ailing Northwest. The aim of KLM's involvement in the takeover was to strengthen its position in the global air transport market. Indeed it could be said that KLM pioneered the field of airline globalization. Initially, KLM's share in Wings Holdings Inc was 11.1 per cent, representing an investment of $100 million. KLM was concerned to keep its stake modest so as not to alarm the US Department of Transportation with this unprecedented foreign interest in one of America's major airlines. Through Wings Holdings, KLM provided Northwest with a $300 million interest-bearing loan in 1990, which could be converted into Northwest shares. KLM then operationalized warrants in March 1991 that brought the total of KLM investment in Northwest to $400 million, representing an interest of 20 per cent.[62] This stake became the basis of a close cooperation between the two carriers that included route allocation, scheduling, code-sharing, and the link-up of the two carriers' computer reservations systems (CRS). The size of KLM's investment showed its determination to forge an alliance that would boost its European market share, while at the same time furthering its share of the North Atlantic market, by linking its services to Northwest's American network. After further financial reshuffling, KLM's direct economic interest amounted to 23 per cent in March 1996.[63]

Initially Northwest continued to fare badly and KLM decided to write off its investment in Wings/Northwest in 1992, in order to keep the operational benefits of the Dutch-American link. At the same time, the partnership with Northwest was intensified by a further US$50 million loan for restructuring the airline after a new liberal air transport agreement between the USA and the Netherlands had cleared some of the American legal obstacles for transnational airline investment and cooperation.[64] In 1993 a number of transatlantic operations were set up as joint ventures under the quality seal 'KLM/Northwest: Worldwide Reliability' that was carried on all the aircraft of the two partners, and expanded to various associated airlines such as the German carrier Eurowings. Eurowings started operating services with mid-size turbo-prop aircraft from Nuremberg, Hannover, Stuttgart, Dresden, Leipzig, Düsseldorf, Dortmund and Paderborn to Amsterdam in 1995, expanding the KLM/Northwest interest in the German market through code-sharing arrangements and participation in frequent-flyer schemes.[65] KLM also increased its interest in the growing market for air transport to Africa by taking a 26 per cent stake in Kenya Airways in 1995, as a first step towards commercial and technical cooperation, and expansion of KLM's globalizing objectives.

In conclusion, KLM's postwar development was heavily influenced by the political setting of the first postwar decade. In the absence of competition from Lufthansa, KLM was able to capture a substantial market share in Europe upon which to build a network. Moreover its home base, Amsterdam's Schiphol Airport, served as the node for the transfer of passengers from abroad to flights that took them to their final destination. This way KLM was able to compensate for the loss of revenue brought about by the collapse of the Dutch colonial empire, upon which its pre-war fortunes had been built. The orientation on the *global* air transport market that this entailed required a pragmatic and commercial approach to the air transport business. In this KLM differed from its European competitors, which continued to interpret their position in the market in terms of their national background. KLM's awareness of the crucial need to forge strategic alliances boosted the airline's international market position, resulting in the unprecedented move to invest in Northwest. In this, and in the subsequent discussions on a full merger with BA, KLM showed its most typical Dutch trait – that commerce precedes the flag.

NOTES

1. Marc Dierikx, *Bevlogen Jaren. Nederlandse burgerluchtvaart tussen de wereldoorlogen* (Houten, 1986), pp. 162–4.
2. Marc Dierikx, *Begrensde Horizonten. De Nederlandse burgerluchtvaartpolitiek in het interbellum* (Zwolle, 1988), pp. 26–33.
3. KLM Accounts and balance sheets, 1920–1939: KLM Board papers, series R-5.
4. C.C. van Baalen, 'De KLM als paradepaardje van een verarmd Nederland (1948–1950)', in: *Politieke Opstellen* dl. 7 (Nijmegen: Centrum voor Parlementaire Geschiedenis, 1987), pp. 9–21.
5. KLM, *Annual Report 1947*.
6. C.C. van Baalen, 'Spitsen als kleurloze minister van Verkeer en Waterstaat', in: P.F. Maas (ed.) *Parlementaire Geschiedenis van Nederland na 1945. Deel III: Het kabinet Drees-Van Schaik (1948–1950). Band B: Anti-communisme, rechtsherstel en infrastructurele opbouw* (Nijmegen, 1992), p. 74.
7. Huub Surendonk, *Onderhoud in beweging. 1921–1996: 75 jaar KLM Technische Dienst* (Amstelveen, 1996), p. 56.
8. KLM, *Annual Report 1962*, p. 9.
9. KLM, *Annual Report 1968/69*, p. 24.
10. H.A. Wassenbergh, *Aspects of air law and civil air policy in the seventies* (The Hague, 1970), pp. 28–32. R.L.M. Schreurs, 'Positie van Nederland in de internationale luchtvaartpolitiek', in: F.A. van Bakelen (ed.), *Teksten vervoerrecht: Luchtrecht* (Zwolle, 1983), pp. 87–90.
11. Written replies to questions asked in Parliament during the Annual Debate on the Budget: *Handelingen der Staten-Generaal*, Second Chamber, 1975–6, V., Noten, 2224–2227. Also: ibid., 1976–7, Bijlagen, Hfdst. XII, nr. 2, MvT, 39–40, and ibid., 1977–8, 14.800, Begroting Hfst.XII, nr.2, MvT, p. 44.
12. Memorandum of the European Commission: *Contributions of the European Communities to the Development of Aviation*, document COM (79) 311, 4 July 1979.

13. For the various political and legal aspects that formed the backdrop to these developments, see P.J. Slot, P.D. Dagtoglou, *Toward a Community air transport policy. The legal dimension* (Deventer, 1989) and Peter P.C. Haanappel, George Petsikas, Rex Rosales, Jitendra Thaker (eds), *EEC air transport policy and regulation, and their implications for North America* (Deventer, 1990).

14. Minutes Board of Directors, 09/01/1940: KLM Board Papers, series R-5.

15. Minutes Board of Directors, 03/04/1951: KLM Board Papers, series R-5.

16. Minutes Board of Directors, 11/12/1952: KLM Board Papers, series R-5.

17. Minutes Board of Directors, 27/11/1951: KLM Board Papers, series R-5.

18. Marc Dierikx, 'Changing Perspectives: the Franco-Dutch competitive experience in intercontinental air transport, 1930–1960', in: *Actes du Colloque International "l'Aviation civile et commerciale des années 1920 à nos jours"* (Paris: Service Historique de l'Armee de l'Air, 1994), pp. 111–43. More extensively: Marc Dierikx, 'Een spel zonder kaarten': KLM-landingsrechten als nationaal belang, 1945–1957, in: D.A. Hellema, C. Wiebes, B. Zeeman (eds.), *Jaarboek Buitenlandse Zaken: Derde Jaarboek voor de geschiedenis van de Nederlandse buitenlandse politiek* (The Hague, 1997), pp. 11–25.

19. Report *Toekomstverwachtingen en groeimogelijkheden van de KLM*, 19/06/1957: KLM Board papers, series R–5 (Supervisory Board).

20. Minutes Supervisory Board, 27/04/1961: KLM Board papers, series R-5.

21. KLM report *Verkenning van het beleid voor de eerstkomende jaren*, 12/12/1962: KLM, DS (RVC), R-5.

22. Minutes Supervisory Board, 28/09/1967: KLM Board Papers, series R-5. Leonard de Vries, *Vlucht KL-50* (Amsterdam, 1969), p. 147.

23. Van der Wal to Van Stapele, 21/02/1969: KLM Board papers, series N-8.

24. Report Vaste Kamercommissie voor Verkeer en Waterstaat aan Staatssecretaris Smit-Kroes, 17/05/1979: HTK 1978–9, Bijlage 15.300 Hoofdstuk XII, nr.55.

25. Handelingen Staten-Generaal, Rijksbegroting 1986, 19 200 Hoofdstuk XII, nr. 2, 63.

26. See: Marc Dierikx, '"Een spel zonder kaarten": KLM-landingsrechten als nationaal belang, 1945–1957', in: D.A. Hellema, C. Wiebes, B. Zeeman (eds), *Jaarboek Buitenlandse Zaken: Derde Jaarboek voor de geschiedenis van de Nederlandse buitenlandse politiek* (The Hague, 1997), pp. 11–25.

27. Marc Dierikx, *Begrensde Horizonten. De Nederlandse burgerluchtvaartpolitiek in het interbellum* (Zwolle, 1988), pp. 116–55.

28. For this and the following, see: Bram Bouwens, Marc Dierikx, *Op de drempel van de lucht. Tachtig jaar Schiphol* (The Hague, 1996).

29. For the problems caused by jet noise around Schiphol, see Marc Dierikx and Bram Bouwens, *Castles of the Air: Schiphol Amsterdam and the development of airport infrastructure in Europe, 1916–1996*, The Hague, 1997.

30. Project Mainport en Milieu Schiphol, *Planologische Kernbeslissing Schiphol en omgeving; deel 1 ontwerp PKB; nota van toelichting* (The Hague, 1993)

31. KLM, *Annual Report 1946*, p. 3.

32. Draft of a Ministry of Civil Aviation report on the operations of BEA and its continental competitors, 02/05/48: PRO, MCA, BT217, p. 68.

33. KLM, *Annual Report 1951*, p. 9.

34. R.E.G. Davies, *A History of the World's Airlines* (London, 1967), p. 500.

35. Report *Toekomstverwachtingen en groeimogelijkheden van de KLM*, 19/06/1957: KLM, DS (RVC), R-5. KLM, Annual Report 1960, pp. 19–20. Annual Report 1965/66, p. 7.

36. KLM, *Annual Report 1983/84*, p. 17.

37. KLM, *Annual Report 1988/89*, p. 24.

38.   KLM, *Annual Report 1956*, p. 13.
39.   KLM, *Annual Report 1957*, pp. 14, 41.
40.   US State Department to Senator Lyndon B. Johnson, 05/06/1956: United States National Archives, Washington DC, State Department, Record Group 59, 611.5694/5-2456.
41.   Marc Dierikx, Changing perspective: the Franco-Dutch competitive experience in intercontinental air transport, 1930–1960, in: *Actes du Colloque l'Aviation civile et commerciale des années 1920 à nos jours* (Paris, 1994), pp. 111–43.
42.   Memoranda of U.S.-Netherlands negotiations on air transport rights, 27 March 1957: U.S. National Archives, State Department files, record group 59, 611.5694/3-2757.
43.   KLM, *Annual Report 1965/66*, pp. 35, 38–9.
44.   KLM, *Annual Report 1973/74*, p. 16.
45.   KLM, *Annual Report 1975/76*, p. 17.
46.   First mention of lease constructions for KLM's expanding fleet is made in the Annual Report of 1969, which indicates four out of five of KLM's DC-8-63s to be leased, plus two out of thirteen DC-9s: KLM, Annual Report 1969/70, pp. 19, 21.
47.   KLM, *Annual Report 1995/96*, p. 18.
48.   Marc Dierikx, 'Bermuda bias: Substantial ownership and effective control 45 years on', *Air Law* 16(1991) no. 3, pp. 118–24.
49.   Agreement about obtaining shares between the State of the Netherlands and KLM, 26 February 1985, reprinted in: KLM, Annual Report 1985/86, 43.
50.   In recent years the title of this publication has changed to *OAG World Airways Guide*.
51.   KLM advertisement in *ABC World Airways and Shipping Guide* (Dunstable, April 1950).
52.   G.I. Smit, R.C.J. Wunderink, I. Hoogland, *KLM in beeld. 75 jaar vormgeving en promotie* (Naarden, 1994), pp. 72–89.
53.   Ibid., 107–25.
54.   Minutes KLM Board of Directors, 03/11/55: KLM Board Papers, series R-5.
55.   Minutes Board of Directors, 03/11/55. Minutes Supervisory Board, 31/01/1957: KLM Board Papers, series R-5.
56.   Minutes Board of Control, 31/10/1957: KLM Board Papers, series R-5.
57.   Cabinet Minutes, 04/05/1962, MR 12: ARA 2.02.05.01, inv.nr. 675. Cabinet Minutes, 23/06/1967, MR 2f: ARA 2.02.05.01, inv.nr. 844.
58.   Press Release of the European Commission IP[91]658, 5 July 1991.
59.   KLM, *Annual Report 1995/96*, p. 26.
60.   KLM, *Annual Report 1990/91*, p. 9.
61.   KLM's managing director Leo van Wijk in an interview 'We hebben nu een betere kans op succes' with *Elsevier* 48(1992) no. 11, 14 March 1992.
62.   KLM, *Annual Report 1991/92*, p. 36.
63.   KLM, *Annual Report 1995/96*, pp. 25–6.
64.   KLM, *Annual Report, 1992/93*, p. 31.
65.   KLM, *Annual Report 1995/96*, p. 16.

1. *La Grande Nation dans tous les ciels*: Air France in the 1950s.

2. Max Hymans, Chairman of Air France from 1948 to 1961.

3. The twin-engined Caravelle was an early success for the
French civil aircraft industry.

4. Paris Orly airport in 1963.

5. Lunch on board a Concorde, 1995.

6. Outback service: BEA Island Rapide landing on the beach at Barra, Scotland, in 1953.

Picture of a man in a hurry

Be where you want to be faster **FLY B·O·A·C**

"Supreme jet comfort all over the world"

BRITISH OVERSEAS AIRWAYS CORPORATION

7. Speed was the theme of BOAC advertising at the beginning of the jet age in 1960.

8. Sombre faces as the Comet 4 is handed over to BOAC by de Havilland in 1958.

9. The swinging sixties: a BEA Trident.

10. Concorde was always a patriotic undertaking: BA advertisement, 1979.

The warm, friendly, loveable Germans invite you to fly with them.

Maybe that will be our image in a hundred, or a couple of thousand years. At least, we hope so.

But right now, people love our machines far more than they love us. After all, we're known more for technical abilities than for gay, carefree attitudes.

And there's one time we're glad this is so: when a crew and mechanics are making an airplane ready for flight.

Anyone who takes such good care of you in the sky can't be all bad here on earth.

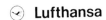 **Lufthansa**

11. Lufthansa turned the Germans' traditional image to its advantage: advertisement, 1963.

12. Aircraft production at the East German VEB Flugzeugwerke in 1959. On the right is the production line of the IL-P14; in the foreground the East German P-152 jet airliner.

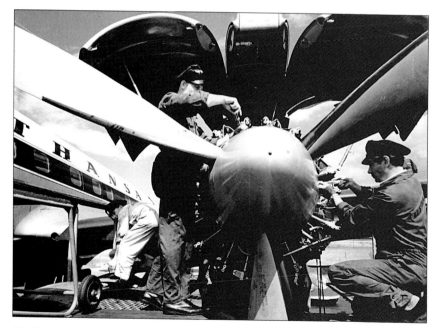

13. Because of its maintenance requirements, the Lockheed L-1049G Super Constellation was called 'Super Complication' by Lufthansa mechanics: servicing one of its Curtiss Wright turbo-compound engines at Frankfurt Airport in 1957.

14. Skirts and profits rose steadily until 1970: Lufthansa stewardess uniforms from 1968.

15. Lufthansa was the first airline to introduce Boeing 747s as cargo jets: taking off from Hong Kong in 1987.

16. Open to the world: the KLM girl of the 1960s.

17. Service on board a KLM DC-6 in the early 1950s.

18. The cosmopolitan airline: a KLM DC-8 honouring Sir Isaac Newton in 1961.

19. Loading cargo into a Boeing 747-300 combi.

20. Global partnership: KLM ground staff working on a Northwest Airlines DC-10.

21. Alitalia's first flight from Turin to Rome, 5 May 1947.

22. For its short- and medium-haul routes, Alitalia used French Caravelles with their distinctive triangular windows.

23. Pope Paul VI was fond of flying: disembarking from an Alitalia aircraft when he was still Archbishop of Milan.

24. The merger of LAI and Alitalia in October 1957.

25. National pride ran high when Pope John Paul II travelled to Warsaw on board a LOT aircraft in 1987. He was welcomed by LOT's General Director, Jerzy Slowinski.

26. A LOT Lisunov Li-2P, the Soviet copy of the DC-3. at Warsaw's Okecie airport in the late 1940s.

27. LOT made repeated attempts to introduce Western aircraft. In the 1960s three Vickers Viscounts were purchased.

28. The Ilyushin Il-18, workhorse of LOT's eastern European routes until their replacement with Western aircraft after 1989.

29. On 26 October 1958 Pan American inaugurated its Boeing 707 jet service with a musical fanfare.

30. A proliferation of commuter airlines in the 1980s created a market for small transports like Fairchild's Metro II.

31. The air freight business proved profitable enough to encourage major airlines like American to configure Boeing 707s (*foreground*) and Boeing 747s (*background*) as dedicated freight haulers.

6 Alitalia and Commercial Aviation in Italy

Amilcare Mantegazza

1 POLICY

1.1 Pre-war origins of Italian aviation

As part of Mussolini's drive to develop a major state-owned industrial sector in Italy, the flagship aviation company Ala Littoria was created in 1934. Mussolini's aim was to remedy the economic instability of the early 1930s. The established policy of purely financial rescue of major industries was replaced with one of massive direct state intervention in the industrial sector, a practice that was to characterize economic development in Italy for the next fifty years. On the advice of Mussolini's economic adviser, Alberto Beneduce, the *Istituto per la Ricostruzione Industriale* (IRI) was set up and given control of the industrial shares that private banks specializing in long-term loans to industry had accumulated in the last decade of the nineteenth century, when they financed the industrial development of the country. It has been calculated that IRI came to possess 21.5 per cent of the total Italian share capital and since it was majority shareholder in 85 per cent of the companies in which it had an interest, its influence actually extended to 42 per cent of Italian share capital.[1]

Mussolini's foreign policy was to project Italy as a great power with colonial pretensions, so even before the rearmament of the late 1930s, an indigenous Italian aircraft industry and a substantial air force had been built up. The Ministry of Aeronautics and the Italian Air Force had been created in 1923. The Ministry of Aeronautics had mainly military responsibilities, but also administered civil aviation, including route assignment to airline companies and route monitoring. In addition communications, weather forecasting and the management of Air Force airfields all had civil dimensions.[2] As a consequence of military spending there were no less than seven major Italian aircraft design and construction companies in the 1930s, as well as four companies producing aircraft engines. The Italian Air Force had a thousand planes, more than sufficient for peacetime purposes, and a series of aeronautical ventures and spectacles projected an image of Italian air superiority, although this was more propaganda than truth. Trans-Mediterranean and transatlantic flights were actively encouraged by the Minister of Aeronautics, Italo Balbo, a favourite of Mussolini, until Mussolini appointed him governor of Libya.[3]

The newly created Ala Littoria, which modelled itself on other European flag-carriers, was born out of an amalgamation of the Società Aerea Mediterranea (SAM), the Società Italiana Servizi Aerei (SISA), the Società Anonima Navigazione Aerea (SANA), the Società Anonima Aero Espresso Italiana, the Società Adria Aero Lloyd and the Società Nord Africa Aviazione. SAM had been created in 1928 by the Ministry of Aeronautics under Balbo. SISA was owned by the naval shipyard and shipping company Cosulich of Trieste, and flew the Turin–Trieste route and between Trieste and locations on the Dalmatian coast. SANA, owned by the naval shipyard Piaggio of Genoa, operated western Mediterranean routes between Italy, France, Spain and Libya. Aero Espresso Italiana was controlled by the Banca Commerciale Italiana and flew the routes Bari–Athens–Rhodes and Bari–Athens–Istanbul. All these companies received state aid in the form of fixed subsidies and also per-kilometre subsidies. In 1932 they owned between them 69 aircraft, 42 of which were Dorniers (built in Italy) and Junkers, the rest being flying boats built by Cantieri Riuniti dell'Adriatico and the Società Italiana Aeroplani Idrovolanti (SIAI). Only one company was excluded from the amalgamation: the Società Anonima Aviolinee Italiane. This remained the property of Fiat, a private concern and the largest automobile and engineering company in Italy. Fiat continued to receive per-kilometre subsidies for operating the Milan–Rome and Milan–Trent–Vienna routes.[4] Ala Littoria's shares were wholly owned by the Ministry of Finance and the airline still received per-kilometres subsidies, but no longer fixed subsidies, from the Ministry of Aeronautics. Average per kilometre subsidies dropped from 15.34 lire per kilometre in 1934 to 10.48 lire per kilometre in 1937–38, as a result of increasing efficiency.[5]

At the time there was a close relationship between military and civil aviation. Civil pilots were trained by the Italian Air Force and were generally highly experienced. Many were veterans of the pioneering trans-Mediterranean and transatlantic flights promoted by Balbo. As a result of the corporate policies adopted by Mussolini, all personnel employed in civil aviation belonged to one trade union, called *Gente dell'aria* (People of the Air) and they enjoyed the privilege of a national work contract.[6]

In 1934, the companies that formed Ala Littoria had flown a total of 3 815 206 km. In 1937–38 Ala Littoria aircraft flew 17 294 431 km and later the service expanded still further. The company penetrated into Europe and the Mediterranean, establishing routes to Spain, Portugal, France, Germany, Austria, Hungary, Czechoslovakia, Palestine, Greece and Tunisia. However, as with other European airlines, the main weight of its services was concentrated between metropolitan Italy and the Italian colonies. In 1938 around 71 per cent of air kilometres were flown between Italy and Albania, Rhodes and the Dodecanese, Libya, the Horn of Africa and Ethiopia – the latter conquered in 1935.[7]

In line with the Fascist policy of autarchy, Ala Littoria replaced the planes it had inherited from the private airlines with Italian aircraft. The Italian aircraft industry designed and built a range of flying boats and land planes using the technology then widely available in Italy. At the end of the 1930s the Ala Littoria fleet consisted of around a hundred aircraft. The most-used were FIAT G-12s and G-18s, designed by Giuseppe Gabrielli, Caproni Ca-101s (used in Africa), Ca-133-148s and Ca-308 Boreas, designed by Cesare Pallavicino, SIAI S-71s, S-73s, S-87s, S-82s and S-83-79s. The flying boats were Macchi C-94s and C-100s, designed by Mario Castoldi, SIAI S-66s designed by Alessandro Marchetti, and Cant. Z-506s and Z-509s, designed by Filippo Zappata for Cantieri Riuniti dell'Adriatico.

However despite the variety of Italian types, a number of problems became evident, for example Italian companies proved incapable of building engines as reliable and powerful as the Rolls-Royce Merlin, the Pratt & Whitney 1830 or the Wright 1820-G2. The Italian aircraft builders had been producing engines for the Italian military for decades and their mentality was very different from that of the American companies. They also supplied a customer – Ala Littoria – that covered its costs with state subsidies. Moreover autarchy made it impossible to use foreign know-how to inspire Italian design, as Giuseppe Gabrielli of Fiat wanted to do.[8]

By 1938, Ala Littoria was planning the inauguration of a Rome–Tokyo route, but this never came about. Some trial flights were also made to South America with a view to establishing a Rome–Buenos Aires route and follow the pioneering example of Balbo in the early years of the decade. In 1939, however, South American routes were taken away from Ala Littoria and assigned to the airline Linee Aeree Transatlantiche Italiane (LATI), which had been specifically created to develop the long-haul South American business. This airline started services a few months before the outbreak of the Second World War, carrying mail, freight and the occasional passenger until the war rendered operations impossible. The managing directors of LATI were Bruno Mussolini, who was killed in a plane crash in 1941, and the Atlantic crossing pioneer Attilio Biseo.[9]

When war broke out, Italy possessed a network of civil air routes that was the fourth largest in Europe. As hostilities intensified the two companies ceased operations, and their aircraft and facilities were taken over by the military.[10]

1.2 Creation of Alitalia in 1957

After the war Ala Littoria was wound up as a demonstration of the demise of the past regime, and in 1946, on the basis of two agreements that facilitated the entry of foreign airlines into the Italian sector, two mixed public/private airlines were set up with Italian and foreign capital. These were Linee Aeree

Italiane (LAI) and Aerolinee Italiane International (Alitalia). Trans World Airlines (TWA) owned 40 per cent of LAI, and British European Airways (BEA) held the same share in Alitalia. The state owned the remaining 60 per cent of both. The foreign capital in these companies served, in part, to meet the terms of the Armistice, which forbade Italy to operate autonomously in the aviation sector. Annexed to the agreements were route tables, which allocated the international network to Alitalia and the national network to LAI. A share of about 16 per cent of Alitalia's capital was later given to a group of Italian private firms, the main one being FIAT. Similarly 20 per cent of LAI was given to FIAT, Bastogi, an established Italian holding, and Piaggio, the engineering and aeronautical company. Later the Italian government authorized various private airlines to operate along with the mixed ones.[11]

The background to the TWA and BEA involvement in the Italian companies was that TWA wanted to break the monopoly of Pan American Airways (Pan Am) on American international routes. It claimed access to Paris and Rome and to Middle Eastern and Far East routes. This provoked a BEA reaction. BEA was eager to keep a foothold in Italian aviation because it wanted 'to check United States civil aviation expansion in Europe', to 'provide a market for British civil air transport aircraft' and 'to preserve a sphere of British influence in the Continental air transport industry'.[12] TWA and the British came to an agreement in which the international routes were parcelled out so as to safeguard the latter's empire network. In allocating route concessions the Aviation Ministry gave Alitalia the international routes and LAI the domestic network; the private companies were allocated regional networks, to which foreign destinations were added in some cases.

The public capital in both LAI and Alitalia was given to IRI, as a normal industrial shareholding. After the war, IRI was no longer the instrument of a dictator but depended for its survival on the will of constantly changing governments. IRI's interest in air transport would seem to have originated both from industrial inclination and political agreement. The potential conflict of interest within IRI, between shipping and shipbuilding on one hand, and commercial aviation on the other, was never realized because IRI's investment in the two airlines was initially modest and the advantage of international air travel over passenger shipping was not appreciated.

The privately owned airlines that had been allowed to operate after the war were soon swallowed up by two mixed ownership companies, and by 1953, when LAI absorbed Fiat's Aviolinee Italiane, only LAI and Alitalia remained. By operating separately, these two airlines were only able to exploit the postwar boom in civil aviation to a limited extent. They were too small to compete effectively with the other major airlines that had been operating over Italy since 1946–7: Pan Am, TWA, BOAC, BEA, Iberia and KLM.[13]

The merger between Alitalia and LAI in 1957 was concluded quickly. Alitalia was able to absorb LAI without difficulty, because TWA accepted Alitalia's and IRI's proposals for the purchase of its share. By this time TWA was a major international airline and the Italian domestic market no longer held much interest for it. The new Alitalia, which was backed by IRI, could now claim to be the Italian flag-carrier.

When the merger took place the public sector of the Italian economy, which was already large, was about to undergo further expansion. The Ministry for State Shareholdings was created and given control of IRI and other new quasi-public authorities. As far as transport was concerned, in 1957 it had been clear for a number of years that aircraft would replace passenger ships as the means of transatlantic travel and, while IRI managed the Italian shipping crisis with state aid, it strengthened its grip on the more promising aviation business.[14]

Many people who were part of Italian aviation at the time favoured the merger and formation of a national airline under the control of IRI. Indeed, the desire for a national airline under IRI control goes back to a plan drawn up in 1941 by Bruno Mussolini. In this plan Ala Littoria had been assigned to IRI with the approval of Donato Menichella, the managing director of IRI.[15] Most of the management of LAI and Alitalia, who had previously been with Ala Littoria or LATI, also favoured the unification of the two companies under IRI; such an aim was part of their business culture. The support of IRI and of the politicians was necessary to achieve this aim.[16]

The architect of the unification was Bruno Velani. An engineer by training, Velani had been a promoter of the original Alitalia in 1946 and since 1949 its general manager. In the years between the wars Velani had been with Ala Littoria, first in charge of the central European network, then concerned with France, Greece and Albania, and later East Africa. He became technical and operations manager in 1939, but then returned to the Air Force, only to be appointed managing direction of Ala Littoria in 1943. He remained general manager of Alitalia until 1957, when he became managing director. Velani had a good technical knowledge of the aircraft industry and he also put to good use his acquaintance with the military and administrative structure of the country, which remained largely unaltered after the fall of Fascism. From 1949 to 1960 the president of Alitalia was count Niccolò Carandini, an ex-ambassador to London. Carandini was an able representative of Alitalia in the international milieu, especially in confrontations with BEA.[17]

While IRI had always been the dominant, although not majority stockholder in Alitalia; in LAI the leading role had always been taken by TWA and the other private partners, and Alitalia only gained a majority stake in LAI a few months before the merger. In Alitalia-Linee Aeree Italiane, as the new company was called, the IRI holding exceeded 50 per cent.

BEA and FIAT had taken an active part in the management of Alitalia at the beginning. However, these and other private shareholders had not fully participated in the increases in Alitalia's share capital, so that by 1956 FIAT and the other private partners had an 8.65 per cent stake instead of their original 16 per cent and the British stake had dwindled to 30 per cent from the initial 40 per cent. After the merger the British share in the new Alitalia was around 9 per cent. The agreement between the Italian Government and the British was due to expire on 8 June 1956, but was extended for a further five years so that BEA could participate in the Alitalia share capital increases and thus maintain its stake at 30 per cent. However BEA did not cover all capital increases and when the agreement finally expired in June 1961, BEA's stake in Alitalia was only 5.4 per cent or 1350 million lire.[18]

Subsequently Alitalia's share base was enlarged several times although IRI's controlling position was unaffected. In 1963 and 1968 preference shares were offered to the public, which brought Alitalia's capital to a face value of 50 billion lire. Since 96.2 per cent of ordinary stock and 54.9 per cent of preference shares were held by IRI, its control was untouched. Buyers of the preference shares were FIAT and some investment trusts. In 1985, during a favourable period for the Italian stock market and for Alitalia's profitability, the capital of the company was increased from 280.8 billion lire to 421.2 billion lire, and IRI offered for public subscription 15 per cent of the capital. In 1986 ordinary Alitalia shares were quoted on the stock market.[19]

The ownership of Alitalia remained substantially unchanged throughout the 1970s and 1980s, during which time IRI underwrote the capital increases and covered the losses incurred by the airline. In 1994 IRI possessed 89.3 per cent of the ordinary shares, 78.9 per cent of the preference shares, and 82.5 per cent of the 'savings' shares (non-voting preference shares).[20]

1.3   State intervention in Italian air policy

Italian air transport policy was theoretically the concern of three different departments within the government: the Civil Aviation Authority or *Civilavia* and the Ministries it came under: Air Defence until 1963, thereafter Transport, IRI and the Ministry of State Shareholdings, and the Ministry of Foreign Affairs. In fact many factors conspired to leave the real decision-making to Alitalia, particularly since none of the departments possessed sufficient know-how or authority.

For bilateral agreements between countries, the Italian Foreign Ministry was naturally involved, but only after Alitalia itself had paved the way. The policy of the Ministry of State Shareholdings was expressed indirectly through IRI; and it was IRI, as major shareholder, that held the purse strings and exercised great influence over the management of Alitalia. It was IRI that selected the top managers and kept track of events through IRI repesentatives on the Alitalia board. Alitalia's performance was monitored mainly from the financial and administrative points of view, with the aim of harmonizing its activity with that of IRI's other interests and programmes. The organizational structure of the IRI group was such that policy-making was delegated to a sub-holding for each sector. However there was no civil aviation sub-holding and furthermore IRI had no personnel with competence or experience in aviation. Therefore Alitalia drew up its own strategic plans. It is noteworthy that there was never any divergence between Alitalia policy decisions and those of the IRI aerospace sector. As we shall see, both the airline and IRI's aerospace companies had links with the American Douglas company until the 1980s.

The Ministry of State Shareholding had a policy that state intervention in industry should have a strong social dimension, and for some time after it was formed one of its main objectives was to eliminate the economic differences between the north and south of Italy by encouraging the development of industry in the south. This policy was passed on to IRI and on again to Alitalia. Helped by government subsidies, Alitalia actively promoted air travel in southern Italy during the 1950s and 1960s.

However this policy had numerous drawbacks. Under pressure from local interest and user groups, Civilavia had made it a condition of granting route concessions that the airline should serve districts with difficult surface transport access and provide flight frequencies that recognized the needs of users. The fares were also low, so low in fact as to justify Alitalia opposing, and Civilavia not granting, profitable routes to other airlines, because Alitalia needed a monopoly on these services to recoup the losses sustained on the *'social'* routes.[21]

In 1957 Alitalia inherited the operating subsidies originally granted to LAI's domestic routes. These subsidies were later renewed and extended to new routes. In 1973 new regulations were drawn up for the payment of these subsidies, in which it was specified that they would be granted only to routes not operating at a profit, but recognized as being in the public interest in social terms or for tourism. It was the connections between Rome and the cities of southern Italy and the Italian islands (small summer tourist resorts like Lampedusa or Sardinia and Sicily) that were the main recipients of subsidies. The annual contribution amounted to 500 million lire up until 1963; after this date it was raised to 750 million, and later increased again.

When in 1963 Aero Trasporti Italiani (ATI) was set up as a subsidiary wholly owned by IRI and Alitalia, to run the southern Italy network, most of the government subsidies received by Alitalia for domestic routes were passed on to ATI. In addition to these, ATI received funds from other sources such as Cassa del Mezzogiorno and the European Community. When ATI was reabsorbed by Alitalia in 1994, it was estimated that it was receiving about 40 billion lire annually, consisting mainly of EEC contributions for maintaining routes in the south and islands, and fiscal relief.[22]

Civilavia regulated airline activity via the issue of route licences and the approval of fares and timetables. Civilavia also vetted agreements between the national airline and foreign carriers. However, as a regulatory body it was handicapped by its subordinate position to the Ministry of Defence, where its role was that of a purely technical body. Later on, when Civilavia was put under the Ministry of Transport, some time was to pass before the bureaucracy, up to that point grappling with railway and motor transport problems, was able to learn the rules of civil aviation and effectively stand up to the monopolistic attitudes of Alitalia.[23] Civilavia did eventually demonstrate a certain amount of independence from Alitalia in allowing more than one airline to operate domestic routes and although these companies only represented minimal competition for the flag-carrier, Alitalia always 'discouraged' the growth of competing airlines. Civilavia granted operating licences on several occasions to companies other than Alitalia. ITAVIA, an airline with private capital and the necessary strong political support, was one of these. It operated from 1960 to 1981 with some success on several domestic routes and even one or two international routes, for example Turin–Geneva. The demise of ITAVIA was largely the result of the disintegration of one of its Douglas DC-9s over the Mediterranean island of Ustica for reasons that are still unexplained.

Nevertheless the fact that Alitalia was hostile to competition and continued to enjoy various forms of state support made life very difficult for any competitor.[24] Until 1972 Alitalia used government-owned airports free of charge and in large part air traffic control facilities were also free. Air traffic control was the responsibility of the air force, even after Civilavia was transferred to the Ministry of Transport, and was separated from the military only in 1980, when the Azienda Autonoma Assistenza al Volo (ANAV) was created. This body was separate from Civilavia and had a legal status not suited to efficient operation in a modern environment. The advantages to Alitalia of its special status were counterbalanced by its overall inefficiency in providing services.[25]

2   AIRCRAFT PROCUREMENT

Immediately after the war, financial resources were not available to the aviation industry to help it compete against the Americans and the British – a task considered technically difficult and politically inappropriate. The aircraft builders, who were already facing financial ruin due to non-payment of German and Italian government contractual obligations, and who had fallen further behind technologically during the conflict, had the additional problem of re-tooling their factories. And instead of encouraging Italian airlines to buy Italian, the Ministry of Aeronautics purchased a batch of 32 war-surplus Douglas C-47s (DC-3s).[26]

Most of the airlines set up in 1946 did not hesitate to use these aircraft, which the Ministry offered at favourable prices. LAI and most of the independent airlines took this line, while Alitalia opted to buy Italian and British aircraft. The shareholders Fiat and BEA persuaded the Alitalia board to buy Italian Fiat G-12s and SIAI S-95s for use on medium-range flights, and a number of British Avro Lancasters refitted for passenger traffic (Lancastrians) for use on international routes, all of which were fitted with war-surplus Bristol Pegasus engines. This turned out to be a poor decision, as the aircraft were unsatisfactory and in any event inferior to those being used by the foreign airlines which soon started operating regular services to and from Italy.[27]

In April 1948, during a heated meeting of the Alitalia board, the shortcomings of the Alitalia fleet were recognized and the decision taken to sell it off and obtain new aircraft as soon as possible. The decision to buy four second-hand DC-4s as replacements on international routes was taken in October 1949. To thwart the attempts of the British shareholders to make Alitalia buy Canadair aircraft (in Sterling) it was necessary to convince them that the only way to obtain the necessary loan from the Export Import Bank (Eximbank) at favourable interest rates, was to buy American. Thus the choice of DC-4s became more or less obligatory.[28]

From that point on American aircraft, and Douglas in particular, became the backbone of the Alitalia fleet. Eximbank proved to be an important link between Alitalia and American aircraft builders, since the airline was always short of capital, and from 1950 onward regularly took loans from the bank. However, there are other reasons why Alitalia became attached to Douglas, related both to its product range and the easy terms it offered, and to the fact that within Alitalia there were forces favouring Douglas.

The easy financial terms that Douglas granted the airline were important. When DC-6Bs were being purchased, for example, Alitalia not only obtained a loan from Eximbank (at 3.5 per cent interest) but also from Douglas itself. The aircraft builder was also generous to Alitalia on other occasions, for example at the beginning of the 1960s with the purchase of DC-8 jets, and again in the 1980s, to regain Alitalia's loyalty after it had started buying Boeing types. Furthermore, since there were Douglas aircraft to fit different market niches, there were advantages from maintenance and pilot training points of view in having an all-Douglas fleet. Alitalia's connections in Italy also favoured Douglas. For example Fiat, an important Alitalia shareholder, had been on good terms with Douglas since soon after the war, when it obtained the Ministry of Aeronautics contract to convert wartime C-47s to civil use. In addition Finmeccanica, the leading engineering company in the

IRI fold, became a Douglas supplier in 1960. It would appear that negotiations to purchase aircraft went on in parallel with discussions to place 'compensatory' orders with IRI engineering companies.

The rivalry with LAI in the years prior to the merger also had an influence on aircraft-purchasing by pushing Alitalia to take decisions in contrast to those taken by LAI. There was also the experience of the unsuitable Lancasters, which made Alitalia wary of British aircraft and hence more pro-American.

In 1953 Alitalia bought three medium-range Convair 340s to use on the European routes the airline had abandoned in 1950. In 1954, Alitalia started replacing the DC-4s with new DC-6Bs and in 1955 four DC-7Cs were ordered.[29]

Table 6.1   Alitalia fleet 1946–56

	1947	*1950*	*1955*
SIAI SM-95	4	8**	–
Avro Lancastrian	4	4	–
Fiat G-12*	4	–	–
Douglas DC-4	–	4	1
Douglas DC-6B	–	–	4
Convair 340	–	–	4
TOTAL	12	16	9
Italian manufacture, %	67	50	–
British manufacture, %	33	25	–
American manufacture, %	–	25	100

* Leased from Ministry of Aeronautics
** 3 SM-95s were leased.
Source: Alitalia, Annual Reports.

LAI's aircraft procurement took an opposite course to Alitalia's. Although its DC-3s were out of date, until the 1960s they were well suited to most Italian airports. In order to operate over the international sector of its network, which included Europe and countries bordering the Mediterranean, LAI placed orders in 1953 for British Vickers Viscount 785s. This proved to be a sound decision as the turboprop Viscount was the best aircraft then available for medium-range traffic. For service on North Atlantic routes, LAI was persuaded by TWA, a long-term Lockheed customer, to buy the Super Constellation L-1049. TWA was part of the Howard Hughes empire and Hughes was a convinced supporter of Lockheed as well as a believer in the merits turboprop aircraft as against jets.[30]

It would seem that throughout most of the 1950s neither Alitalia nor LAI seriously considered buying jet aircraft. The investment required would have been large and their respective major partners either did not provide the aid required (IRI) or pursued other options (TWA). In Alitalia's case, since it had been excluded from the prestigious North Atlantic route, jets were considered an unnecessary luxury for its long-range traffic which mostly consisted of emigrants to South America.[31] Both airlines also felt that turboprop aircraft had a competitive edge over jets, at least over short and medium distances, moreover with the exception of the French Caravelle there were no short-range jet aircraft on the market at the time. Lastly there was the fact that just prior to the merger, Alitalia had bought piston (DC-7) and turboprop (Viscount) aircraft.[32]

Nevertheless, despite the success of the Viscount, it became clear from 1959 onwards that jets were superior to propeller-driven planes,[33] and when Alitalia and LAI merged in 1957, they made immediate plans to purchase jets in order to keep up with other airlines. The newly created airline had a fleet of 36 aircraft of six different types, involving excessive maintenance and inventory costs. It was fortunate that the new Alitalia managed to extricate itself from the LAI order for the Super Constellations, which would have increased the fleet diversity even further. The imminent arrival of long-haul jets meant that Alitalia could not ignore them any longer. The race was on among the world's airlines; in 1957 508 jets were ordered, 484 four-engined and 24 twin-engined.[34] The crucial decisions were taken by Alitalia's management in 1957 and 1958, first the Douglas DC-8 was chosen and then the Caravelle. These orders marked a decisive turning-point for Alitalia because from then on it became a modern airline and fairly soon after its fleet consisted entirely of jet aircraft. The timing of the change from propeller-driven to jet aircraft was also fortunate since the 1960s were another decade of strong growth in air traffic, the annual rate being only slightly below 15 per cent every year from 1960 to 1969. Alitalia took advantage of this to boost its international market share and break into new sectors.[35]

Alitalia preferred the DC-8 over the Boeing 707 mainly because by 1960 Douglas had equipped the DC-8 with Rolls-Royce Conway engines, before Boeing's decision to use the same engine in the 707. The Conway was a fan-jet and consumed about 25 per cent less fuel than the original pure jet engines, it also had more power and gave the aircraft a considerably longer range. This was essential for the Rome to New York route, which was a thousand miles longer than London to New York.

Furthermore Douglas agreed to deliver DC-8s in June 1960, just in time for the Olympic Games being held in Rome that year, and two more late the same year, allowing the company to use its new aircraft on the important North Atlantic service.[36] Between 1960 and 1966, 14 DC-8s and 21 French Caravelles were delivered to Alitalia, so that in 1966 the airline's fleet included 35 operational jets. At the same time, the number of Viscount turboprops also increased from nine to 16. The first Caravelles went into

Table 6.2 Alitalia fleet 1957–95

	1957	1962	1967	1972	1977	1982	1987	1992	1995
Propeller aircraft									
Douglas DC-3	12	4	–	–	–	–	–	–	–
Convair 340 Metropolitan	6	–	–	–	–	–	–	–	–
Vickers Viscount 785	6	14	14	–	–	–	–	–	–
Douglas DC-6	3	3	–	–	–	–	–	–	–
Douglas DC-7C	2	5	–	–	–	–	–	–	–
Jet Aircraft									
Douglas DC-8/43 passenger	–	9	14	11	–	–	–	–	–
Douglas DC-8/62 passenger	–	–	2	8	8	3*	–	–	–
Douglas DC-8/62 cargo	–	–	–	2	2	1*	–	–	–
Sud Aviation Caravelle	–	14	21	18	–	–	–	–	–
Douglas DC-9/32 passenger	–	–	7	33	33	19	8**	–	–
Douglas DC-9/32 cargo	–	–	–	2	2	–	–	–	–
Boeing 747 passenger	–	–	–	5	5	4	7	14	11
Douglas DC-10/30	–	–	–	–	8	–	–	–	–
Boeing 747 combi	–	–	–	–	–	3	4	4	4
Boeing 747 cargo	–	–	–	–	–	1	1	1	1
McDonnell Douglas MD-11	–	–	–	–	–	–	–	6	8
Boeing 767	–	–	–	–	–	–	–	–	4
Airbus A300/B4	–	–	–	–	–	8	8	–	–
Airbus A321	–	–	–	–	–	–	–	–	10
Boeing 727/200	–	–	–	–	–	18	–	–	–
McDonnell Douglas MD-80	–	–	–	–	–	–	21	74	90
Aerospatiale ATR 72	–	–	–	–	–	–	–	–	4
Aerospatiale ATR 42	–	–	–	–	–	–	1	11	9
TOTAL	29	49	58	79	58	57	50	110	141

* To lease
** In leasing 13 DC-9/32 and 1 MD-80 + 4 DC-9/32 leased.
Source: Alitalia, Annual Reports

service with Alitalia in 1960. The early deliveries, notwithstanding the late placement of orders, were possible thanks to the generosity of Air France which released aircraft earmarked for itself, and is an indication of the good relationship that Alitalia had with the French airline, born out of the Air Union discussions.[37]

In 1965 Alitalia signed an order for 28 Douglas DC-9/30s, all of which were delivered in 1970. At the same time, the long-range fleet was also increased. An updated version of the DC-8 (the series 62) was introduced, which was also available in an all-cargo version. The series 62 was a very long-range aircraft, capable of covering up to 5 000 nautical miles, or flying non-stop from Italy to Brazil. Fourteen of these DC8s were put into service. As noted previously, the Douglas–Alitalia relationship was further strengthened after 1966 with an agreement by which Aerfer, a company controlled by Finmeccanica, became a subcontractor for the production of DC-9/30 fuselage panels.[38]

Alitalia placed its first order for Boeing 747s jumbo jets in 1966, at a time when the charter flight phenomenon was reaching its peak. Alitalia took the view that the jumbo would help them meet the threat posed by the charter companies. The idea was that because the unit operating costs of jumbos were lower, blocks of low cost tickets could be sold to package tour companies. However it proved impossible to segregate normal tariff passengers from tourist passengers, as planned, when the Boeing 747 was introduced, and as a result there was a tariff war, aggravated further by a concomitant fall in traffic volume.[39]

Immediately after the introduction of wide-bodied aircraft there was a marked slowdown in the growth of international air transport, so that the average annual growth rate was only 8.8 per cent during the 1970s, almost half that recorded in the 1960s.[40] At the same time the wide-bodied jets brought a major increase of capacity. To meet the envisaged increase in demand, Alitalia placed orders in 1970 for a further three Boeing 747s and four Douglas DC-10s, as well as one Douglas DC-8/62 and two Douglas DC-9s. However, the following year, the 747 order was reduced to one, while the number of DC-10s was doubled. The DC-10 was Douglas's answer to the 747 and offered innovative solutions in terms of engine layout. Alitalia had an input to the design of this aircraft, as part of the European buyers' consortium ATLAS, while Finmeccanica built important components. The intercontinental version bought by Alitalia had a capacity between that of the DC-8 and the Boeing 747 which permitted Alitalia to use the aircraft in several sectors. In 1973, Alitalia put four Douglas DC-10s into service, and in 1975 the number rose to eight, these aircraft remaining part of the fleet until 1982.[41]

The rise in the costs of passenger transport on routes where the 747 was employed, was partially offset by the impetus these aircraft gave to cargo traffic. The almost circular cross-section of the 747 fuselage permitted half to be devoted to passenger seating and the other half to cargo, which was loaded by means of an entirely mechanized fibreglass loading unit designed by Boeing itself. Alitalia began to exploit these characteristics of the 747 in the 1970s, when Italian exports to the world underwent considerable expansion, as 'Made in Italy' became fashionable. Many luxury products such as designer clothes and shoes were sold to the USA and the Far East, and Pininfarina car bodies went to General Motors, much of this cargo being carried by Alitalia. As a consequence, Alitalia's freight traffic reached almost 20 per cent of total traffic, compared to the 10 per cent of the 1960s.[42] By

Amilcare Mantegazza

1974, Alitalia had equipped itself with a long-range fleet more than sufficient to meet its needs. However its medium- and short-range aircraft were ageing and inadequate, so in the same year seven Boeing 727s were bought for medium-range operations, with two further 727s added in 1978 and four more in 1979.

In 1978 and 1979, Alitalia made more aircraft procurement decisions, both for the short to medium and the long-haul sectors. A further six Boeing 727s were purchased as well as eight Airbus A300s. The latter was the first of the European aircraft to be included the Alitalia fleet. The Airbus A300 was well suited to many of the routes that Alitalia operated, such as Central Africa and the Persian Gulf, which were half way between medium and long-range flights.

Partly because of the suspension of the Douglas DC-10's airworthiness certificate in 1979, Alitalia accepted a Boeing offer in which five Boeing 747s (in service since 1970/71) were taken back in part exchange for eight long-range 400-series 747s, which were capable of flying from Rome to Australia non-stop, and which included passenger, combi and cargo versions. Ten more aircaft were delivered by 1980 (three Boeing 727/200s, three Airbus A300s, and three Boeing 747s), ten in 1981 (three Boeing 727/200s, three Airbus A300s, and four Boeing 747s) and two in 1982 (one Airbus and a Boeing 747). Over the same period eight DC-8/62s and four DC-9/32s were sold off, and five old 747s were returned to Boeing. Between 1984 and 1986, the long range fleet increased by four Boeing 747s.[43]

In 1982 McDonnell Douglas won back Alitalia as a customer for its short-range aircraft by offering to buy and resell Alitalia's Boeing 727 fleet in exchange for an order for twenty MD-80s. The MD-80 offered various advantages over the 727, such as two-man crews. Moreover being a stretched version of the Douglas DC-9, it fitted in well with the numerous DC-9/30s in the Alitalia and ATI fleets, and was consistent with the long-standing relationship between Douglas and Alitalia.[44] During the mid-1980s McDonnell Douglas was developing the long range MD-11, projected for delivery in 1990, and Alitalia ordered six combi versions of the aircraft. These were supposed to be 'very long range aircraft able to carry the maximum payload [200 passengers plus 30 tons of cargo] more than 11,000 km non-stop'. Alitalia intended to use this plane to 'develop new intercontinental medium-volume markets, offer faster and more frequent connections on its established routes and to encourage the development of its cargo business'.[45]

After four years of good economic results and increasing traffic volume, Alitalia drew up a programme in January 1989 which called for an investment of 2400 billion lire over the four years to 1992; this was approved by the IRI board. For its long hauls Alitalia planned to buy another Boeing 747 in addition to the twelve already in service. The 747 was to be delivered the following year together with first two of the six MD-11s. Alitalia planned to purchase another five of these in 1993–4. Also planned was the acquisition of four medium-range Airbus A300s, three to be delivered in 1989 and one in 1990. For its short-range needs 23 McDonnell Douglas MD-80s were ordered, three ATR-42s and ten 100-seater Airbus 321s.[46] After the Gulf War in 1991 Alitalia's worsening economic diffciulties forced changes in this purchasing plan and to reduce debt many aircraft were sold or leased.

3 ROUTE NETWORK

Before the Second World War the greater part of Italy's network of international routes had served Mussolini's African Empire. Subsequently the direction of route development was determined by geography, history and political expediency arising from Italy's post-war status as an occupied nation, as well as economic considerations. Furthermore the plans of two major players at the time, TWA and the British airlines BOAC and BEA, had to be taken into account. TWA had secured a foothold in Italian aviation after the war, being particularly interested in Mediterranean routes to the Middle East and Far East. In this context LAI was assigned the role of a feeder, channelling Italian traffic into TWA's international network. Similarly BOAC wanted to safeguard its routes to South Africa and Egypt.

Italy is a Mediterranean cross-roads between Europe and Africa. For flights from the Atlantic, Italy is the last stop before the Middle East and this was important as long as aircraft range was limited. For centuries Italy has attracted tourists and in recent times become a destination for mass tourism, including pilgrimages. The industrialization of Italy in the nineteenth century prompted many agricultural workers to emigrate, so Italian communities sprang up in many parts of the world. General and religious tourism was thus augmented by the flow of people of Italian descent from all over the world. These developments, as much as Italy's economic miracle in the 1950s and 1960s, was an important factor in the growth of Alitalia's traffic. The growth of Italian industrial activity both inside and outside Italy involved increasing numbers of businessmen and these people represented a significant proportion of Alitalia's passengers.

3.1 International routes

In the early 1950s, having had to abandon most of its European routes, Alitalia concentrated its efforts on routes to South America, the Horn of Africa and some parts of North Africa. These routes, especially to South American, represented the backbone of Alitalia's international network throughout the 1950s. In Brazil and Argentina in particular, but also in Uruguay and Venezuela, there were large Italian communities, and a certain amount of business traffic originated there as the longest-established Italian immigrants became part of the business élite in these countries. In the 1950s these countries also attracted a mass of emigrants simply in search of a better life, to whom Alitalia offered instalment-plan tickets. In Africa, although Ethiopia and Libya were no longer Italian colonies, Italians continued to work there and Somalia was made an Italian mandate in the postwar period. While North Africa, the Middle East and the Horn of Africa continued to be important destinations in the Alitalia network, national airlines in competition with Alitalia sprang up in 1950s; and during the 1960s a worsening political climate often rendered Alitalia's continued presence impossible.

Routes to the main European countries were gradually expanded after 1952, however, the first addition to Alitalia's routes outside Europe came in 1956 when its South African service was inaugurated in competition with BOAC and South African Airways.

North Atlantic routes

When Alitalia and LAI were merged, in addition to the domestic market, the latter airline contributed an important sector to the Alitalia network on the North Atlantic, which during the 1960s was second in size only to the US domestic market and was certainly the fastest growing. Eventually this sector came to dominate Alitalia's international operations, overtaking South America. During that decade the North Atlantic routes became major earners for Alitalia, although towards the end of the period competition from the charter companies cut into profit margins. The New York and Boston routes (direct from Milan and via London), acquired by the merger, were augmented in 1960 by a service to Chicago with a stopover at Montreal. In 1964 the airline started a direct Milan–Chicago service, and in 1966 a Milan–Lisbon–New York route.

During the 1950s and 1960s, Alitalia improved its position *vis-à-vis* other airlines operating the North Atlantic routes, raising its share from 1.3 per cent in 1957 to 5.5 per cent in 1964. Thereafter, however, no matter what efforts were made to improve services, Alitalia's market share remained unchanged, in line with average growth. From 1966 Alitalia's plan was to increase the number of destinations within the USA. Atlanta, Dallas and Los Angeles were included as part of an aim to 'offer services that formed efficient connections between other Alitalia sectors, flights to and beyond the Pacific, and new connections to Asia via the countries of Eastern Europe'. But this plan was frustrated by the refusal of the American authorities to concede additional destinations. Alitalia regarded this as unjust because bilateral agreements between the USA and European countries, including Italy, which dated back to 1948, permitted American airlines to operate 'fifth freedom' traffic between European countries, yet the same concession was not given to European countries within the huge American domestic market.

In addition, there was considerable growth of US charter companies (known as supplementals) on transatlantic routes during the 1960s, competing with the regular carriers.[47]

After repeated attempts to bring about changes in the bilateral agreement between the USA and Italy, Alitalia finally ceased to adhere to it. Protracted negotiations followed until 1970 when a new agreement was signed, guaranteeing Alitalia additional stops at Detroit, Philadelphia, Atlanta and Los Angeles (the latter inaugurated in 1984). This led Alitalia to boost its capacity on the North Atlantic routes in a fruitless attempt to raise its market share, at the expense of the other airlines. At the beginning of the 1970s, when the Boeing 747s were introduced, the transatlantic routes within the Alitalia network accounted for almost 40 per cent of the available seat/kilometres. However the increase in capacity after wide-bodied aircraft were brought into service was not matched by a comparable rise in demand, and the North Atlantic routes became heavy loss-makers.

Present network structure

To combat this unfavourable situation, Alitalia carried out a wide-ranging reorganization in 1974. The biggest change was that the transatlantic routes were cut to 27 per cent of total capacity. A similar capacity was allocated to Europe, the Mediterranean and the Middle East, which constituted the airline's specialist sector. The remaining 46 per cent of capacity included routes to the Far East and Australia, West Africa, South Africa, South and Central America and domestic connections. The economic weight of the North Atlantic and Europe-Mediterranean-Middle East sectors still determined whether the balance sheet was in the red or in the black.

Economic ties between the USA and Europe are very close; there are also strong economic links between these countries and North Africa and the Middle East. The fact that Alitalia's main operations are within these relatively integrated economic units has meant it has been unable to exploit geographical phase differences within economic cycles. Thus during the 1970s and 1980s demand on the North American and European routes followed similar trends, although it is true that the Mediterranean region tended to differ. Demand remained slack up to 1982 on the North Atlantic and up to 1983 in Europe. Alitalia sought to manage this situation by cutting stopovers, reducing frequencies, and where possible substituting aircraft. For example, between 1977 and 1984 Boeing 747s were replaced with Douglas DC-10s on Canadian routes. On the other hand, following the pruning of USA routes in 1974, these were activated again in 1978 by changing the configuration of the Boeing 747s to increase their capacity. For the rest of the 1980s air traffic grew, although there was a lull in 1986. The growth lasted till 1990 when Iraq invaded Kuwait, and resumed a year later after the Gulf War had ended.

Alitalia's participation in the development of air transport to the Far East was only marginal and the airline managed to attract only a modest amount of business in relation to its overall capacity. However throughout the 1970s and 1980s this sector was characterized by a higher load factor (between 60 and 70 per cent) than any other route and consequently remained profitable. Alitalia's traffic to the southern hemisphere remained small over this period. The airline sought to expand this area in view of the fact that the bulk of its operations were concentrated in the northern hemisphere and were consequently highly seasonal, as Italians and the Europeans in general tend to take their holidays in June, July and August.

Following US deregulation and the liberalization of civil aviation in the European Community in the 1980s, the ability of airlines to attract traffic depended more on route structure and this became an important factor in competition between airlines. The Alitalia network did not develop according to the 'hub and spoke' model typical of American airlines, principally because the geographical position of Rome discouraged it; instead Rome attracted medium-range traffic from the Mediterranean area and long-range traffic to and from the Middle East and Far East. The Alitalia network therefore concentrated more on direct connections, so that it largely avoided feeding other European hubs, but the airline was susceptible to passive 'sixth freedom' loss of passengers, especially on long range multi-stopover routes. Furthermore many European airlines have adopted aggressive penetration policies, supplying their own hubs by flying directly to many of the smaller Italian airports.[48]

3.2 Airport–airline relationships

In the period between the two world wars, Italian legislation placed civil aviation under the control of the military authorities and military airports also handled civil traffic. After the Second World War, it was obvious that military airports could not handle both military and civil traffic. In addition, some of these airports, most notably Rome's Ciampino airport, were too close to built-up areas to permit the runaway extensions and terminal expansions necessary to accommodate the increasing size of aircraft and rapid expansion of passengers. The Italian government did not introduce any new legislation, but proceeded with the inadequate policy of only expanding the airports it owned. Because new initiatives did not come from the centre, municipal authorities and other local interests (for example chambers of commerce) stepped in, which resulted in the building of numerous small airports. In a report commissioned in 1971 by the Italian Lower House (Chamber of Deputies) to investigate this problem, the director general of Civilavia (the civil aviation regulatory body) admitted that the situation was 'rather diversified' in that there were 'airports built and operated by government; airports built by government but operated by private interests; airports built and operated by consortia; airports built and run by private companies set up by public bodies; and airports run by consortia or by public bodies'.[49] As a result of this situation it was impossible to promulgate a nationally integrated policy for airports and airport management. In 1972, following the Lino report, which detailed the serious shortcomings of Italian airports and air traffic control, the government allocated 200 billion lire for modernization.[50]

When in 1963, Civilavia was finally separated from the Ministry of Defence and became part of the Ministry of Transport (despite the fierce opposition of the military) not all services were transferred to civil authorities and the Ministry of Defence retained responsibility for air traffic control. This situation ended in 1979 with the creation of an autonomous agency for air traffic control (called ANAV) which reported to the Ministry of Transport.

After the War the international airports at Rome and Milan assumed a central role in air transport in relation to their key positions in the Italian economy. For many years, therefore, Alitalia has maintained two home bases, Milan and Rome, the latter at Fiumicino being the company's main base. Fiumicino, the airport which replaced Ciampino, was opened in 1960 in time for the Rome Olympics and was run by the state (at a loss) until 1974. In 1974 a limited company Aeroporti di Roma (AR) was created to run the two airports of the city (Fiumicino and Ciampino); its main shareholder was IRI. In 1983 Alitalia bought control of AR and although the company's efficiency was well below the European standard, from 1984 it began making profits. For the occasion of the World Cup football competition in 1994 Fiumicino was finally connected by rail to the centre of the city although the frequency and level of service left something to be desired. In 1995 Alitalia sold its shareholding in AR back to IRI for 400 billion lire, as it was in need of cash.[51]

At Milan, a limited company controlled by the Milan city authorities runs both Linate, just outside the city, and Malpensa, about 40 km west of the city. Notwithstanding the fact that both airports have been run by a single management, it has not been possible to solve the problem of achieving their balanced utilization. At present medium range European flights are at saturation point at Linate, while intercontinental flights use Malpensa. The airlines prefer Linate to Malpensa largely because of the lack of transport between Malpensa and Milan. While Linate can be reached by public transport from the centre of Milan in a matter of minutes, Malpensa relies on very busy roads. Although the restructuring of Malpensa and the construction of a fast rail link to Milan's city centre are recognized as necessary before traffic can be transferred from Linate, no one is willing to assume responsibility for the work, which has been at the advanced planning stage for some time.[52]

The Milan and Rome airport systems have other problems too. They are inefficient compared to their major European counterparts, both from the point of view of organization and from the use of modern technology. Because of this inefficiency they have a lower number of air transport movements (ATMs) than the standard in Europe and America and this prevents them from being transformed into hubs characterized by large numbers of connecting flights. Alitalia is in part responsible for this situation, since until recently it divided its investment between Rome and Milan, and was unable to improve the efficiency of either sufficiently to make them more attractive than other European hubs. Since 1990, Alitalia has lost a significant amount of North Italian traffic to other European airlines, estimated to be in the region of 2.5 million passengers.[53]

4 OPERATING AND FINANCIAL PERFORMANCE

LAI and the original Alitalia were airlines of only modest size. Following the 1957 merger, there followed a decade of strong growth, and Alitalia became the sixth largest airline in Europe. The increase in the volume of operations went hand in hand with a long run of profitability so that from 1959 to 1971 the airline was continually in the black.

4.1 Alitalia and LAI 1947–57

As long as Alitalia and LAI divided the Italian civil air transport sector between them, they were unable to compete effectively with the larger airlines on the major international routes. In certain areas their routes also overlapped and they were in direct competition with each other. LAI's operating conditions were such that its aircraft flew considerably fewer hours than did the aircraft of the larger companies, and hence service frequency was also lower. The smallest aircraft flew half the hours of the same aircraft in the Pan American, TWA, BEA and KLM fleets; the larger ones a quarter less time. Furthermore the organizational structure was poorly developed relative to the demand. There were also problems with air crews: the older pilots had to be retrained to fly new, larger aircraft and air crew training took longer than that of other categories of employee. Alitalia's operations were similarly disadvantaged. This situation not only made it difficult to attract passengers from rival airlines, but considerably increased costs, although Alitalia's financial situation was better than LAI's, which had a 5.1 billion lire debt arising from the purchase of aircraft.[54]

Table 6.3 Alitalia's financial results (million lire at current prices), fleet capacity and passengers carried

	1947	*1950*	*1953*	*1956*
Gross income	175	2 582	4 034	8 547
Expenses	304	2 842	3 764	8 498
Profits or *Losses*	*130*	*260*	269	49
Seat/km available	10 213 752	117 111 383	167 730 080	397 384 749
Ton/km available	1 595 315	16 272 216	19 844 213	46 148 859
Flight hours	2 345	14 843	12 422	22 128
Km flown	621 108	4 538 580	4 198 335	8 127 286
Passengers carried	10 306	24 089	36 563	116 394
Pass./km carried	5 483 911	60 650 019	108 097 755	217 077 269
Ton/km carried	578 501	7 979 927	12 764 244	25 542 164

Source: Alitalia, Annual Reports of years cited.

In 1956 Alitalia's traffic, which included a small number of domestic routes, in addition to its international operations, was centred on Rome. The main markets were Central and South America – 45 per cent of the company's business – together with Europe and North Africa/Middle East (26.3 per cent and 22.8 per cent respectively). In the same year profits were just under 50 million lire on an income of 8.5 billion lire; with 1.5 billion lire budgeted for depreciation, and capital of 2.5 billion lire. Long-term debts, arising from renewal of the fleet, had reached 2.3 billion lire.[55]

The less encouraging financial condition of LAI probably influenced TWA's decision to detach itself from its Italian interests. More decisive, however, was the fact that Carandini and Velani had the support of IRI, and hence of the Italian government. TWA did not want business partnerships with governments. On the other hand IRI came out of the deal rather well, since it had to pay only 420 million of the 1688 million lire due to the American airline, according to the 1946 agreement, to liquidate its interest in LAI. The remainder was offset against four Super Constellations (Lockheed L-1049/A) that were handed over by LAI to TWA.[56]

4.2 Alitalia's expansion, 1958–74

Table 6.4 illustrates the extent of Alitalia's transformation and growth following the merger. On the supply side, between 1958 and 1969, the number of kilometres flown increased fourfold, seating capacity went up nine times, and ten times in terms of available tonne-kilometres. In the same period, the number of passengers carried increased four and half times, passenger-kilometres by five times and tonne-kilometres by three. The increase in capacity occurred over a remarkably short period. In 1957 the size of the company grew as a result of the merger and then jets started replacing piston aircraft from 1960. In terms of size and cruising speed, jets provided a level of performance almost double that of piston-engined aircraft. Among International Civil Aviation Organization (ICAO) carriers, the average number of seats per aircraft was 53 in 1959 and 101 in 1969; average speed was 345 km/h in 1959, and 570 km/h in 1969. Already in 1962, the available seats per aircraft and average speed on Alitalia European routes were 72.6 and 544 km/h respectively; the corresponding figures for domestic and international routes were 55 and 308, and 126.3 and 654 respectively.[57]

It has to be said that other airlines grew at a similar rate over the same period. Worldwide growth in commercial aviation was rapid after the end of the war and continued at a very sustained rate (around 15 per cent per year in volume) for more than 25 years. The pivotal market of this long growth phase was the North Atlantic, especially Europe-USA.[58]

Table 6.4　Alitalia financial results (millions of lire at current prices), fleet capacity and passengers carried

	1947	1952	1957	1962
Capacity offered	1 192.737	15 889.737	71 295.108	579 921.832
Load carried	578.501	11 446.123	40 712.120	300 080.198
Load factor %	48.5	72.1	57.1	51.7
Number of passengers	10 306	17 692	221 271	2 077.571
Revenue	174	3 090	13 582	83 322
Costs	304	2 920	13 577	82 267
Profits or *Losses*	*130*	169	5	1 055
Number of employees	293	419	3 073	7 288
Labour productivity	4 070	37 923	23 200	79 572

	1972	1977	1982	1987	1992
Capacity offered	2 152 599.375	2 579 750.200	2 847 886.622	3 432 534.555	
Load carried	1 133 501.620	1 559 835.704	1 718 883.199	2 310 113.827	3 773 000.0
Load factor %	52.7	60.5	60.4	67.3	65.3
Number of passengers	5 415.446	6 733.380	7 258.485	9 149.946	
Revenue	285.290	872.673	2.520.130	3.563.547	5.427.368
Costs	291.516	861.550	2 518.791	3 496.645	5 429.185
Profits or *Losses*	*6.266*	11.122	1.338	66.901	*16.817*
Number of employees	14.822	16.694	18.734	18.453	18.828
Labour productivity	145.230	154.531	152.017	186.014	

Source: Alitalia Annual Reports, various years.

It is clear that the merger with LAI enhanced Alitalia's performance. An organization and route network of sufficient size was created to permit a policy of expansion, despite the tough international competition which, according to the company's management, had up to that point smothered Italian aviation. Alitalia's management was also able to exploit the fact that the new company was now the sole national carrier. It targeted the Italian communities dispersed throughout the world in a bid to penetrate international markets; it also promoted Italy as a major tourist destination, and as a country whose economy was growing rapidly. In addition the hold on the domestic market provided the airline with a secure and growing customer base.

Alitalia was particularly aggressive on scheduled European routes, expanding its services following renegotiation of bilateral accords and continual adjustment of pool agreements. Thus, the airline projected a higher profile in areas where it was already present and muscled into some new markets for the first time. In order to affirm its presence on the North Atlantic and at the same time put pressure on the USA, Alitalia occasionally undercut the tariffs fixed by IATA which gave rise to major controversy in 1969.[59]

Although it acquired a majority interest in the Turin-based 'aircraft rental and services' company Società Aerea Mediterranea (SAM) in 1961, Alitalia's involvement in the charter sector was limited. During the 1960s, when mass tourism to Italy had not yet taken off, this was perhaps understandable. In any event the company was undergoing rapid expansion and did not have the resources to dedicate to this sector. Furthermore Velani was hostile to the charter business, believing it represented a 'danger' to scheduled airlines.[60]

It is significant that Alitalia's image took shape in the 1960s and was influenced by the 'economic miracle' taking place in Italy at the time. The airline's logo and colours date from 1969. Alitalia, cheap cars and motorways were symbols of Italy's new industrial economy. Migrants from southern Italy, who travelled by train to the cities of the north to find work, drove home in their new cars to visit their families, or flew home with Alitalia. In 1969, Alitalia launched 'a sweeping campaign to change its image from the traditionally nationalistic, to a more modern world-wide appearance'. Promotion such as this 'revealed a new, more aggressive, drive by Alitalia in world airline competition'. The old bow-and-arrow symbol and the old three-colour paint scheme of the national flag was replaced by a modern-looking green and red 'A', on a white background. Of modern appearance too was the glass and concrete skyscraper which served as the company headquarters in Rome after 1967. Italian fashion also made its appearance in the lime-green uniforms designed for Alitalia hostesses.[61]

Alitalia was profitable from 1959 to 1970, the period corresponding to its rapid growth phase. Athough its profits were always modest, they were present for too long a period to be the product of balance sheet manipulations and they demonstrate that fare levels and traffic volumes were more than sufficient to cover costs. Alitalia enjoyed the favourable factors that affected industry as a whole during these years: exchange rate stability, low oil prices, and reduction of operating costs due to the introduction of jets. The latter was the main reason why available tonne-kilometre (ATK) costs were cut by more than 37 per cent. However there were additional factors peculiar to the Italian airline. For example, the modernity of Alitalia's organizational structure and the pioneering role it played in the use of computers. The first contract between Alitalia and IBM, which had an important subsidiary in Italy, was in 1962, and this was soon followed by the setting up of internal administration and personal reservation systems (ARCO). However sales and advertising costs, and also costs relating to the installation and operation of the computer network of directly dependent Alitalia agencies in Italy and abroad, were unusually high.[62] The factor which contributed most to cutting costs in the 1960s was the flexibility of Alitalia's organization, which included the subcontracting of maintenance and overhaul work, and resulted in lower overhead expenses and fewer ground employees than its main competitors. Another advantage was that salary levels in Italy were below those of other European countries, while the productivity of the aircraft (usually American) was equal to that of other airlines.[63]

It is important to note that proceeds from the sale of the company's piston-engined aircraft made a significant contribution to earnings during the 1960s, when the company sold off five Douglas DC-7Cs, eleven Douglas DC-6s, six Convairs, four Douglas DC-3s and 16 Vickers Viscounts. This is indicative of the existence of a lively second-hand market for aircraft related to the formation of many new airlines. Between 1960 and 1966 investment in Douglas DC-8s, Sud Aviation Caravelles and Vickers Viscounts amounted to 97 billion lire; a further 40 billion was invested in other fixed assets. Additional expenses were the pilot training school, whose costs were not considered as fixed expenditure. These investments were financed by the issue of bonds for 10 billion lire, the raising of 75 billion lire internally, the non-distribution of profits, and recourse to bank loans, which were readily available in the 1960s, both in Italy and in the USA. Eximbank, for example, provided loans at 6 per cent.[64]

That Alitalia's profits were linked more to exceptionally favourable circumstances than to any underlying strength, emerged during the early 1970s when the jumbo jets started coming into service, and the North Atlantic, to which Alitalia was heavily committed, became a significant loss-maker. Oil price rises, inflation, and a fall in demand threw Alitalia's balance sheets into disarray. Its difficulties in the early 1970s were further aggravated by the replacement of the management team under Carandini and Velani, who had successfully led the company for 20 years. A leadership battle ensued between Cesare Romiti and Umberto Nordio. Eventually Nordio emerged as victor and remained head of Alitalia until 1987. Cesare Romiti gained his management experience with the shipping sector of IRI and was Velani's candidate (on the advice of Mediobanca, it would seem) to take over Alitalia. Nordio was appointed to the board of directors in 1973 and was also from the shipping sector; after a few months he induced Romiti to resign.[65]

4.3 Alitalia's difficulties, 1973–95

When Nordio took over in 1973 things were particularly difficult. Alitalia had been making losses for three consecutive years, while the even balance of 1971 had only been achieved by adjusting certain items to take account of the devaluation of the lira against the dollar. The oil crisis in 1973 produced a shock, but the ensuing price rise was not limited to oil, and a general inflationary trend set in that affected all sectors of the Italian economy. Italy's inflation rate was in double figures for a considerable part of the 1970s and was only brought under control in the 1980s. This occurred despite the fact that in 1978 and again in the early 1980s there were further sharp rises in the price of oil. In 1978 the process of airline deregulation began in the US domestic market and this had international repercussions, one of which was to weaken the role of IATA in determining tariffs. Another effect was to trigger a major round of renegotiation of bilateral agreements, and the adoption of more liberal ground rules.[66]

In December 1973 Alitalia's management decided on a three-pronged policy to correct the company's worsening situation: fleet renewal, rationalization of services and a limit to new staff recruitment.[67] However, in spite of these measures, Alitalia was in the red for three more years and only in 1977 did profitability return. In 1978 a further increase in oil prices caused new losses and Alitalia did not see profits again until 1983. Another four years of profitablity were followed by a new and serious run of losses after 1988.

The steps taken in 1973 represented a sharp departure from the policy of expansion that had characterized Alitalia's management up to that point. They reflected the need to increase efficiency and in particular to reduce operating costs – or at least keep them under control. This was an important new objective which should have borne fruit in the succeeding years, but both external and internal pressures conspired to frustrate this objective. It was during the 1970s that relations with the trade unions were conditioned by a general climate of social unrest in Europe. This unrest was particularly intense in Italy for reasons connected with the country's recent economic development. Not only did wages keep pace with inflation in this period, but some groups of workers managed to improve their position relative to others. Many groups in the air transport industry obtained significant benefits and these were maintained and even increased during the 1980s. An internal factor which further obstructed Alitalia's drive to efficiency was its flag-carrier status. This brought with it a guarantee of state support and a monopoly of certain routes and, sheltered under this umbrella, Alitalia executives underestimated the trend toward market liberalization in Europe in the 1980s. Not even Nordio's international managerial experience was sufficient to break the cultural straitjacket which bound the thinking of many Alitalia executives. The integration of the services previously handled by subcontractors also reveals the lack of vision at Alitalia. If integration formed part of the policy of expansion begun in the 1960s and justified then by the economies of scale it yielded, when it was put into effect in the 1970s expansion was no longer on the agenda and there were no longer compelling reasons for the policy.[68] Lastly Alitalia suffered from the crisis affecting state shareholdings. This required a fundamental rethinking of the Italian model of state ownership that had been in place since the 1930s. A major aspect of this model had been its role as an instrument of social policy, but in effect it was concerned with maintaining a system of patronage based on social and electoral consensus. This resulted in frequent management appointments to state holdings such as Alitalia of people without the necessary business qualifications. Such appointments were made in Alitalia at the end of the 1980s and coincided with the airline's greatest losses.

4.4　Labour productivity and labour relations

Labour productivity, calculated on the basis of available tonne/kilometres per employee, rose during the 1960s thanks to the introduction of jet aircraft.

Table 6.5　Alitalia personnel levels and labour productivity, 1957–87

	Staff	Ground	Flight	Flight personnel productivity[*]	Total productivity[*]
1947	293	238	55		4 570
1952	419	?	?	?	37 923
1958	4 074	3 427	647	243	23 200
1962	7 288	6 153	1 135	510	75 472
1967	10 399	8 778	1 621	716	111 758
1972	14 822	11 756	3 066	702	145 230
1977	16 694	13 301	3 393	760	154 531
1982	18 734	14 760	3 974	716	152 017
1987	18 453	14 255	4 198	817	186 014
1992	18 828	13 573	5 255		

[*] measured in available tonne/kilometres per employee
Source: Alitalia, Annual Reports of years cited.

The number of employees at Alitalia rose from 4074 to 12 431 between 1958 and 1969. However, the increases in capacity and traffic volume were even greater than the rise in staff numbers so there was a significant increase in labour productivity.

In the first few years following the switch to jets, Alitalia had difficulty in finding pilots trained to fly them and there was no agreement with the Italian military in this area. After an initial group of pilots had been trained by Douglas in the USA, Alitalia set up its own training centre in 1961. However this did not prevent problems from arising in 1965 when the shortage of pilots forced the company 'to restructure certain sectors of the medium-range network, and reduce the volume of flights over the entire long range network'. Since the cost of the training school was very high, Alitalia was relieved to reach an agreement with the Italian Air Force in 1968, by which the Air Force shared the cost of the school and ex-military pilots were channelled into the commercial sector.[69]

Starting in 1969 and continuing throughout the 1970s, Alitalia was affected by major labour conflicts. The cause of these clashes with the trade unions was often the issue of contract renewal. In addition to Alitalia and ATI, air traffic controllers and airport service workers were also involved. Trade union membership was high in the Italian economy and together with the mood of militancy that characterized the 1970s this brought considerable gains to the Italian working population. Through the mechanism of the *scala mobile* spending power lost through inflation was rapidly and automatically restored, while working conditions and social services were also improved. Alitalia's problems came mainly from those employees with the greatest industrial power, who were able to exploit the situation to a greater extent than other workers. Flight personnel in general, and pilots in particular, had the power to totally paralyse the airline and often used or threatened to use that power. Pilot training required at least 18 months and considerable expense, so pilots were a scarce and valuable resource. Furthermore pilots' expectations derived from comparison with their colleagues in other airlines, rather than with general economic conditions in Italy. Thus Italian pilots followed the example of the Americans and formed their own organization, which during the 1970s demanded pay rises which produced condemnation in the press. In fact Italian pilots' earnings in the 1970s were not as high as some claimed and had not reached the level of their American or even European counterparts. According to Nordio, Italian pilots earned between 11 and 35 million lire a year in 1975 ($17 000–$53 500 at the 1975 rate of exchange) depending on experience and grade; it seems that seniority was more important than in other airlines in determining remuneration level.[70] Only in the following decade did the cost of Italian labour reach European levels, and between 1988 and 1992 the earnings of an Italian pilot were around the average, and often above the average for the Association of

European Airlines (AEA) companies. Recent studies have brought to light restrictions in their contracts that were mainly responsible for the low productivity of Alitalia pilots.[71] The renewal of the Italian pilots' contract in 1994 gave rise to a long struggle that resulted in the sacking of the airline's newly appointed boss. It would seem that private assurances of a substantial pay rise in exchange for industrial peace had been subsequently withdrawn by management.[72]

5 STRATEGY

5.1 Collaboration and competition with other airlines

The agreements between TWA and BEA and BOAC which gave rise to LAI and Alitalia were the first examples of international cooperation directly affecting Italian airlines. The links with the British had a greater influence on the choice of aircraft than the development of routes in this period. LAI bought Viscounts and later so did Alitalia, but these purchases were overshadowed by purchases from the Americans. TWA's outlook was independent and competitive and the company severed its links with Alitalia when it was nationalized. The British airlines on the other hand retained an interest in Alitalia until 1961.

The move towards European coordination in the 1950s was more as a political initiative than something sought by the airlines themselves. The earliest collaboration was in 1951 and was the subject of a European Council debate. There was talk of creating a consortium of European companies under the aegis of a supranational aviation authority, to run services within a common European airspace. However, the desire of most countries to keep their own national flag-carriers prevented implementation of these ideas. For many years, the only European organization concerned with air transport was the European Civil Aviation Conference (ECAC), which had the limited purpose of monitoring the development of civil aviation in Europe, and promoting coordination and balanced growth in the sector.

Then in 1957, Air France, Lufthansa, Sabena, and Alitalia started negotiations to merge their operations and give this cooperation legal form as an organization to be known as Air Union. The initiative was a result of the enthusiasm generated by the signing of the Treaty of Rome and the founding of the Common Market. Fear of overcapacity following the introduction of jet aircraft in the 1960s also played a part. For Alitalia, it was important to keep pace with European carriers like Air France. In essence, Air Union was a proposal for the joint operation of international services on the basis of predetermined traffic quotas. Until 1959 negotiations remained at the airline level, but later political factors became more important and these called for an international convention to supersede agreements between airlines. In 1965 two documents were drawn up, one the Air Union statute, prepared by the airlines, the other a convention drawn up by the governments. However there were many fundamental issues on which agreement could not be reached and in the end these differences caused the whole Air Union idea to be abandoned.[73] Alitalia was committed to collaborative initiatives at its highest level, particularly during the negotiation phase with the airline companies, even though Velani doubted that such a wide-ranging accord could be reached. Although Alitalia did not hesitate to obtain early delivery of some French Caravelles by supporting the French position over Air Union, it was careful not to become dependent on French aircraft. The protracted length of the Air Union negotiations also hindered the reaching of an accord, since over the years the objectives and negotiating power of the individual airlines changed markedly and in fact Alitalia was one of those which developed fastest. An important sticking point over Air Union concerned quotas and the methods used to determine them. Alitalia, unlike the other airlines, felt that the reference to a common European airspace would be compatible with Article 7 of the 1944 Chicago Convention only if the legal form of Air Union were that of a single company. The entry of political representatives into the negotiations rendered everything more difficult and probably doomed Air Union to failure. Discussion also centred on the intergovernmental committee's powers of intervention, on national policies, tariffs, admission of new member airlines and relations between existing members.[74]

While multilateral agreements such as Air Union, had difficulty in getting off the ground, bilateral pooling agreements between European airlines were common and well supported. Pool agreements existed between the wars, but after 1945 they increased in significance.[75] Alitalia entered into pooling agreements with all the main European carriers. Analysis of these agreements shows that Alitalia's position *vis-à-vis* its European competitors changed from one of weakness to increasing commercial power, and this power was supported by the tougher line taken by the Italian aviation authorities. Relations with the British airlines clearly illustrates this. Early on, BEA had managed to 'to establish a network of services through Italy to the Middle East with full traffic rights', and also 'fifth freedom' rights. By 1958 the imbalance in the Anglo-Italian agreement had become so severe that the British government felt obliged to award the Italians 'fifth freedom' rights in London en route to the USA. The imbalance was emphasized by the fact that in the pooling agreements signed in the 1950s, there were no limitations on revenue transfers to the company that won more custom. When the Anglo-Italian agreement expired in 1965, Civilavia (with Alitalia's agreement) threatened not to renew the fifth freedom conceded to BEA unless it obtained a decrease in the frequency of BOACs flights through Rome to

South Africa. In the pooling accord reached in 1966, not only was no transfer limit introduced, but there was also a capacity premium as well as the exclusion of the first 5000 single sector paying passengers from the pool in favour of Alitalia. Only in 1983 did pool agreements limit the amount of money that could be transferred between the two airlines.[76]

From the 1960s onwards, Alitalia had to face growing competition from charter airlines, some of them independent, others affiliated to scheduled European airlines. The charter companies were getting a growing share of the tourist traffic from Northern Europe to the Mediterranean, including Italy. In 1968, for instance, non-scheduled services within Europe carried 9.6 million passengers. Of these, 5.4 million used independent charter companies and 4.2 million used charter companies affiliated to the regular airlines. Although in 1961 Alitalia took over the charter carrier Società Aerea Mediterranea (SAM), it was not successful in developing SAM to a level at which it was able to compete with other European charter airlines and in 1971 SAM's operations were incorporated into those of Alitalia itself. As Alitalia's annual reports show, results for charter operations were not good and charter flights represented only a small percentage of the airline's business.[77]

5.2   Collaboration and competition with other modes of transport

Throughout the 1950s, air traffic within Europe increased at a slower pace than in the USA and one of the reasons why Europeans travelled less by air was that the Continent was already served by a comprehensive railway system. In Italy the railways are state-owned and although during the Second World War the rail network suffered severe damage, after the war they were the primary means of transport for goods and passengers. The percentage of Italian car ownership was low and the road network could not cope with a major volume of traffic. In 1945 most Italian trains were still steam-powered, and although the network was soon to be electrified, rail investment was at a lower level than that for roads. In addition, a tariff policy was introduced to keep passenger fares low, so that rail travel was cheap at the expense of rail profits and investment in the sector.

In the second half of the 1950s, Italy embarked on a major road construction programme. Alongside the production of the earliest economy cars, IRI started building an extensive motorway network. Only recently have the Italian railways started to alter course. On the main lines they have been using the high-speed train, the *Pendolino*, which can compete with air travel. This train can reach 250 kph, but unlike other high-speed systems has not required complete modernization of the track and uses lighter rolling stock. The Milan–Rome high-speed link is proving competitive to air travel, because of the unsatisfactory road and rail links between Fiumicino airport and Rome city centre, and the problem of fog affecting the northern airports in the winter.[78]

5.3   Globalization and strategic alliances

Since the late 1980s Alitalia's performance has deteriorated. Losses followed losses from 1988 onwards. In the five years from 1990 to 1995 losses were close to 1000 billion lire. Over the same period net indebtedness also increased, reaching almost 2000 billion lire. The recovery programme Alitalia agreed with the government and the trade unions has so far not been able to put Alitalia back in the black. Despite personnel cuts, the sale of company property and subcontracting of some of its operations, Alitalia remains a sick patient. Alitalia's poor performance can be mainly attributed to the difficulty it has in overcoming past problems, and of adopting a competitive business philosophy more in line with the aviation policy of the European Community. In the last two years the winds of liberalization have made themselves felt even in Italy. One or two private airlines have been licensed to operate on the most remunerative routes, notably the Milan–Rome stretch once dominated by Alitalia, resulting in a reduction in tariffs.

As regards external growth, the purchase of a 30 per cent share in the Hungarian flag-carrier, Malev, has brought little benefit to Alitalia. Similarly, the agreement with the American airline Continental has failed also to deliver significant benefits.

NOTES

1.   The calculation is in G. Gualerni, *Storia dell'Italia industriale. Dall'Unità alla Seconda Repubblica*. (Milan, 1994) p. 154 note 5.
2.   Touring Club Italiano–Regia Aero Club d'Italia, *Annuario della aeronautica*, 1932 (Milan, 1932).
3.   The following designed and built aircraft: Fiat group, Caproni group (also engines), the engineering group Breda; the shipbuilding groups Cantieri Riuniti dell'Adriatico of Venice and Piaggio of Genoa; and the aeronautical companies Società Italiana Aeroplani Idrovolanti (SIAI) and Savoia-Marchetti. There were many other smaller companies also (see. J. W. Thompson, *Italian Civil and Military Aircraft* (Fallbrook, Ca. 1963). Furthermore Isotta Fraschini, Piaggio, and Alfa Romeo built aero-engines.
        For a history of the Regia Aeronautica (Italian Royal Air Force) and the pioneering/spectacle flights of Balbo see G. Rochat, *I. Balbo aviatore e ministro dell'aeronautica 1926–1933* (Ferrara I. Bovolenta, 1979) (where on p. 125 is reported the 1000 efficient planes figure) and F. Minniti, 'La politica industriale del Ministero dell'aeronautica. Mercato, pianificazione, sviluppo', Part I, *Storia contemporanea*, no. 1, 1981, and Part II in *Storia contemporanea*, no. 2, 1981.

4. Founded in 1928 and controlled by the state, the Società Aerea Mediterranea (SAM), absorbed Transadriatica in 1931, which at least at the beginning was backed by Junkers and flew services between Venice, Münich, Vienna and cities on the Adriatic. Later it also absorbed Aero Lloyd, which managed the Albanian network.

    For more on commercial Italian aviation see: *Annuario della aeronautica* 1932 op.cit.; *Ali sul mondo* (Florence, 1990); R. Abate, *Storia dell'aeronautica italiana* (Milan, 1974); G. D'Avanzo, *I lupi dell'aria* (Rome, 1991); *Le ali della rondine. The origins of Italian commercial aviation* (Catalogue of exhibition: *Le ali della rondine*, Rome, 7–26 May 1992) (Roma, 1992). However the history of Italian commercial aviation in the years of Fascism is still to be written, and much of the documentation is buried in the archives and publications of the Ministry of Aeronautics.

5. Per kilometre subsidies are specified in *Lettera* of 8 May 1937 No. 23346 at the Ministry of Finance, in the Central State Archive (Archivio Centrale dello Stato [ACS]): Min. Aeronautica, Gab. 1937, b. 35.

6. Touring Club Italiano, *Annuario* op. cit., pp. 490–4, and ACS, Min. Aeronautica Gab. 1940, b. 107.

7. In ACS, Min. Aeronautica Gab. 1937, b. 35.

8. On aircraft of the *Ala Littoria* see (in ACS) 'Segreteria particolare del duce (SPD)', Carteggio ordinario (CO), b. 157, Note on the work of Comandante Bruno [Mussolini] for the new organization of civil aviation, Rome 4 October 1941. Also 'Le ali della rondine', op. cit. and G. Gabrielli, *Una vita per l'aviation* (Milan, Bompiani, 1982) (From the end of the First World War the Fiat designer was Celestino Rosatelli, but in the 1930s Giuseppe Gabrielli, began to emerge).

9. On the birth of LATI, see (in ACS), Min. Aeronautica, Gab. 1939, b. 43 and 1940, b. 99 and SPD, CO b. 157; also: B. Delisi, *La linea atlantica dell'Atlantico Sud. Il contributo dell'Ala Littoria*, in *Le ali della rondine* op. cit.

10. See G. Alegi, *L'aviazione commerciale in Italia. Problemi e prospettive di ricerca*, in *Le ali della rondine*, op. cit. and R. Gentili, *L'aviazione commerciale in Italy*, in *Ali sul mondo*, A.I., 1990.

11. This account of the rebirth of Italian commercial aviation is based on contemporary aviation magazines, such as *Alata*, published weekly in Milan, and newspapers; published memoires, such as that by P. Piacentini, *SOS dell'aviazione civile italiana. Verità e documenti* (Roma 1947); the Executive Committee Minutes (ECM) of the Alitalia Board; and a few documents from Alitalia's archives such as a copy of the agreement between the Italian Aircraft Ministry and BOAC, dated June 1946. Act n. 88, 4 September 1946 entitled Italian state or IRI to participate in founding national and international transport corporations with the contribution of Italian individuals or corporations of state consent. IRI has based its flag-carrier status upon this Act. See G. Martini, *Il servizio di trasporto aereo di linea* (Milano, Giuffrè, 1976) pp. 55–60 e S. Tomasino, *I trasporti aerei* (Milan, 1971).

12. The intentions of TWA are stated in *Alata* 1946 and those of BOAC-BEA in the Memorandum of P.C.F. Lawton, BEA's General Manager, 12/11/1951 in BEA Board Paper 55. Information on the decisions of the Italian authorities is also drawn from *Alata*.

13 . For the foreign airlines flying to Rome and other Italian cities in 1946–7 see *Alata*, no. 5, November 1945, no. 3, March 1946, nos. 4–5, April–May 1946, no. 6, June 1946, nos. 8–9, August–September 1946.

14. A brief account of the development of the public sector of the Italian economy can be found in M.V. Posner and S.J. Woolf, *Italian Public Enterprise*, (London, 1967).

15. Donato Menichella was one of the founding fathers of IRI and also a prominent *commis d'etat*. After the war, he became governor of the Bank of Italy. On IRI and other postwar matters, see Posner & Woolf, *Italian*, 1965.
16. The proposals of Bruno Mussolini can be found in ACS, SPD, CO, b. 157 Proposte op.cit.
17. On Velani and Carandini, D'Avanzo, *I lupi* op. cit., pp. 194–6; 391–401; 456 (footnote 77).
18. Memoranda by Financial Controller R.L. Weir: 13 September 1954, in BEA Board Paper 87, 14 April, 1961, and 22 June 1961 in BEA BP 60.
19. On Alitalia shares on the stock market see: Ricerche e studi SpA (R&S), *Alitalia-Linee Aeree Italiane (Gruppo IRI)*, (Milan, 13 November 1973) pp. 50–63 and *R&S*, various years.
20. This information can be found in Alitalia Annual Reports, various years, and also in *R&S*, op. cit, various years.
21. Regarding the system of state participation see: P. Saraceno, *Il sistema delle imprese a partecipazione statale nell'esperienza italiana*, (Milano, 1975); and *Il sistema delle partecipazioni statali*, L. Pennacchi ed., (Bari, 1980).

    A discussion of Alitalia and Civilavia policies regarding the national network up to the middle of the 1970s, can be found in the essay 'Criteri ottimali di determinazione delle tariffe ed uso ottimale delle risorse nel trasporto aereo interno', in G. Tucci, *Trasporto e regolamentazione economica. Struttura e riforma nel trasporto aereo*, (Milan, 1987), pp. 187–230.
22. For a certain period 200 million lire went annually to subsidize Somali Airlines. See IRI Archives, Pratica 404, f. 100 'Accordi e convenzioni dal 1946 al 1991'. The 40 billion lire is quoted in a press release, in *La Repubblica*, 11/5/1994. Figures of direct government subsidies for Alitalia between 1960 and 1968 have been estimated by Tomasino – 1960: 1722 million lire: 1965: 968 million lire; 1968: 700 million lire, in Tomasino, *I trasporti* op. cit., p. 124, table 3.II. On ATI see: D'Avanzo, *I lupi dell'aria*, op. cit., pp. 680–4.
23. The shortcomings of Civilavia were a lively issue in official documents concerned with civil aviation such as the Chamber of Deputies (*Camera dei deputati*), report *Problemi delle gestioni aeroportuali in Italy. Indagine conoscitiva della X Commissione permanente*, no. 11, (Rome 1972). See also Air Press, *Rapporto sull'aviazione in Italy* (Rome, various years). It is possible to find all the main information on Italian civil aviation in Air Press periodical publications.
24. In Air Press, *Rapporto sull'aviazione in Italy* (various years), it is possible to find yearly operating performance of ITAVIA, until its network was absorbed by Alitalia in 1981. A brief history of ITAVIA is in G. D'Avanzo, *I lupi dell'aria*, op. cit., pp. 743–52. On pp. 752–4 there is also information on the other private airline, Alisarda (originally owned by the Aga Khan and later known as Meridiana, operating alongside Alitalia). This company flew between mainland Italy and Sardinia.
25. When in 1963 Civilavia was transferred to the Ministry of Transport, communications services remained the responsibility of the Ministry of Defence, until 1980 when Law 23/5/1980 No. 242 created the autonomous company ANAV.
26. The result of this was that many aircraft companies had to close (for example, the Caproni group and the aeronautical sections of Breda, Cantieri Riuniti dell'Adriatico, and SIAI.). Only the aeronautical section of Fiat, l'Aermacchi and some factories owned by IRI remained open. In the 1950s the Ministry of Defence again started placing orders on behalf of the armed forces both in Italy and abroad with NATO. In 1959 Giuseppe Gabrielli designed the Fiat G-91 aircraft which was adopted by NATO forces.

C–47s were bought by the Ministry of aeronautics for $640000 and resold or given gratis to the independents. LAI paid $5000 for each of its 15 C–47s plus spares. Source, D'Avanzo, *I lupi* op.cit., p. 449, note 53.

27. The discussions leading up to these decisions are in Alitalia Executive Committee Minutes (ECM), 1946.

28. The discussions are recorded in Alitalia ECMs of 19 and 21 April 1948 and the decision to buy Douglas DC–4s from a Peruvian company are in ECM 11 October 1948, 4 February, 20 March 1950.

29. Alitalia annual reports, various years.

30. R.W. Rummel, *Howard Hughes and TWA* (Washington: Smithsonian Press, 1991), pp. 189–229 (chap. 13 and 14) and N. Dietrich and B. Thomas, *Howard: the Amazing Mr. Hughes* (Greenwich, Conn. 1972).

31. F. Serena, 'Scelte economiche e politica commerciale', in *Vent'anni Alitalia*, (Roma, Edizione speciale di *Freccia Alata*), 1967, p. 107.

32. Alitalia, annual reports 1955 and 1956.

33. R. Miller and D. Sawers, *The technical Development of Modern Aviation*, (London, 1968) p. 153ff.

34. The figures of orders for jets in ITA, Le Transport aérien commercial de 1960 à 1970. Perspectives pour la nouvelle décennie, (Paris, 1971), table. on p. 91.

35. For the annual growth rate of 15 per cent in the 1960s, see table. A, based on ICAO data, attached to Nordio hearing, in Camera dei deputati, *Situazione dell'aviazione civile in Italia. Indagine conoscitiva della X Commissione permanente*, (Rome, 1976), p. 367.

36. On the choice of the DC–8 equipped with Conway, interview with engineer Luigi Di Giorgio, June 1991 (Di Giorgio shared with top management responsability on the choice of aircraft) and D'Avanzo, I lupi, op. cit., table on p. 619; about the delivery before time of the DC–8, D'Avanzo, I lupi, op. cit., pp.662–4.

37. On Caravelle delivery, see Di Giorgio interview op. cit.

38. G. D'Avanzo, *I lupi op. cit.*, p. 676; B. Catalanotto C. Falessi, *1969–1989: I vent'anni dell'Aeritalia*, (Milan, 1989) pp. 60–1.

39. See Di Giorgio interview, op. cit.

40. See footnote 36 and V. Ruggiero, *Il trasporto aereo europeo* (Napoli, 1984), p. 35.

41. 1970s growth overestimate is in Alitalia *Annual Report 1970*. About DC–10s, see Di Giorgio interview, op. cit.

42. Di Giorgio, and Alitalia, annual reports.

43. Di Giorgio and Alitalia, Annual Reports. In 1974 the fleet was constituted of 30 long range aircraft: five Boeing 747s, six Douglas DC–10s (that would become eight the following year) eight Douglas DC–8/62s and 11 Douglas DC–8/43s; as well as 49 short to medium range aircraft: 33 Douglas DC–9/32s and 16 Caravelles.

44. Alitalia, *Annual Report 1982*.

45. Alitalia, *Annual Report 1986*.

46. *Mondo AZ*, 1989, January, p. 9.

47. 'The Americans have one huge advantage. Air treaties signed after World War II let them fly between many European cities: Paris to London and Brussels to Munich, for example. Generally they can't haul passengers within a country. But their border-hopping privileges mean they can build feeder networks inside Europe, while European airlines have no such rights within the U.S.', S. Toy and J. Rossant, 'The Carnival is Over', *Business Week*, 9 December 1991, p. 18. For a French view on the subject: Jacqueline Dutheil de la Rochère, *La politique des Etats-Unis en matière d'aviation civile internationale* (doctoral thesis, Paris, 1971).

48. C. Leporelli, 'Strutture e gestione delle reti di trasporto aereo: strategie di impresa e indicazioni per la politica pubblica'; C. Leporelli and C. Zamboni, 'Competizione e struttura di rete nel trasporto aereo: un'analisi empirirca del mercato a medio raggio'; M. Ajmone Marsan and F. Padoa Schioppa Kostoris, 'Concorrenza tra sistemi di rete nel mercato del trasporto aereo: un'analisi dei percorsi diretti e indiretti', *Struttura di mercato e regolamentazione del trasporto aereo*, F. Padoa Schioppa Kostoris, ed., (Bologna, 1995), pp. 27–71; 203–50; 251–320.

49. Camera dei deputati, *Problemi delle gestioni aeroportuali in Italia. Indagine conoscitiva della X Commissione permanente,* (Rome, 1972), p. 13.

50. The 200 billions increased to 351 in 1975, but were still few compared to the provisions of the 'Lino' report which amounted to 1000 billion lire, Air Press, *Rapporto 1974–1975 sullo stato dell'aviazione in Italia*, (Rome, 1975), pp. 89–93.

51. D'Avanzo, *I lupi* op. cit., pp. 795–6 and 801.

52. R. Bianchi, 'Gli aeroporti milanesi: il ruolo della SEA', *Rivista milanese di economia*, no. 15 (July-September) 1985, pp. 75–85 and D'Avanzo, *I lupi, op. cit.,* pp. 798–9.

53. M. Ajmone Marsan and F. Padoa Schioppa, 'Concorrenza op.cit., in *Struttura di mercato*, op. cit., pp. 263–6.

54. Alitalia, annual report, various years.

55. The data on routes come from Alitalia annual reports, various years.

56. The financial report of the merger in IRI Archives, pratica 404, fasc. 32.

57. Quoted from *Vent'anni Alitalia*, op. cit., p. 32. For data on the capacities and average speeds of European ICAO airlines in ITA, *Le transport aérien commercial de 1960 à 1970. Perspective pour la nouvelle decennie*, Paris 1971, pp. 49–59; for data on Alitalia, see Air Research Bureau, *Document ARB/318*, Paris 1964, Appendix 3.

58. See footnote n. 36.

59. L. Doty, 'U.S. Route Policy Irks European', *Aviation Week & Space Technology, 14 April 1969 for more on the campaign to lower tariffs to the USA.*

60. P. Ray, 'The Impact of Tourism', *Flight International,* 10 December 1970. Page 900 reports that at a symposium organized by the Institut du Transport Aerien (ITA) in Paris, the founder and President of Alitalia, Bruno Velani claimed that 'a very high proportion, if not all, of the supplementals' traffic was composed of tourists taken from scheduled services... In the summer of 1969', he claimed, 'supplementals carried 39.7 per cent of US traffic on New York–Rome; on New York–Amsterdam the proportion was 67.4 per cent; and reached the "incredible" figure of 68.2 per cent on New York-Frankfurt'. He also suggested that the charter carriers' 'dangerous' activities should be restricted.

61. 'Alitalia begins image change campaign', *Aviation Week & Space Technology*, 17 November 1969, p. 43.

62. See Alitalia *Annual Report 1962* and following years; V. Berlen, 'Le telecomunicazioni e le prenotazioni', *Vent'anni Alitalia, op. cit.* pp. 87–96.

63. In *Ricerche e studi SpA 'R&S'*, SpA Alitalia-Linee Aeree Italiane (Gruppo IRI) (Milan, 1973, Unpublished paper), On p. 40 it is maintained that Alitalia had a small number of ground employees because it subcontracted many services.

64. In *'R&S'*, SpA Alitalia op. cit., pp. 32 and 37, the role of Eximbank during the 1950s and 1960s is emphasized when loans covered only 40–50 per cent of aircraft price.

65. D'Avanzo, *I lupi del cielo*, op. cit., pp. 690–9.

66. Rigas Doganis, *Flying Off Course*, (London: 1985), pp. 45–57.

67. See Nordio hearing, in Camera dei deputati, Situazione dell'aviazione civile cit., pp. 87–8.
68. 'Lo svilupppo della flotta Alitalia. La manutenzione degli aerei Alitalia', *Corriere della Sera*, 9 June 1981, and *1947–1987 Quarant'anni Alitalia la città del volo*, (Roma, Alitalia, s.d.). During the 1970s Alitalia took over responsibility for several activities that had previously been contracted out. These were carried out in the 'Zona Tecnica' at Fiumicino airport and included maintenance of all types of aircraft and engines, pilot training using flight simulators (the flying school was based in Sardinia), and meal preparation. The building complex not far from the airport was enlarged to accommodate the computer and telecommunications systems. In addition, Alitalia's participation in technical agreements, such as ATLAS, meant an increase in the number of technical and skilled employees. 1979 saw the initiation of the substantial expansion of the 'Zona Tecnica' at Fiumicino, which company literature referred to as 'Città del volo' (flying city). On maintenance costs, 650 billion lire in 1995, see 'Alitalia, competitività nella manutenzione', *Il sole-24 ore*, 2 June 1995.
69. In Alitalia, annual reports 1962–68: with the aim of developing its own training programme, Alitalia bought four Macchi MB-326/D trainers and flight simulators.
70. Nordio hearing, in Camera dei deputati, *La situazione dell'aviaizione civile op. cit.*, p. 90.
71. M. Bianco and T. Piccirilli, 'La produttività del personale navigante tecnico: un confronto tra le principali compagnie aeree europee', in *Struttura del mercato op. cit.*, table. 17, p. 346.
72. The story was in all the Italian newspapers between August and October 1995.
73. In IRI Archives, minutes of 1958, 1959, 1960 and 1961 Air Union meetings in F. 54; and F. Santoro, *La politica dei trasporti della Comunità economica europea*, (Torino, UTET, 1974), pp. 384–416; and L. Asperges, 'Il trasporto aereo nella politica della CEE: storia di un fallimento', *Automobilismo and automobilismo industriale, (1972, pp. 330–57, nn. 5–6.*
74. Pro-memoria dal quale risulta la posizione di Alitalia nei riguardi di Air Union, 11/11/1961, in IRI Arch. F. 54.
75. The pool is an agreement between two or more airlines operating on the same route, on the basis of which, to avoid harmful competition and stabilize costs, all or part of the revenue is put into a joint fund. See Doganis, 'Flying Off Course, op. cit., pp. 29–33.
76. Quotations from letter from H.E. Marking to J.F. Dempsey (General Manager, Aer Lingus) dated 1 March 1966, in Ministry of Aviation, EB/14/040, Review of Anglo-Italian Air Services Agreement 1964–65 and Pooling Accords, texts 1965 and 1983.
77. ITA, *Le transport aérien commercial de 1960 à 1970. Perspective pour la nouvelle décennie* (Paris, 1971, pp. 49–59).
78. T. Flink, 'La rivitalizzazione di un settore maturo: le ferrovie ad alta velocità in Italia', *Annali di storia dell'impresa*, no. 8 (1992), pp. 419–50.

7 LOT: Connecting East and West in Poland

Joanna Filipczyk

The Polish national airline LOT resumed activities almost eighteen months earlier than its rivals in western Europe. A temporary operations base was established in 1944 in Lublin in central eastern Poland, the seat of the Moscow-controlled Polish Committee that was to play a major role in governing the Soviet-liberated zones of the country. In August 1944 LOT started two domestic services: from Lublin north to Białystok, and south to Rzeszów and Przemyśl. Despite the priority given to mail services in cooperation with the civil aviation division of the Military Aircraft Command, LOT's two single-engined Polikarpov Po-2 aerial taxi biplanes managed to carry 4811 passengers before the end of 1944. However, it was only on 16 March 1945, with the end of military control over its operations, that LOT could formally recommence its activities.

1 POLICY

In the socialist concept of the economy that was established in Poland after 1945, air transport occupied a different position to that in Western Europe. It was established essentially as a service branch of the government, and considered as a public utility rather than in terms of a luxury international transport system. This had important consequences for the Polish government's policy towards LOT.

1.1 Origins

In the years after 1918, the period in which air transport companies were founded in the leading western European countries, the newly declared Republic of Poland was in turmoil and conflict. Such conditions were ill-suited for founding a high-risk venture in air transport, despite the poor shape of the new country's surface communications. Reflecting the political importance of the Franco-Polish alliance after the First World War, the first scheduled service to and from Warsaw was provided by the French Compagnie Franco-Roumaine de Navigation Aérienne in April 1921. Poland granted the airline a subsidy in the form of free fuel to the value of one million francs per year.[1] The establishment of a genuine Polish airline did not take place until 1922, when local oil industry interests combined the German Junkers group to form an airline called Aerolloyd Warszawa. The new company started services on 5 September 1922 on two domestic routes, Warsaw–Lvov, and Warsaw–Gdańsk, expanding its network to Kraków the following year, and to Vienna in April 1925.

In that same spring of 1925 Aerolloyd Warszawa severed its German ties and became an all-Polish joint stock company, Aerolot. Strengthening its capital base, Aerolot prepared for further expansion of its route network and for competition from a second Polish aviation company, Aero T.Z., which was founded in Poznań in 1925 and aimed to operate services between Warsaw, Poznań, and Berlin.[2]

The Polish government was slow to react to these developments and take an interest in the new mode of transport. It was not before 1928 that increasing aeronautical activity led to the adoption of a Polish Aviation Law by Parliament, the Sejm, and a Civil Aviation Office was created within the Ministry of Communication. Thereafter, however, the government targeted civil aviation as a means of rapid transport in a large country that suffered from an underdeveloped surface infrastructure and this brought profound changes in the Polish aeronautical field. The Civil Aviation Office took a leading role in bringing about a merger between Aerolot and Aero T.Z. resulting in the founding of a new national carrier, Polskie Linie Lotnicze, known under its acronym LOT, on 28 December 1928. To allow for a degree of managerial freedom, LOT was organized as a limited liability company. The initial capital stood at 8 million złoty (US$1.2 million), divided into 100 shares. Following the German model, it was intended to divide the shareholding between the government, which was to take a 60 per cent majority, and a number of municipalities. This financial division turned out to be unrealistic as it quickly became clear that it would not be possible to extract 3.2 million złoty from the country's already overburdened cities. The state was forced to expand its shareholding to 86 per cent and only three other parties participated in the venture: the regional government of Silesia (10 per cent), and the cities of Bydgoszcz and Poznań (2 per cent each).[3]

Next to continuing the services operated by Aerolot and Aero T.Z., the new national airline opened two new routes in January 1929. Having invested in the new venture, the city of Bydgoszcz was rewarded by inclusion in LOT's domestic network. The Silesian industrial centre of Katowice was also connected. During the 1930s the focus of LOT's operations shifted from domestic to international air services. In line with Polish policy at the time, LOT's operations created a broad axis of north–south services connecting the Baltic and the Mediterranean regions through Warsaw. Extending the Warsaw–Lvov route further south, Bucharest was included in the network in 1930, and the following year, LOT reached Sofia and Salonika. In the second half of the 1930s LOT's services included Athens and Lydda (Palestine). The rich city of Beirut was included in 1938 and the same year a second southern route connected Warsaw with Budapest, Belgrade, Venice and Rome. Northward, a LOT service to Vilnius (northeastern Poland), Riga and Tallinn was started in 1932, and extended to the Lithuanian capital Kovno and Helsinki. Strangely LOT gave little attention to developing services to *western* Europe, other than maintaining a service to Berlin, operated in pool with Deutsche Lufthansa. Services from Warsaw to Paris and London were operated by Air France and British Airways Ltd., LOT's only other western route was to Copenhagen, with scheduled flights beginning in 1938. No services to the Soviet Union were offered. In all, LOT carried 35 400 passengers and just over a thousand tons of cargo in 1938, making it the largest of the eastern European airlines, just ahead of the Czechoslovak carrier, CSA.

The German attack on Poland in September 1939 brought a violent end to LOT's activities. In the subsequent four weeks of fighting and the Luftwaffe's bombing campaign against Warsaw, all of LOT's aircraft were destroyed.

1.2 Government aims

At the end of the Second World War, Poland was in even greater turmoil than it had been after the First. The Soviet troops that occupied the country made no effort to leave after it was liberated from the Germans and a Soviet-dominated provisional government took over in Warsaw. Under this provisional government the preparations were made to remodel Poland according to Stalin's views. A revival of the privately owned LOT of pre-war days was out of the question. However because of the unsettled circumstances and the severe damage that the prolonged fighting had caused to Polish surface transport, air transport was of vital importance to the provisional government. Its first act was to place the airline under state control on 6 March 1945. The decree was published in *Polish Monitor*, number 1, item 1, after which the new LOT was established on 16 March.

LOT reopened its domestic services as soon as it had the aircraft to do so and the first of three circular routes around the larger liberated cities of the new Poland was initiated on 30 March 1945: Warsaw Łodz–Kraków–Rzeszów–Lublin–Warsaw, Warsaw–Łodz–Poznań–Katowice–Łodz–Warsaw, and Warsaw–Olsztyn–Gdańsk–Bydgoszcz–Warsaw. Establishing direct connections to the former German cities that were now under Polish sovereignty, was also essential and on 14 May 1945 a service between Warsaw and Gdańsk (Danzig) was inaugurated, followed by a service to Wrocław (Breslau) on 18 June 1945.[4]

Even before the new People's Republic of Poland was officially declared LOT's affairs had become government business. LOT's first postwar task was to provide transport for officials and generally speed up the business of government. By the end of 1945 LOT offered regular domestic services to seven cities. LOT's perceived function was reflected in the annual government subsidies that ensured its survival, the use of military airport facilities and close cooperation with the transport branch of the Polish Air Force.

Although LOT liked to see itself as a genuine airline, evidenced by its IATA-membership and choice of aircraft, essentially it remained a service branch of the government and its development was heavily influenced by politics. At the 20th Congress of the Communist Party of the Soviet Union in February 1956, Khrushchev publicly denounced Stalinism, and indicated that in different countries there might be different 'roads to communism'. In the wake of this shift in the Soviet party line, strikes and riots broke out in Poland in June 1956, putting pressure on Poland's United Workers Party to chart a more *national* course. Serious tensions arose, leading to a stand-off with the Soviet Union in October and the appointment of Poland's nationalist communist Wladyslaw Gomulka as First Secretary of the Party. Gomulka embarked on an ambivalent course towards relaxing the Party's political and economic hold on the country.[5] LOT profited from these changes and was allowed to raise prices on its domestic routes to recoup more of its operational costs, and to develop more international services to western Europe. However, the Polish policy towards bilateral air traffic agreements tended to serve defensive economic objectives and Poland demanded safeguards from foreign airlines to protect LOT's monopoly of ticket sales in Poland. Rather than earning foreign currency from foreign customers, LOT became an instrument for saving foreign currency by restricting Polish travellers to the national airline that offered tickets in złoty. LOT's role was not only to provide air transport, but also serve as an instrument for acquiring and preserving foreign currency – a role which changed little throughout the 1960s and 1970s.[6] The airline's development was dictated by traffic goals which were set by the government. This situation only began to change in the 1980s, after new legislation on state-owned enterprises introduced the concept of managerial responsibility for LOT's financial performance, as well as for the meeting of specified traffic goals.

Although LOT had been reorientating itself to western standards of airline operation before 1989, the dramatic events of that year marked a clear turning point in the airline's history. After a decade of Solidarity-led social and political unrest that the Jaruzelski government was unable to quench, and economic stagnation that no reform plan could cope with, the government of the United Workers Party crumbled in the aftermath of the June 1989 elections. In August of that year the electoral breakthrough of the opposing Solidarity movement opened the door to a complete transformation of Poland's political and economic system, which fundamentally affected LOT.[7] The new government neither could nor wished to maintain LOT as a protected branch of government. Consequently in the following years, LOT was manoeuvred onto the runway of privatization.

1.3 Changing attitudes to regulation

In the first years after 1989, LOT prepared for a return to western business practices. On 14 June 1991 a large majority of the Sejm voted to gradually transfer LOT's ownership from the Polish State to the private sector. Under the new law, LOT was to become a joint-stock company, in which shareholding would be divided on a more or less equal basis between the Ministry of Transport and Maritime Economy, and the private sector. Originally, the act had stipulated full privatization, but parliament considered this too extreme. A revised bill was endorsed by Poland's prime minister, Hanna Suchocka, on 16 November 1992. The following month the airline was reorganized as a joint stock company, PLL LOT SA, with a capital of 2072 billion złoty (approximately US$149.5 million). In the run-up to the floatation of LOT on the Warsaw stock exchange, 100 per cent of the stock was held by the State Treasury. It was the first step on the road to partial privatization along the lines of other European flag-carriers. To guide the airline on its new course, LOT's commercial director Jan Litwiński, was promoted president and chief executive officer in 1993.

To achieve the new ownership structure the government set aside three years, although in December 1994 the Council of Ministers decided to extend the deadline until December 1996 to allow more time for restructuring, and particularly for divesting LOT's non-core businesses such as ground handling.[8] The government announced it would hold on to 51 per cent of LOT's shares to meet the requirements of substantial ownership and effective control,[9] while 29 per cent of the shares were to be made available to private investors, in the search for whom the government turned its eyes to the western airlines.[10] The remaining 20 per cent were to be offered to the airline's employees at a preferential discount. In 1996 the percentage of shares to be offered to private investors was raised to 34 per cent, to enhance LOT's attraction for prospective investors, with a mere 15 per cent remaining for employee shareholders.[11]

2 AIRCRAFT PROCUREMENT

Like its western counterparts, LOT struggled with problems of aircraft procurement. In the period after 1945 however, its decisions were compounded by political and financial factors that were even more complex than the circumstances troubling its competitors in the West.

2.1 Propeller aircraft

After severing its ties with Junkers in 1928, LOT first relied on six single-engined Fokker F.VIIA aircraft for flying operations. These were soon augmented with three-engined Fokker F.VIIBs that were licence-built in Poland by Plage & Laskiewicz for LOT and for the Polish Air Force, and with several types manufactured by Podlaska Wytwòrnia Samolotów that elaborated on the Fokker designs. In 1934 the LOT fleet reached 33 aircraft, and thereafter a programme of modernization was undertaken, focusing on American Lockheed L-10 Electras (seven) and L-14 Super-Electras (eight), and Douglas DC-2s (two). A single Junkers Ju-52 brought the total strength up to 18 aircraft by the end of 1938.[12]

LOT soon managed to acquire a sizeable fleet of transport aircraft once peaceful conditions had been restored in 1945. The strategic importance that the Soviet Union attached to Poland helped the Polish government plead its case in Moscow for the release of transport aircraft from Russian stocks. In 1946 LOT already possessed 39 aircraft including nine Douglas C-47s and 25 Soviet-built copies known as the Lisunov Li-2. While these converted military transport planes were to form the mainstay of LOT's fleet until 1960, they were replaced on the more prominent international routes by purpose-built twin-engined Ilyushin Il-12s, of which LOT acquired five in 1949. Although no bigger than the Li-2s, the Il-12 did offer slightly improved operating characteristics.[13] Like the other eastern European carriers, LOT used these aircraft, of which production totalled around 3000, on its domestic routes. To equip LOT's few international services was more of a challenge. Medium-range Russian aircraft were nearly impossible to obtain and as Poland had no dollars with which to buy American models, the only option was to shop for aircraft in western Europe. Typically LOT ended up operating five French four-engined Sud-Est SE.161 Languedocs on its international routes. These 33-seaters had started life as Bloch 161s in 1937 and were already obsolete when LOT followed Air France and Iberia and brought them into service in 1947. Their unsatisfactory operating economics caused LOT to withdraw them in 1950.[14] With the shift in LOT's objectives to domestic operations in the Polish six-year plan of 1950–6, the Languedocs were replaced with Il-12s and the improved Il-14P when these became available in 1955. Thus, by the mid-1950s, LOT possessed a fairly homogeneous fleet of Soviet twin-engined aircraft. The domestic routes were operated with the ageing Li-2s, and the international routes were flown with the Ilyushins.

2.2 Jets or turboprops?

At this stage LOT's problem was how to respond to the development of jet aircraft, given its meagre foreign earnings. Like BEA, LOT was an exclusively short- and medium-haul carrier and it looked at jet propulsion with scepticism. A medium-haul passenger jet was under development in the Soviet Union, and the first of this generation, the twin-engined Tupolev Tu-104 – derived from the Tu-16 bomber – made its first flight on 17 June 1955.

Unlike the Czech carrier, CSA, which bought six of the 70-seat Tu–104As for their services to Paris, London, and other western cities, LOT decided to stick to piston-engine technology in anticipation of Soviet turboprop designs such as the four-engined Ilyushin Il-18 that promised better seat-kilometre economy than the jets[15] However as the Il-18 would not be available until 1960, LOT had to decide what to do in the meantime. Was it necessary, in order to face competition from Aeroflot and CSA, to acquire a stop-gap aircraft that would offer a better standard of comfort and performance than the Il-12s and Il-14s? As LOT began to attach more importance to revenue-earning international services to western Europe at the end of the 1950s, it decided to go for comfort rather than for speed. Consequently LOT bought three second-hand Convair CV-240s from SABENA which acted as a stop-gap until the Ilyushin Il-18 turboprop became available. LOT received three of these in 1961, followed by five more after 1964.[16]

Table 7.1 LOT

	1946	1951	1956	1961	1966	1971	1976	1981	1986	1991	1996
Propeller-driven											
Polikarpov Po-2	5	–	–	–	–	–	–	–	–	–	–
Lisunov Li-2/C-47	25	24	30	14	2	–	–	–	–	–	–
Douglas C-47	9	6	5	–	–	–	–	–	–	–	–
Ilyushin Il-12B	–	4	3	–	–	–	–	–	–	–	–
Ilyushin Il-14P	–	–	6	12	12	9	–	–	–	–	–
Convair CV-240	–	–	–	4	–	–	–	–	–	–	–
Vickers Viscount 804	–	–	–	3	1	–	–	–	–	–	–
Ilyushin Il-18	–	–	–	3	8	8	9	9	9	6	–
Antonov An-24W	–	–	–	–	10	15	17	17	16	11	–
Jet engined											
Tupolev Tu-134	–	–	–	–	–	5	5	5	–	–	–
Tupolev Tu-134A	–	–	–	–	–	–	5	7	7	7	–
Antonov An-26*	–	–	–	–	–	–	12	12	–	–	–
Ilyushin Il-62M	–	–	–	–	–	–	6	7	7	7	–
Tupolev Tu-154M	–	–	–	–	–	–	–	–	2	11	–
Boeing 767-200	–	–	–	–	–	–	–	–	–	2	2
Boeing 767-300	–	–	–	–	–	–	–	–	–	1	2
Boeing 737-400	–	–	–	–	–	–	–	–	–	–	4
Boeing 737-500	–	–	–	–	–	–	–	–	–	–	6
Alenia ATR-72	–	–	–	–	–	–	–	–	–	–	7
Total	39	34	44	36	33	37	54	57	41	45	21

Source: LOT Polish Airlines
* On lease from the Polish Air Force

Like western airlines such as KLM, LOT favoured fuel economy over speed in its choice of aircraft. Yet LOT carried this preference one step further than its rival airlines in the Eastern Block by ordering not only the Il-18, but also buying three second-hand Vickers Viscount 804 turboprops from British United in 1962. In fact, in the early 1960s LOT was the only airline from the Eastern Block to operate Western aircraft. The Viscount order showed LOT's appreciation of cost-control, as the airline was gradually released from national planning structures.[17] Unfortunately, within a short space of time LOT lost two of its Viscounts in crashes, probably because of insufficient training with western equipment. A month after its arrival, one went down at Warsaw's Okęcie airport and the second crashed in Belgium in August 1965. For its short-haul routes LOT replaced these and other losses with the twin-engined Antonov An-24 turboprop, which was introduced on its routes to Kiev and Leningrad in April 1966. In all, LOT bought 17 Antonovs, making the An-24 the most numerous type in its fleet.[18]

In the long run the general trend towards jets became irresistible and when Tupolev's new twin-engined Tu-134 became available in 1966, LOT ordered two to be delivered in 1968. They replaced the An-24s on the longer routes, such as those to Madrid and Amsterdam. The remaining Viscount was sold to New Zealand in 1967. Over the next ten years LOT's fleet of Tu-134s was gradually increased to twelve aircraft.

In 1970 Poland and the USA reached a bilateral air services agreement that offered opportunities to LOT to explore the American market. After initial differences had been resolved over controlled ticket sales in Polish currency, LOT was given operating rights in 1972 on the Warsaw–New York route, next to Pan American Airways, although the Americans put some limitations on its frequencies. This meant that LOT now had to buy long-range aircraft if it wished to operate the route and it bought two Ilyushin Il-62M aircraft in the Soviet Union.[19] Their number was gradually increased to nine aircraft by 1980, as LOT's intercontinental network grew with routes to Montreal in 1976, and destinations in Asia. However, LOT was unhappy with its Ilyushins and while its continued expansion prompted the order of 14 of the more versatile Russian Tupolev Tu-154Ms in 1986 (these jets could be used in a mixed configuration for passengers and cargo), LOT also actively investigated the possibility of buying western aircraft.

2.3 Acquiring western aircraft

Even before the final collapse of the Polish system in 1990, LOT was heading for change. In 1986 the airline had replaced the last of a long line of political appointees as directors with Jerzy Słowiński, whose business orientation took the airline away from its traditional focus on government-designated transport targets and aligned more with the commercial operations of Western airlines. As a first step, three different Western aircraft – the Boeing 767, the Airbus A310, and the McDonnell Douglas DC-10 – were considered as a replacement for the Ilyushin Il-62s. A year later, LOT decided on the 767, Boeing being chosen over the Airbus and the DC-10, because the 767 had a better range with full payload and because Boeing was able to provide the fastest delivery. Two 767s had been ordered by the Romanian carrier TAROM, but the deal had collapsed as a result of political difficulties between the USA and Romania. To reduce the risk to the foreign creditors that LOT needed to obtain the aircraft, a complicated lease arrangement was negotiated with a syndicate of 17 banks, led by Citicorp, to enable the speedy acquisition of a single 208-seat 767-200ER for LOT's transatlantic routes. In April 1989, LOT's first Boeing arrived in Warsaw. It was followed by a second leased 767-200ER later that year, and two larger (249-seat) 767-300ERs in 1990 and 1995. The aircraft replaced the Ilyushin Il-62Ms and made good business sense as the Boeings not only had a better payload, but also saved LOT nearly 50 per cent in fuel costs over the Russian aircraft.

In the years after 1989, LOT made a complete changeover from Soviet aircraft to Western equipment. This change was dictated by the superior economic characteristics of the Boeings, as well as the measures taken by the ECAC member states to discourage the use of noisy aircraft at Western European airports after 1988.[20] In 1990 LOT decided to replace its Antonov An-24 turboprops with Western aircraft.[21] The choice was between the Canadian de Havilland Dash-8 and the French-Italian Alenia ATR-72 turboprops, and the British BAe-146 jet. The idea of buying the British jet was soon abandoned because of its higher operating costs and the choice narrowed to the two turboprops. LOT faced the question of whether it should go for a smaller plane, characterized by lower operating cost-per-flight (Dash-8), or a bigger aircraft that would have lower cost-per-seat (ATR-72). After some negotiating and consultation with the government, LOT chose the 64-seat ATR-72 as the Franco-Italian consortium provided a more attractive package of compensation orders for the Polish aviation industry. As in the case of the Boeing contract, LOT arranged a leasing agreement for the ATR-72. A contract was signed for a 12-year lease covering 77 per cent of the value of the aircraft, with a purchasing option at the end of the contract. It was arranged by the Crédit National de Paris and the Banque National de Paris. The first ATR-72 arrived in Warsaw in August of 1991 for use on LOT's domestic and short-haul routes to the newly independent ex-Soviet republics.

Because of LOT's narrow operating margins, fleet standardization was of key importance. Through the American Bankers Trust and with guarantees from the American EXIM Bank, LOT was able devise the lease of four 147-seat Boeing 737-400s for its denser medium-haul routes, and six 108-seat 737-500s for its other European routes in replacement of the Tu-134 and

Tu-154.[22] As with the Boeing 767s, a 12-year lease period was agreed upon, covering 85 per cent of the value of the aircraft, LOT providing the remaining 15 per cent itself. As the actual owner of the aircraft, a 'special purpose vehicle' was created called Marta Leasing, conveniently located in the tax haven of the Cayman Islands. Deliveries of the new Boeing 737s began in December of 1992, and within a year LOT had withdrawn all of its Tu-154s from scheduled services.[23]

Initially LOT wanted to sell its Russian aircraft to China and it had a favourable transaction worked out, but the deal fell through when the Russian manufacturers proved unable to service the aircraft and refused to sign a contract with the Chinese. In its search for other buyers, LOT ended up negotiating a transfer of its Ilyushins and Tupolevs to the former Soviet republics. With the collapse of the Soviet Union, these new countries were left with an air transport infrastructure but no aircraft, as Aeroflot had withdrawn as many planes as it could to Russia. The LOT sale helped such newly formed airlines as the Belarus carrier Belavia, Air Ukraine, Lithuanian Air, Estonian Air, and Latvian Air, to literally take off. The transactions took some time to be completed, because the former Russian republics could not pay at once, and special ways had to be found to complete the financing of the deal. As evidence of the problems encountered, seven unsold Il-62s and three Tu-154s were still parked on the tarmac at the LOT maintenance base at Warsaw's Okęcie airport in June 1996, along with a single Il-18.

3 ROUTES

In the first decade after 1945, LOT's route network reflected the political realities of the Cold War. More services to western European capitals were inaugurated from the mid-1950s, when the airline expanded its network on a more commercial basis. Reaching out to Polish communities abroad featured prominently and would continue to do so after the changes of 1989, when LOT adapted its network to the new international environment.

3.1 Network development

The network of air services that LOT developed in the 1930s had a general north–south pattern with a predominantly international emphasis. After the Second World War, the first foreign services that LOT opened were to Berlin in May 1946, to Paris and Stockholm in July, and to Prague in August. LOT's initial activity reflected political conditions and the Polish airline focused on the eastern European capitals that had come under the Soviet sphere of influence. In 1947 scheduled flights to Budapest and Belgrade were started, with Bucharest following in 1948, and Sofia and Moscow in 1955.

Despite international politics, and aircraft procurement difficulties, LOT was also allowed to start a limited number of services to the West. During much of the 1950s, LOT and the Czech carrier CSA were the only airlines from the Eastern Bloc to operate such services.[24] LOT opened routes to Copenhagen in 1948, Brussels in 1949, and Vienna in 1955. The emphasis remained, however, on providing domestic air transport, and this developed comparatively fast. In the first half of the 1950s new services were opened between Warsaw and Rzeszów, and between Gdańsk and Szczecin. There was also fast growth in the area of services for agriculture and forestry.

Further international expansion was directed towards opening more routes to western Europe for which Berlin was the staging point. For many years LOT and CSA were the only Eastern European carriers providing services between East Berlin and the West. LOT's services began under the directorship of Andrzej Skala (1957–9), after the airline took delivery of its Ilyushin Il-14 aircraft in 1956. The following year LOT was able to start new routes to Athens and Tirana. In 1958 the first scheduled LOT flights landed in London and Zurich, with Amsterdam included in the network in 1959, and Rome in 1960. By that time, LOT and CSA had lost the duopoly on air services from Eastern Europe to the West, as Malev from Hungary and TAROM from Romania joined in. LOT's international expansion proceeded erratically as it followed the pattern of Poland's economic and political relations abroad. Between 1960 and 1965 LOT's propeller-driven planes first appeared among the jet aircraft that increasingly dominated the airports of Frankfurt, Helsinki, and Cairo.[25]

As its number of four-engined Ilyushin Il-18 turboprops doubled to eight in the mid-1960s, the LOT network was extended to Beirut, Milan (1966), Vilnius, Leningrad and Kiev (1968), Istanbul (1969), Madrid, Geneva and Nicosia (1970).[26] In the domestic market, flights to Bydgoszcz and Katowice were resumed and new direct connections to Katowice, Kraków, Wrocław and Rzeszów were opened as the ageing Ilyushins were replaced by the Antonov An-24s. Poland's two main ports, Szczecin and Gdańsk, were also connected.

From 1972 onwards a more liberal approach to Poland's international civil aviation and the acquisition of Ilyushin Il-62 aircraft, enabled LOT to open a service to New York – gateway to America's sizeable Polish community. New York immediately became LOT's most important international destination, although bilateral differences over reciprocal market access forced LOT to operate under the guise of charter flights until 1973. Five years later the New York route accounted for nearly a third of LOT's total international traffic. Flights to western Europe were even more important and carried around 40 per cent of LOT's international traffic. By contrast, low fares made traffic between Comecon partners unrewarding and LOT tended to neglect the East European routes.

In the 1970s LOT also began expanding into the Middle East, and following the principle of socialist political allegiance, Baghdad was included in the network in 1972, followed by Damascus (1973). Foreign air transport increased by 37 per cent in 1972, and 30 per cent the next year. In 1975 Benghazi, and Lyon – France's second largest city – joined the LOT network. In 1976 Montreal and Bangkok were connected, the latter a way-station for travellers between Poland and the Polish immigrant community in Australia.

If such services reflected Poland's general political interests, this did not mean that competition was completely absent among eastern Europe's airlines. In 1978 CSA started operating between Bratislava and Warsaw, competing with LOT's Warsaw–Vienna services. As the LOT service was to a western capital, LOT operated it under the IATA tariff structure and load factors suffered when CSA drew off passengers with the much lower EAPT fares agreed between the airlines of the Comecon countries (see below). The Warsaw–Bratislava service facilitated cheap travel to Austria, as a bus waiting at Bratislava airport took passengers across the Czech border to Vienna for as little as five dollars. The CSA service led to a serious dispute between the supposedly fraternal LOT and CSA airlines, and was ultimately suspended in 1982.[27]

After a fairly static period in the 1980s, the LOT network underwent a dramatic change with the collapse of Eastern Europe's socialist states, and the crumbling of the Soviet Union in 1991. LOT now followed the example of western airlines and shaped its network and timetables as 'spokes' based on LOT's 'hub' at Okęcie Airport. The region east of Poland, covering an area of 22 400 square kilometres inhabited by 270 million people, offered an enormous potential market for air transport. Moreover the 1.5 million Poles living beyond Poland's eastern border could now be more easily targeted by LOT. Even before the Soviet Union's disintegration, the total number of international passengers from this region had reached 14 million (1990), 151 000 of whom travelled to Poland on LOT, against 100 000 on Aeroflot. The following year, with the end of the Soviet Union, this business collapsed when the artificial EAPT rates were abolished and replaced by IATA's dollar-based fares. As a result, LOT's share of passenger traffic dropped to a mere 34 000 and Aeroflot's to 72 000. For the Polish civil aviation authorities this was the cue to recapture the eastern market and expand LOT's presence there. Bilateral air transport agreements were signed accordingly with Belarus, Estonia, Lithuania, Latvia and Ukraine, while air transport agreements were concluded with Armenia, Azerbaijan, Georgia, Russia and Uzbekistan. In 1992 new services were opened, joining Warsaw with former Soviet cities like Vilnius, Kiev, Lvov and Minsk – all with sizeable Polish minorities. That LOT had correctly judged latent Polish patriotism among these communities, showed in the fact that these routes were the fastest growing of its eastern network; statistics for 1992 showed a 68 per cent increase in passenger embarkations on its eastern routes to 57 229.[28]

With new western aircraft entering service, LOT expanded its eastern network with a service to Riga in June 1993 and Tallin in July 1994, while also developing its traditional connections to Moscow and St Petersburg, in the expectation that the Russian market would be profitable. Accounting for 40 per cent of LOT's income from the eastern network, the Russian market provided by far the largest proportion of its revenue. Within that market LOT identified receipts from business traffic (25 per cent of the total in 1993, 30 per cent in 1994) as the most important segment. Operating results improved further after the introduction of business class seating on LOT's ATR-72s in the summer of 1994.[29]

Privatization and rising living standards gave LOT a chance of growth despite increased competition from carriers in the new republics. However, western airlines were also testing the market in the east and in the early 1990s their more effective marketing, better access to CRS and their ability to offer worldwide connections made the west Europeans formidable rivals; Lufthansa carried five times LOT's passenger volume on its eastern routes, Air France three times more and British Airways twice as many. Conflicts ensued. In 1993 air traffic between Poland and Britain was temporarily suspended because of difficulties over market access, while there were also disputes with KLM over the Dutch carrier's capacity between Warsaw and Amsterdam and its policy of feeding eastern traffic into its transatlantic services at Amsterdam.

These developments prompted LOT to put more effort into improving its share of the market from the former Soviet Union. Compared to the preceding year, LOT's profits from the eastern routes rose by 50 per cent in 1994, with the share of LOT in this market rising and the idea of a Warsaw hub materializing. Providing easy transfer in Warsaw for traffic from Belorussia, Lithuania and the Ukraine, to Europe, the USA, Canada and the Far East, now became an important part of LOT's business strategy.[30]

3.2 The airport environment: Okęcie

To make Warsaw a major air transport hub, LOT needed the full cooperation of the airport authorities in Warsaw. Their willingness to embark on a programme of rapid expansion and modernization at Okęcie was vital, because the airport had long been neglected.

After the Second World War, the increase in air transport resulted in the reconstruction of a number of airports across Poland. As early as March 1945 LOT was able to open branches in Lublin, Kraków and Rzeszów, followed by Poznań, Katowice, Olsztyn, Gdańsk, Wrocław and Szczecin before the year was out. Within the airport reconstruction programme, Warsaw received most attention. Okęcie airport, originally intended as a temporary facility, opened on 29 April 1946. For a period of 15 years, LOT was put in charge of its administration and operated its facilities, although these were also used by military aircraft. Ground handling was also entrusted to LOT. Conditions during these postwar years were extremely difficult and aircraft maintenance had to take place in the open air. At the end of the 1950s airport and airline operations were separated and LOT surrendered its jurisdiction over Okęcie to the Ministry of Transportation and Maritime Economy.

As was the case with LOT itself, the first *six-year plan* for Okęcie was not ready before 1950. It was part of the modernization and extension of airports under LOT's administration and was geared to facilitate the expansion of the airline. Two main runways were built, which were eventually extended to 3690 and 2800 meters. However with funds for airport improvement lacking, Okęcie's 'temporary' facilities continued to be used until 1969, when a new terminal building, designed to handle up to 700 000 passengers per year, was completed. The new terminal was a typical example of the ineptitude of Poland's central planning process, for in the year it opened, Okęcie already handled over 850 000 passengers and the number would nearly double over the next three years.[31] Not surprisingly a low standard of service characterized Okęcie throughout the 1980s, both with regard to airside procedures and passenger movements. Complicated customs proceedings made boarding time considerably longer than in the West. Nevertheless, traffic volume doubled between 1985 and 1989, although movements through Okęcie collapsed in 1990 with the disintegration of the centrally planned economy and the accompanying 30 per cent fall in Polish living standards. International passenger movements fell by 30 per cent, domestic movements by 80 per cent.[32]

Following the example of phased privatization of state enterprises in branches of the economy where competition was limited, airport operation was reorganized and entrusted to a separate public corporation by act of parliament on 23 October 1987. This was in line with the policy of the Ministry of Ownership Transformation to gradually return Polish Railways (PKP) and LOT to private ownership. It was, at the same time, intended to speed up technological improvements in air traffic control, an area in which Poland lagged behind both western and eastern European standards.[33] Thus Okęcie became a part of the Przedsiebiorstwo Państwowe Porty Lotnicze (PPL), a state-owned enterprise that operates the airport facilities on a leasehold basis. While LOT is the most important customer of PPL, providing 60 per cent of its annual turnover, PPL and LOT are separate enterprises. Cooperation is close and LOT and PPL jointly operate a ground-handling company called LOT Ground Services (LGS), which replaced direct ground handling by LOT in August 1992.

While LOT was undergoing restructuring and renewal of its fleet in the late 1980s, Okęcie was also modernized. The old 1969 terminal had become seriously overcrowded, handling three million passengers per year by the end of the 1980s, over four times the number for which it had been designed. In 1986 the government embarked on an expansion program for which large-scale investment was necessary; estimated costs ran up to DM 300 million. After tenders had been received from a number of international construction companies, a contract for the new terminal building, a cargo centre and a catering facility was awarded to the German Hochtief company in May 1990, even before all the financial aspects of the project were settled. The finance was to be raised by the airport itself, with additional assistance from Citibank in Frankfurt. 50 million ecus in funds for the airport's technical equipment was borrowed from the European Investment Bank in Luxembourg within the framework of a financial agreement of December 1992 called 'Poland – Modernization of the Airport in Warsaw'. The agreement required that PPL seek managerial assistance from a western airport organization and for this purpose the Irish Airports Authority Aer Rianta was chosen for a period up until the spring of 1995. The contract with the Irish included strategic advice on the drawing up of the Okęcie Airport business plan for 1993–1997.

On 17 June 1992 the new passenger terminal was opened, thus enabling LOT to expand its *hub* operations at Okęcie.[34] Terminal 1 was equipped with eight passenger-loading bridges and designed to have an annual capacity of 3.5 million passengers. With its striking triangular floor plan and lilac paintwork, the new terminal reflected LOT's raised expectations of service, comfort, and speed of handling. At the same time, Polish trade and tourism was also on the increase and Warsaw's new airport encouraged a sense of competition with neighbouring countries which were also changing their airports into regional air traffic centres. As a result Okęcie moved up to a median position in the 1994 survey of 30 European airports by European Data & Research Limited.

4 PERFORMANCE

The political circumstances under which LOT operated between 1945 and 1989 affected the operational performance of the airline and set it apart from its western counterparts. Nevertheless LOT's reorientation towards commercial operations preceded Poland's general political and economic transformation in 1989.

4.1 Operating and financial

Like other pre-war airlines, LOT did not manage to break even before 1940. For its financial survival, it depended on government subsidies, which amounted to 6.5 million złoty (US$1 million) in 1938, the last full year of operations. Total revenue stood at 28 per cent of total costs, a reasonable performance, comparable to that of Britain's Imperial Airways.[35]

In the postwar period, LOT became a service branch of the government. A network of domestic routes was established for transporting official passengers and mail, and was operated irrespective of the financial results.[36] Losses were covered by government subsidies. One of the few reminders of LOT's pre-war days, was its continued membership of IATA which was in itself peculiar. It signified LOT's reluctance to become isolated from the IATA rate-setting and clearing-house structures which it had joined in 1931. In practice, however, LOT flew only a few IATA routes in this period.

Despite the emphasis on five-year planning that was characteristic of the political economy of all socialist states that followed the Soviet model, a medium-term plan for LOT's development was not produced until 1950. The plan covered the period 1950 to 1956 and provided for short-haul domestic, and medium-haul international services along routes that served political interests.[37] However, a shortage of aircraft and the limited possibilities for international services between countries of the newly formed Comecon bloc meant that the plan had to be redrawn before the year was out. As a branch of government, LOT's activities were modelled on the Soviet Aeroflot airline. This meant that LOT's prime function was within Poland, and its attention was redirected to domestic services, the provision of special transport for large forestry projects, and other aerial activities like crop-dusting. International expansion came last in its list of priorities and accounted for no more that 35 per cent of LOT's total effort in 1955.

Table 7.2   LOT productivity, 1946–95

	Passengers (1000s)	Cargo (tonnes)	Available ton/km (000)	Revenue ton/km (000)	Pass. load factor	Overall load factor	Total Staff	Revenue ton/km prod. per employee
1946	54 489	368.1	2 537	1 397	65.9	55.1	–	–
1951	113 176	1 062.1	5 187	3 509	65.7	67.6	–	–
1956	199 565	3 333.3	10 054	8 350	79.6	82.5	1 424	5 864
1961	201 632	4 048.1	35 772	14 283	61.5	56.1	1 571	9 092
1966	493 717	9 813.1	69 549	35 544	52.7	54.3	2 756	12 897
1971	1 085 723	12 576.4	132 377	71 734	58.9	55.5	3 513	20 420
1976	1 560 773	21 860.3	330 306	174 495	61.4	53.1	4 830	36 127
1981	1 711 126	11 703.1	408 633	236 305	67.8	57.8	6 118	38 625
1986	1 810 524	9 427.1	458 291	285 291	72.3	62.3	6 132	46 525
1991	1 208 000	10 804.2	713 000	349 996	67.9	49.1	6 227	56 206
1995	1 839 000	17 100.0	898 400	488 000	69.6	54.3	4 161	117 280

Source: LOT Polish Airlines

After 1956 LOT was allowed to raise fares on its domestic routes to recoup more of its operational costs, and open new services to western Europe. As the majority of foreign airlines were not allowed to sell tickets in Poland, providing reciprocal air services became the main driving force behind LOT's international expansion.[38] To offset the higher domestic fares, seasonal rates were introduced in 1958. With increased awareness of costs, the organizational and economic structure of the airline was reorganized in 1958 to improve its financial results and the balance of LOT's activities was shifted towards the operation of the more profitable international routes. In 1960 these already accounted for 64 per cent of LOT's total capacity offering and 38 per cent of passenger numbers – a rapid expansion that brought problems in the field of aircraft procurement. In the first half of the 1960s, lack of suitable aircraft became an obstacle to further growth.

A more serious hindrance was the agreement by the aviation authorities of Bulgaria, Czechoslovakia, the German Democratic Republic, Hungary, Poland and Romania, at an international conference in Budapest in March 1963, to adopt a common system of passenger fares for citizens of the participating states on all services between them. With fares set at 50 per cent of the standard IATA rates, the agreement also provided for uniform prices for aircraft servicing. The system was introduced in April 1963 and named Jedinyi Awiacjonnyj Pasażirskij Tarif (Common Air Passenger Tariff), or EAPT. Uniform air cargo rates, known as Jedinyi Awiacjonnyj Gruzowoj Tarif (EAGT), were also agreed, although Poland did not sign this later convention. Regional cooperation between the six eastern European countries was organized in a multilateral forum, known as the '6-pool', which existed until the end of the 1980s. Poland's participation in the EAPT tariff structure was disastrous for LOT. Fares were dictated by socialist ideas about the international division of labour that underlay Comecon's price mechanism. Common rates also acted as a brake on competition between the various Comecon flag-carriers.[39] Each airline's targets were defined in terms of annual transport quotas set on the basis of national economic planning and results were measured in terms of transport performance only. Neither market share nor financial performance were considered to be relevant indicators of an airline's development.

The EAPT structure was fine-tuned in the so-called Berlin Agreement, signed by Aeroflot, CSA, Interflug, LOT, Malev, Tabso, Tarom, and the associated airlines of Mongolia, in the context of the Comecon Standing Committee on Air Transport in October 1965.[40] Under this agreement technical assistance had to be provided on an exchange basis between carriers, and was aimed at furthering cooperation between the participating partners in technical and trade matters, as well as in settlement of payments.[41] But if an East European counterpart of IATA was envisaged, this never materialized. Moreover, the calculation of air transport fares and rates bore little resemblance to the actual cost of providing services. Compared to west European and IATA fares, air tickets were extremely cheap.

In consequence, LOT's financial returns nose-dived to an operational loss of 270.7 million złoty in 1966, thereafter experiencing a gradual recovery. LOT continued to show a profit year after year, thanks to huge government subsidies that made the airline appear to be profitable, even though the Polish government returned considerable sums to the state treasury after the airline's accounts had been closed. In this fashion a flow of money rotated from the national budget to LOT and back again.[42]

With these low fares in place, air transport developed swiftly in the second half of the 1960s. LOT even expanded its activities through a blocked space arrangement with KLM for Polish passengers to transfer in Amsterdam to KLM's transatlantic services.[43] Cargo services were also developed, using the four-ton cargo capacity of the Ilyushin Il–18. East Berlin, Frankfurt, Amsterdam and Moscow served as LOT's cargo destinations for transporting agricultural products and smaller industrial components.[44] Economic reforms within the socialist countries and more stable East–West relations contributed to further growth. Nevertheless, the total number of passengers carried by LOT and the other Comecon airlines remained modest when compared to the West. In 1972 the total air travel of the seven Comecon countries stood at 4.8 million passengers.[45] By way of comparision, the three main traffic centres in northwestern Europe – London, Paris and Frankfurt – registered over ten times that number.

In Poland, the political and economic reforms under the new government of Edward Gierek, who had replaced the increasingly repressive Gomulka as party leader in 1970, contributed to LOT's increasing traffic, even though passengers remained predominantly businessmen and government officials. Not before 1981 were ordinary Polish citizens allowed to keep their passports at home. Access to one's own passport had to be applied for, and took at least six weeks, after which an equally long waiting period for visa applications could be expected. For this reason LOT had difficulty developing its home market. When the airline showed a profit in 1970 of 245.7 million złoty (209.8 million złoty in 1971) most of the money flowed back to the Treasury through taxation, although the airline was allowed to keep some portion for self-finance.[46] In the early 1970s the Polish economy boomed and the country's growth rate ranked among the third highest in the world.[47] In 1972 foreign air travel went up by 37 per cent although domestic air transport was in deep crisis. After a continuous period of rapid growth between 1966 and 1973 in which domestic passenger numbers increased more than threefold, the rapid rise in fuel prices that followed the Arab–Israeli conflict of 1973 forced LOT to raise its fares drastically. This resulted in a plunge in passenger numbers in 1974 and necessitated cost savings. The airline's dependency on subsidies to offset the inherent problems of operating a short-haul network increased, but because of the subsidies, LOT recovered 93 per cent of its costs on domestic routes by 1978. Domestic losses were cross-subsidized with profits from services to Western countries. These profits came from the large number of foreign passengers; in terms of foreign air transport per capita, Poland actually ranked last among the Comecon members with a mere 30 passengers per 1000 Polish inhabitants in 1978.[48]

The actual state of LOT's finances was hidden since its accounts were arranged in such a way that made it difficult to determine what was actually going on at the airline. Although, thanks to EAPT rates, a ticket from Warsaw to Moscow in 1989 cost the same as the taxi fare from Okęcie airport to the suburb of Ursus, LOT showed small profits throughout the 1980s (see Table 7.3), as more finance-based management techniques were introduced. LOT was allowed to earmark part of its profits for investment and other purposes such as the construction of employee housing. Low fares resulted in relatively high passenger numbers. In 1989 LOT carried over 2.3 million passengers, but the number decreased by 48 per cent over the next two years when the new national progressive income tax coincided with LOT's introduction of more economic fares.

Table 7.3 Financial performance of LOT, 1979–95

Year	Total revenue (million złt)	Net profit (million złt)	Net loss (million złt)	Net profitability (net profit as % of revenue)
1979	7 100	1 200	–	16.9
1980	9 000	1 900	–	20.9
1981	10 000	2 400	–	24.0
1982	8 600	500	–	5.8
1983	13 900	1 600	–	11.9
1984	22 900	3 700	–	16.2
1985	31 900	5 100	–	16.0
1986	40 000	4 900	–	12.1
1987	59 600	5 300	–	8.9
1988	114 500	10 900	–	9.5
1989	454 400	132 700	–	29.0
1990	2 663 000	–	7 800	−0.3
1991	3 671 100 ($344.7 mln)	–	470 900	−12.8
1992	4 946 600 ($356.9 mln)	–	84 100	−1.7
1993	6 623 300 ($360.3 mln)	42 000	–	0.6
1994	8 977 454 ($18.8 mln)	20 882 ($0.9 mln)	–	0.2
1995	10 268 760 ($38.8 mln)	60 168 ($2.5 mln)	–	0.6

Source: LOT Polish Airlines
* No realistic złoty/dollar conversions can be given prior to 1991.

This new course necessitated a radical business reorientation, which LOT managed to implement within a short period of time. Under its new director, Bronisław Klimaszewski, the airline's results improved from a 471 billion złoty ($44.2 million) net loss in 1991, to a 63 billion złoty ($3.4 million) profit in 1993, despite a 72 per cent deterioration of the exchange rate for the Polish currency, as well as replacement of LOT's Russian aircraft with the leased Boeings and ATRs that raised the airline's debts to US$ 500 million. At US$ 300 million the order for the nine 737s was the biggest in LOT's history. However the new aircraft brought enormous improvements in the airline's financial results.[49]

4.2 Labour productivity

A common characteristic of all eastern European airlines was their relatively high staffing levels and low productivity (see Table 7.2). The socialist conception of air transport as a branch of government and a public utility, meant that there were no restrictions on employment. Initially more cost-conscious than its Comecon neighbours, LOT managed to keep its staffing levels below those of comparably sized airlines like CSA and Interflug. Operating the turboprop Il-18 and An-24s instead of Tu-104 jets, LOT's staffing level remained relatively stable until the mid-1960s at about 60 per cent of CSA's. Nevertheless, when LOT introduced its own jets in the mid-1960s, it found that the increasing complexity of the aircraft, and the further growth in its services and fleet size (see Table 7.1), made an expansion of the workforce inevitable. It also led to an increase in the airline's bureaucracy that made effective control of LOT's staff numbers more difficult. By the mid-1970s, LOT's labour force stood at the same level as that of CSA, though it remained well below that of the East German Interflug.

In the following years, escalating staff numbers posed an increasing financial burden on the airline. With 7296 employees, LOT had the largest workforce of the eastern European carriers in 1990 and this presented a serious drain on the airline's resources at a time when LOT wanted to restructure its operations on the basis of costs and productivity.[50] At the beginning of 1991 drastic treatment could no longer be postponed and a broad staff reduction plan was initiated. As a result employee numbers fell by 42 per cent between 1991 and 1994. Staff cuts on this scale reflected LOT's changed outlook on the air transport business. In April 1993 a new organizational structure was introduced to simplify business procedures, facilitate the decision-making process and improve the flow of information within the company. In addition, a yield management strategy was implemented.[51] These changes preceded the adoption of western-style financial and accounting techniques in 1994 that aimed to improve cost control.[52] Combining changes in aircraft with changes in business mentality, LOT also discovered the advantage of broadening its marketing strategy from passengers to cargo transport, especially on long-haul routes. Having replaced its Soviet equipment with Boeings, LOT not only achieved a substantial reduction in operating costs, but also gained a surplus capacity on its passenger routes that could be marketed for cargo. This was especially attractive on routes to the USA. LOT's Boeing 767s could take up to seven tons of cargo next to their normal passenger load, compared to less than a single ton for the Tupolevs and Ilyushins. To market this capacity, cargo activities were concentrated in a new department in 1995: the Office for Cargo and Mail. It combined the efforts of the Operating Department, which serviced customers, the Marketing Department, which handled problems with trade, the Department of Tariffs and Regulations, which covered pricing, the Office of Rented Flights that dealt with charters, and the Finance Department, responsible for overall supervision.[53]

LOT was also successful in separating its ground handling from the core business of air transport, which reduced its financial obligations and provided a strong incentive to upgrade the quality of ground services. In July 1992 an independent subsidiary was created called LOT Ground Services (LGS). Starting capital stood at 50 million złotys (approximately US$3.6 million). Though PLL LOT initially held all shares in the new venture, 49 per cent of LGS was subsequently sold to Poland's national airport company PPL. LGS began operations in September 1992 under American management, made possible by a three-year contract with AMR, the parent company of American Airlines.

LOT was now under way towards strengthening its position among European airlines. Nevertheless, it was taking considerable risks in expanding its share of the market, both in its intensive utilization of aircraft and in its financial situation. Leasing its entire fleet put a burden on the airline's ability to produce profits. With LOT's partial privatization in train, the airline needed to reduce its annual debt payments and in November 1995 LOT announced it had concluded a refinancing agreement for its three leased Boeing 767s. A conglomerate of five Japanese banks had agreed to buy out the original American lessors and conclude a new ten-year leasing agreement for the aircraft. For the debt-laden LOT, this was an important step as it reduced annual dues by more than five million dollars. When the Polish parliament named a special panel to expedite LOT's privatization in August 1995, and even considered amending the 1991 legislation to enable the sale of the government's majority stake in the airline, LOT's attraction for investors began to grow.[54]

4.3 Marketing and service

Even before the political changes of 1989, LOT's approach to marketing was already more developed than most other Polish companies. Nevertheless, promotion of air transport had in the past not received the professional attention it deserved. LOT had relied on high-street campaigns that advertised its services with posters, as a result of which brand recognition of the LOT product was poorly developed.

LOT's first step was to initiate a large-scale advertising campaign in the Polish newspapers, *Rzeczpospolita* and *Gazeta Wyborcza*. Radio commercials were also undertaken, as was poster advertising, while trucks with LOT slogans were used to take the airline's crane symbol beyond the immediate environment of the airport and into the streets of the larger cities. Advertising concentrated on focusing the public's attention on the company's transformation, and stressed LOT's commitment to creating a *'dom poza domem'* (home away from home) for its passengers. It was necessary to break with the old image of LOT as an airline with old-fashioned aircraft and a fossilized organizational and operational system. For the more experienced traveller, a frequent-flyer programme was introduced in 1991 to boost customer loyalty.[55] This was necessary since LOT's share in the Polish air travel market had sunk below 50 per cent since 1989.[56] Taking its campaign one step further in 1994, LOT launched television commercials which focused on the dynamics of aviation and enticed first-time customers to board LOT aircraft with special promotional offers. At the same time the airline's long history was utilized in the slogan 'tradition is commitment'.

5 STRATEGY

Before 1989 most of LOT's strategic cooperation with other airlines had taken place within the framework of the '6-pool' that emerged when the EAPT and EAGT rates were introduced in 1963. At the beginning of the 1990s, LOT was faced with the need for a radical reorientation on the international market. If a relatively small carrier like LOT wished to survive in the increasingly deregulated and competitive international air transport environment, it would need to strengthen its position. To achieve this, LOT followed a twofold strategy.

5.1 Collaboration and competition

On the one hand, LOT embarked on various forms of cooperation with new airlines that emerged from the dust caused by the collapse of the old socialist order. Several partners were found as new air transport enterprises were created in the former republics of the USSR. Apart from the sale of surplus equipment on special terms, cooperation included training in return for tacit commercial agreements to transport the new airlines' passengers through Warsaw, where they continued on LOT's European and intercontinental flights. Consequently, LOT adjusted its timetables to provide attractive transfer connections in Warsaw. This cooperation provided LOT with improved access to the new national markets in the CIS (Commonwealth of Independent States).

LOT and its eastern partners applied common fares and rates for their flights, so that, for example, passengers might fly from Chicago to Warsaw according to the LOT tariff and then continue on Belavia to Minsk at the same rate. Services between Warsaw and Kiev, and between Warsaw and Lvov, were operated under a pool arrangement with Air Ukraine and LOT had its own blocked seats on Air Ukraine flights. Nonetheless, LOT's returns – and service – continued to suffer from restrictive capacity clauses in these bilateral arrangements. In the cooperation with Lithuania and Ukraine, equipment also became an issue. LOT's ATR turboprops were often too small for the traffic, but capacity limits prevented the Poles from using one of their Boeings. As elsewhere in the air transport business, capacity agreements and inter-airline deals on route schedules, determined the pace of market growth.

Against a background of hub-building in east European capitals, LOT cooperated with the Slovak carrier Tatra Air after the break-up of Czechoslovakia in 1993. In August 1994 Tatra was granted permission to open a new service between Bratislava and Warsaw to feed into LOT's European and intercontinental network. Operations started in January 1995 using Tatra's 35-seat Saab 340B turboprops, which provided speedy travel from Warsaw via Bratislava to Kosice.[57] On some European routes cooperative alliances were also started with Swissair, Austrian Airlines, Lufthansa and CSA.

5.2 Globalization and strategic alliances

Beyond regional cooperation with other east European carriers, LOT's directors realized they needed a strong strategic partnership that would serve the airline's global interests. Facing a deterioration of results and market share on its eastern routes, LOT looked around for a strategic alliance. Although the immediate aim of such a partnership was to attract foreign investment as LOT proceeded towards privatization, the wider goal was access to a well-developed international computer reservations system (CRS). This had become a crucial element in airline marketing and gaining access was imperative for airlines of the ex-Comecon countries seeking a position for themselves in international air transport. Two years after the CRS Amadeus and Galileo, owned by syndicates of the largest western airlines, had become operational in 1989, their market share in global ticket sales was already over 84 per cent.[58]

Another reason for acquiring an ally was the ongoing process of aviation liberalization within the European Community (EC). The benefit to LOT of first-hand experience of the business strategies and operational procedures of a major international carrier were clear. Informal approaches were made to no less than 18 foreign airlines, but only two, American Airlines and Delta, turned out to be interested in developing the kind of strategic partnership that LOT had in mind.

To determine the best partner for LOT, it was decided to use professional consultants American Bankers Trust. The result was a Memorandum of Understanding between PLL\ LOT SA and American Airlines, signed in Warsaw in May 1994.[59] The agreement provided for mutually rewarding cooperation, including code-sharing, coordinated scheduling of flights, cooperation in marketing (including linking their respective frequent-flyer programmes), and integrated ticketing of passenger and cargo traffic.[60] Of these, the agreement on code-sharing was the most important to LOT. It would also prove the most troublesome, since a code-sharing agreement meant that a travel agent's computer screen showed more LOT connections when arranging a booking than the airline actually offered. Code-sharing thus artificially expanded the scope of LOT's services and connections, and thus its potential market share. On the other hand, a general code-sharing agreement such as American Airlines wanted, extended beyond the bilateral relationship between Poland and the USA and held the danger of LOT becoming less visible in the CRS. It was therefore in the interest of LOT to limit the agreement to bilateral services. In the event, the United States Congress refused to approve the agreement between American Airlines and LOT on antitrust grounds, ruling that it would create a monopoly of the Polish-American air transport market to the exclusion of other carriers. Unless American Airline's competitors, such as the KLM/Northwest and Lufthansa/United alliances, were also allowed to enter the virtual reality of air travel booking between Poland and the USA, Congress would veto the deal. This left the Poles, for whom the code-share agreement was of overriding importance, with no option but to open up their market.

Cooperation in airport handling was easier to achieve. American Airlines permitted LOT to use its own terminal facilities in the USA, avoiding the inconvenience of having to go from one terminal to another when transferring between American and LOT flights. The consulting agency of American Airlines' parent company, AMR Services, made a thorough analysis of LOT connections in order to work out the scale of preferential charges and connections. In August 1995 AMR reported favourably on LOT's connections. Thus strengthened, LOT was well-poised to continue as one of eastern Europe's prominent airlines.

In conclusion, it is evident that LOT, as an airline from eastern Europe, suffered from basically the same business and policy problems that troubled the major carriers in the West. In the immediate postwar period there was a close link between the political aims of the government and the route structure that LOT was obliged to maintain. In the 1950s, the problem of reconciling government policy with the need to balance the books, was compounded by a struggle with technology and whether to chose turboprop or jet aircraft. LOT took longer to make the transition to jets than western European airlines and even its partners in the 6-pool. The reason it took so long appears to be rooted in LOT's operational and business strategy, not in government involvement. The EAPT/EAGT fares agreed in 1963 made it difficult for LOT to maintain a semblance of commercial rationale in its operations. In addition, it had to struggle with a stagnant economy in Poland and in the Comecon countries, once it had completed the transition to jets in the 1970s. The change increased LOT's dependency on government subsidies and problems were further aggravated as it discovered it operated less economical aircraft than its western counterparts. In the second half of the 1980s this led to the decision to adopt western aircraft. Thus by the time of the upheval to Poland's political and economic system in 1989, LOT was already realigning its operations toward western business practice. Combining creative leasing arrangements with a general restructuring and a reduction in the workforce, LOT managed to achieve profitability in 1993. The airline's new commercial course was emphasized further after the decision in 1991 to put LOT on the road to privatization.

NOTES

1. Alexandre Herlea, 'The first transcontinental airline: Franco-Roumaine, 1920–1925', in: *From Airship to Airbus: the history of civil and commercial aviation*, vol. 2, William F. Trimble (ed.), *Pioneers and Operations* (Washington DC, 1995), pp. 56–7.
2. Mieczysław Mikulski, Andrzej Glass, *Polski Transport Lotniczy* (Warszawa, 1980), pp. 50–81.
3. Mieczysław Mikulski, 'Komunikacja Lotnicza w Polsce w Latach 1919–1939', in PLL LOT, *LOT Wczoraj Dziś i Jutro* (Warszawa, 1979), pp. 7–32.
4. Mieczysław Mikulski, Andrzej Glass, *Polski Transport Lotniczy* (Warszawa, 1980), pp. 149–88.
5. Peter Calvocoressi, *World Politics since 1945* (London, 1989), p. 135. R.J. Crampton, *Eastern Europe in the Twentieth Century* (London, 1994), pp. 283–7.
6. Marek Żylicz, *International Air Transport Law*, (Dordrecht, 1992) [Utrecht Studies in Air and Space Law, vol. 12] p. 4.
7. Eugeniusz Sobecki (ed.), *Podstawowe Informacje dotyczce Lotnictwa Cywilnego* (Warszawa, 1994), pp. 95–114.
8. Interview with Zbigniew Kiszczak (chairman of the supervisory board) on the problems of privatization, in *LOT Żurawie* no. 15/95, pp. 2–3. Internationale Nederlanden Groep ING Bank NV, *Memorandum Informacyjne Polskie Linie Lotnicze LOT SA: Program Emisji Komercyjnych Weksli Inwestycyjno-Terminowych K.W.I.T.* (Marzec, 1996).

9.   Marc Dierikx, Bermuda bias: 'Substantial ownership and effective control 45 years on', *Air Law* 16(1991) no. 3, pp. 118–24.
10. Zbigniew Kiszczak, 'LOT na Prywatyzacyjnej Ścieżce', *LOT Żurawie* 21/95, p. 7.
11. LOT information on the *Internet*: 'LOT goes public' (http://www.poland.net/LOT/goes.html) 15 April 1996.
12. Jerzy Osiński, Henryk Żwirko, *Biuletyn Informacyjny Lotnictwa Cywilnego – 50 Lat PLL LOT* (Warszawa, 1980), pp. 1–10.
13. Genrikh V. Novozhilov, 'The design of military and passenger aircraft in Russia', in *From Airship to Airbus*, vol. 2, William F. Trimble (ed.), *Pioneers and Operations* (Washington DC, 1995), p. 199.
14. LOT, 'Tabor Ilość według typów 1945–1980' (statistical summary). Kenneth Munson, *Kleine luchtvaart encyclopdie: Verkeersvliegtuigen sinds 1946* (Amsterdam, 1968), pp. 118–119.
15. For a comparison of the technical and economic data of the Tu-104 and the Il-18, see Novozhilov, pp. 203–5.
16. Kazimierz Szumielewicz, *Polskie Linie Lotnicze S.A.* (Warszawa: PLL LOT Zakładowy Ośrodek Informacji Technicznej i Ekonomicznej, 1996), pp. 5–8.
17. Żylicz, *International Air Transport Law*, p. 4.
18. Kneifel, *Fluggesellschaften und Luftverkehrssysteme der sozialistischen Staaten*, pp. 51, 178.
19. Żylicz, *International Air Transport Law*, p. 40.
20. Krysztof Rutkowski, 'Eastern European air transportation: paths to an integrated system of European air transportation', in: *Selected Proceedings of the Sixth World Conference on Transport Research, Lyon 1992*, vol. 1, *Land use, development and globalisation* (St.Just-la-Pendue: WCTR, 1993), pp. 431–42 (p. 434).
21. Piotr Głażewski, 'Leasing Bez Precedensu', *LOT Żurawie* 7/95, pp. 3–4.
22. Barbara Zamow, 'Nowy Boeing 767 w Barnach LOT-u', *Gazeta Transportowa* no. 21/95, pp. 1, 3.
23. Polish Airlines LOT S.A., *Annual Report 1993*.
24. Research data based on analysis of the biannual *ABC World Airline Guides* underlying Marc Dierikx, Peter Lyth, 'Le Développement du réseau européen de transport aérien, 1920–1970: un modèle explicatif', in: Michèle Merger, Albert Carreras, Andrea Guintini (ed.), *Le réseaux européens transnationaux XIXe–XXe siècles: quels enjeux?* (Nantes, 1995), pp. 133–57.
25. Kneifel, *Fluggesellschaften und Luftverkehrssysteme der sozialistischen Staaten*, pp. 50–1. Jerzy Osiński, *Sytuacja Ekonomiczna Transportu Lotniczego 1960–1970* (Warszawa, 1972).
26. Jerzy Woydyłło, 'LOT-em na Wschód', *LOT Żurawie* no. 6/95, pp. 1–3.
27. Jerzy Osiński, Henryk Żwirko, *Biuletyn Informacyjny Lotnictwa Cywilnego – 50 Lat PLL LOT* (Warszawa, 1980), p. 1–10.
28. PLL LOT, 1992 breakdown of scheduled air traffic.
29. PLL LOT, *Annual Report 1994*, p. 11.
30. PLL LOT, *Annual Report 1994*, p. 10.
31. Przewozy w Centralnym Porcie Lotniczym Okecie w latach 1945–1978 (table), in J. Osinski, H. Żwirko, *Raport informacyjny lotnictwa cywilnego. 50 Lat PLL LOT. Wydanie Specjalne 20/80* (Warszawa, 1980), p. 201.
32. Ryszard Stawrowski, Nowy MDL Warszawa Okęcie, in: *Przegląd Komunikacyjny* no. 6/92, p. 21–22. PPL, *Case study: Port Lotniczy Warszawa w dobie przemian ekonomicznych* (undated, 1992).
33. Cf.: Irena Hajduk, Joanna Filipczyk, *The Polish transportation system in integrated Europe* (European University Institute Colloquium Paper DOC.IUE 131/1993 (Col. 28).

34. Jadwiga Sławińska, *Działalność Lotnisk Komunikacyjnych w Polsce w Latach 1991–1995* (Warszawa, 1996), pp. 3–12, 18–20.

35. Comparison based on data published in *L'Aérophile* 46 (1938) no.7, p. 138, and on O.J. Lissitzyn, *International Air Transport and National Policy* (New York, 1942), p. 192.

36. Żylicz, *International Air Transport Law* p. 4.

37. Ibid.

38. Żylicz, p. 4.

39. Biuletyn Informacyjny Lotnictwa Cywilnego, *50 lat PLL LOT. Wydanie Specjalne 20/80* (Warszawa, 1978).

40. Jerzy Osiński, 'Umowa Berlińska o Współpracy Przedsiębiorstw Przewozu Lotniczego Krajów Członkowskich RWPG', *Biuletyn Informacyjny Lotnictwa Cywilnego* 19/65 (Warszawa, 1965).

41. *Biuletyn Informacyjny Lotnictwa Cywilnego* (Warszawa, 20 December 1965). J.L. Kneifel, *Fluggesellschaften und Luftverkehrssysteme der sozialistischen Staaten: UdSSR, Polen, CSSR, Ungarn, Bulgarien, Rumänien, Kuba, Jugoslawien und der VR China* (Nördlingen, 1980), pp. 25–7.

42. Akumulacja Finansowa i Jej Podział, 1965–1971 (table), in: *Transport Lotniczy 1971* (Warszawa, 1971), p. 8.

43. Żylicz, *International air transport law*, p. 7.

44. Kneifel, *Fluggesellschaften und Luftverkehrssysteme der sozialistischen Staaten*, p. 202–3.

45. Zbigniew Landau, V clav Prucha, 'The rise, operation and decay of centrally planned economies in Central-Eastern and South-Eastern Europe after World War II, in: Václav Prucha (ed.), *The System of Centrally Planned Economies in Central-Eastern and South-Eastern Europe after World War II and the Causes of its Decay* (Prague, 1994), pp. 9–37. J.L. Kneifel, *Fluggesellschaften und Luftverkehrssysteme der sozialistischen Staaten: UdSSR, Polen, CSSR, Ungarn, Bulgarien, Rumänien, Kuba, Jugoslavien und der VR China* (Nördlingen, 1980), pp. 246–7.

46. Akumulacja Finansowa i Jej Podział, 1965–1971 (table), in: *Transport Lotniczy 1971* (Warszawa, 1971), p. 8. In 1969 LOT was allowed to keep 92.1 per cent of its 361.7 million złoty 'profit' (brought about, in part, by a 153.8 million złoty subsidy), while in 1970 all but 17.6 per cent (43.2 million złoty) went back to the treasury (against 51.2 per cent, or 107.4 million złoty, in 1971). In *Fluggesellschaften und Luftverkehrssysteme der sozialistischen Staaten*, Kneiffel quotes entirely different profitability figures for 1970: a profit of 160 million złoty, which, he claims, the airline was allowed to keep for self-financing purposes (pp. 51, 116).

47. Calvocoressi, *World Politics since 1945*, p. 145. Crampton, *Eastern Europe in the Twentieth Century*, p. 360.

48. By contrast, Hungary had 117.8 international travellers per 1000 inhabitants, while Bulgaria counted 116.8. The GDR had 69; Czechoslovakia 50.7; Romania 37.7. Osiński and Żwirko, *Biuletyn Informacyjny Lotnictwa Cywilnego*, p. 113.

49. Interview with Jan Litwiński and Zbigniew Kiszczak, Rachunkiem Ekonomicznym posługujemy się coraz skuteczniej, in: *LOT Żurawie* 15/94.

50. David Woolley, 'Eastern European airlines. Changing fleets and philosophies for a world without the wall', *EXXON Air World* 43 (1991) no. 3, 4–7.

51. Polish Airlines LOT S.A., *Annual Report 1993*.

52. Polish Airlines LOT S.A., *Annual Report 1994*.

53. Interview with Krzysztof Ziębicki (LOT's head of marketing), Dobrej Opinii Kupić nie można, in: *Gazeta Transportowa* 21/95, p. 1, 3. *LOT Żurawie* (15/10/1995).

54. Financial information provided by LOT Polish Airlines (undated). LOT information on the *Internet*: LOT Goes Public, 15/04/1996 (http://www.poland.net/LOT/goes.html).
55. Jerzy Woydyłło, Kampania z Rozmachem, in: *LOT Żurawie* 12/95, pp. 1–2.
56. Polish Airlines LOT S.A., *Annual Report 1994*.
57. Jerzy Woydyłło, 'Powitanie na Okęciu', *LOT Żurawie* 7/95, pp. 5–6. Interview with Roman Stoličny (commercial manager of Tatra), in: ibid, p. 6.
58. EEC Commission case no. IV/34.632 *Combination of the Galileo and Covia CRSs: notice pursuant to Article 19(3) of Regulation no. 17.*
59. Jerzy Woydyłło, 'Prosto z Dallas', *LOT Żurawie* 7/95, pp. 1–2.
60. LOT *Żurawie*, 01/04/1995.

8 Airlines, Entrepreneurs and Bureaucrats: the American Experience
Roger E. Bilstein

The story of postwar developments in the United States airline service reflects trends that had become basic to the evolution of air transport during the pre-war era. Active participation by federal agencies launched the airmail service that spawned the first airlines in the 1920s, and federal regulations as well as airmail subsidies evolved as essential features for continuing development. Technological achievements in the 1930s created a crucial legacy for successful postwar airliners. The war itself played a key role in accelerating the role of air transport. While it is true that a fascinating cast of characters filled the roles of bold entrepreneurs before and after the war, the framework of federal programs and legislation represented a continuing pattern of influence and interaction.

Traditionally, airlines in the US were categorized by extent of service offered, revenues, and types of aircraft. For many years after World War II, the largest companies were called *trunk carriers* and comprised the domestic airlines known as the Big Four – American Airlines , Eastern, TWA, and United – along with Pan American Airways as the international star. In 1981 statisticians acknowledged that several companies equipped with smaller aircraft and generally shorter but densely populated routes, generated revenues and passenger miles that entitled them to higher recognition. A redefined category of *major airlines* included those with revenues of US$1 billion or more annually in scheduled service. Dynamic regional carriers like Alaska Airlines and Southwest Airlines had become uniquely successful with revenues that compared to much larger carriers. Consequently, the line-up of major airlines in 1994 included Alaska, America West, American, Continental, Delta, Northwest, Southwest, TWA, United, and USAir. A pair of all-cargo airlines also qualified on the basis of revenue: Federal Express (FedEx) and United Parcel Service (UPS).

A second tier, the *nationals* represented airlines with revenues of $100 million to $1 billion. Although some offered long-haul flights and a few international services, the nationals tended to operate within a particular geographic area. Among the nationals were several airlines who operated as *local service* and *supplemental* (non-scheduled charters) carriers in the era prior to deregulation.

The *regionals* represented the third tier, reporting revenues of $20 million to $100 million. Their services remained largely tied to the carriage of travelers between major cities and smaller towns within the same region. Equipment ranged from planes with 60 seats or more down to transports of 30 seats or less. Lastly there were a variety of cargo carriers, such as Emery, FedEx, UPS, and others, licensed to carry cargo only. The largest cargo operators in the 1990s began as small package carriers that also specialized in overnight delivery of documents. They became integrated cargo carriers, selling door-to-door service by combining traditional airline schedules with the services of a freight forwarder.[1]

Within the confinement of a single chapter, this essay focuses on selected airlines and selected individuals as characteristic of broader trends for the industry as a whole. The preceding chapters have highlighted major developments of the European experience, revealing similarities as well as differences. The American story also reveals some similarities with its European counterparts, although the differences are often striking. One major factor is the sheer size of the American domestic market, representing 3.5 million square miles of territory and a population of more than 250 million people. This king-sized market is served by equally king-sized airlines. In 1994, with foreign routes added, the revenue passenger miles (RPM) for United Airlines totaled 108 million; American Airlines, 98 million; Delta 86 million; and Northwest 57 million. By comparison British Airways, the world's fifth largest – and Europe's biggest – reported 53 million RPM; trailed by Lufthansa with 35 million; Air France, 31 million; KLM, 25 million and Alitalia, 18 million. The top four U.S. airlines therefore represented more than double the total of Europe's top five carriers.[2]

Historically, intense competition for market share has engaged the American operators. This competitiveness had much to do with the way America's airlines evolved, the types of aircraft they ordered, their aggressive marketing and their relationship with the federal government.

1 POLICY

1.1 Origins

For several years after World War I, a number of pioneering air transport companies appeared and vanished in the United States. Although the Post Office Department successfully developed the US Air Mail service, constructing a navigational system and attracting an increasingly air-minded clientele, it remained a government monopoly. Some independent companies focused on passenger travel, but failed due to the lack of suitable aircraft and sufficient subsidies to provide a financial cushion and dependable cash flow. By the mid-1920s, federal legislation had changed the entire picture of American aviation.

Traditionally, the US Post Office used private corporations like the railroads for the transport of mail. Having established the feasibility of airmail service and built a clientele, postal authorities and Congressional leaders crafted the Air Mail Act of 1925. Under contract to the Post Office, private airlines took over airmail operations and these pioneers formed the basis of larger companies that became major US airlines, such as American, Eastern, TWA, United etc. At the same time, a high-level presidential commission submitted its recommendation for federal legislation to promote the development of civil aviation. As a result, the Air Commerce Act of 1926 established an Aeronautics Branch within the Department of Commerce. The new office expanded existing navigational systems and developed new ones, monitored airfield operations, and evolved regulations to certify both aircraft and the pilots who flew them.

All of this enhanced the willingness of insurance companies and lenders to participate in the rapid developments that followed. Individualistic entrepreneurs and builders played a key role in early American aviation progress, but the legislation of federal support in 1925–6 was fundamental. Moreover, the research activity of the National Advisory Committee for Aeronautics and continued military evolution of aircraft made significant contributions to the technology of flight and the increasing capability of civil aircraft for air transport service.[3]

Subsequent legislation during the early 1930s led to a more viable system of independent airlines and the encouragement of larger aircraft more suitable for passenger travel. The Aeronautics Branch of the Department of Commerce deployed radio navigation systems and formalized protocols for an air traffic control network. A major legislative milestone occurred in 1938 with the Civil Aeronautics Act which merged the Aeronautics Branch and several other departments into one entity, the Civil Aeronautics Authority (CAA). The idea was to preserve order in the industry, protecting carriers from destructive competition, and setting rates at levels which were fair to consumers while providing a reasonable return for the airlines. The CAA held powerful mandates over airmail rates as well as passenger ticket prices. Additionally, the CAA determined specific routes, interline agreements, and mergers. A separate agency known as the Air Safety Board covered safety issues and investigated accidents, but was transferred to the CAA in 1940, when new legislation renamed it as the Civil Aeronautics Board, or CAB. The CAB was to play a major role in airline development for the next three decades.[4]

While the CAB and its successors clearly influenced much of the US airline experience, the government did not control the industry. Competition remained fierce, and represented a key factor that not only contributed to the evolution of outstandingly successful airliners, but also shaped the operations of American airlines.[5] In contrast to many European experiences, no single US airline enjoyed a monopoly, received subsidies to offset losses, or was forced to accept particular aircraft for nationalistic reasons.[6]

1.2 Government Aims

There were further legislative changes in the postwar years. The proliferation of airline movements triggered amendments to flight rules and air traffic control around congested airports, especially at night and during bad weather. The introduction of radar in the postwar era helped, but airliners beyond the immediate jurisdiction of airfields retained considerable freedom of movement. In any case, the unprecedented expansion of air travel and the increased congestion of airliners in airspace around major hubs raised the spectre of airliners colliding in the sky.

In fact one of the first major disasters occurred in daylight in comparatively open airspace. In 1956, a United DC-7 and TWA Constellation, both eastbound, took the same flight path at the same altitude over the Grand Canyon in Arizona and collided at 21 000 feet, killing all 128 passengers aboard. Following the Grand Canyon crash, public opinion about airline safety in the US reached a strident level, demanding an overhaul of the responsible federal agencies. A series of Congressional committees concluded that the CAA was buried too far down in the Department of Commerce to have an effective voice in obtaining funding for essential services. Dedicated professionals in the CAA strove to sustain efficient and alert administration of the country's airways, but cronyism and indifference within the Department of Commerce often frustrated their efforts. By 1958, a new Federal Aviation Agency (FAA) was in place, and functioned independently of the Department of Commerce. As chief of the FAA, President Eisenhower appointed Elwood P. Quesada, an ex-US Air Force general with several years of experience as an executive in the aerospace industry and on federal commissions. The Grand Canyon incident also mobilized the CAA to order long-range radar equipment and institute new procedures for surveillance and separation of air traffic. Under General Quesada, the FAA expanded its control systems to encompass the entire national airspace.

Subsequent legislation created the Department of Transportation in 1967. Although alert to the dangers of smothering the initiative of national aviation authorities, Congress nonetheless inserted aviation into the new department. While the FAA's acronym remained the same, its amended status resulted in a change of name to Federal Aviation Administration. The FAA retained broad responsibility for the operation of a nationwide air traffic control system, to assure safe separation of civil and commercial aircraft in all phases of flight. Further, the FAA either strengthened or acquired jurisdiction over a broad range of other aviation safety activities, such as the certification of civil aircraft designs, pilot training and mainten- ance procedures. The CAB meanwhile retained jurisdiction over economic aspects of aviation, such as airline routes and air fares.

With the 1967 legislation another new entity appeared with the National Transportation Safety Board (NTSB). Many aviation experts felt that the CAB, which had responsibility for airline accident investigations, should not be allowed to police itself. Accordingly, the NTSB became responsible for the investigation of all transportation accidents, including civil aviation. Originally an autonomous agency within the Department of Transportation, the NTSB became a completely independent federal agency with passage of the Transportation Act of 1974. NTSB investigations had two goals. One was to determine the cause of an accident, and the other was to make recom- mendations for future safety.[7]

All of this resulted in an enviable safety record. During the 1960s, airline experts pointed out that airline travel in the United States had not only surpassed rail travel in safety, but also represented a safer mode of travel than driving in one's car.[8] However no system is infallible and in May, 1996, a low-fare airline named ValuJet lost a DC-9 over Florida and 110 people died. Subsequent investigation uncovered such serious dis- crepancies in maintenance procedures that the FAA grounded the airline. However in July 1996, the crash of a TWA Boeing 747 over Long Island, New York, killed 230 passengers and, at the time of writing, remains un- explained.[9]

1.3 Attitudes towards regulation

In many ways, the beginning of the end for the CAB occurred when Dwight D. Eisenhower became President in 1952. The new administration was keen to enact a number of reorganization plans that would change the face of a federal bureaucracy shaped by Democratic policies since 1933 when Franklin Roosevelt took office. Although the CAB's stewardship of America's air routes had arguably been a success, and it had agreed to fare increases to cover the introduction of jet airliners, operators became increasingly hostile to the CAB's power over their finances. This friction heated up during the 1970s, as fuel prices rose by more than 200 per cent and labor costs climbed more than 130 per cent. Moreover, the integration of wide-body jets into airline fleets put new pressures on load factors and declining profit margins. Critics charged that the CAB responded too slowly to change in the industry, that the CAB's point-to-point route structure had led to high operating costs, and that it ought to be abolished. Even though the CAB permitted carriers to increase fares, restrict new services that might compete, and limit capacity on major routes so as to develop more profitable load factors, such actions did little to help the airlines' problems. In fact, the traveling public now became incensed at higher fares and frequent problems in finding an available seat.

By 1974, when Gerald Ford inherited the Presidency from Richard Nixon, the sentiment for deregulation was not only growing, but engaged bipartisan support. In the US Senate, Democratic Senator Edward Kennedy chaired the hearings of a Subcommittee on Administrative Practice and Procedure which concluded that airline prices would fall if federal control of competition was ended. Newspaper editorials across the country complained that government regulations were too burdensome on American industry and contributed to inflation. The same issues applied to the airline business. By this time, the CAB itself had reached a similar conclusion. In 1975, a special CAB study concluded that the airline industry was 'naturally competitive, not monopolistic', and stated that the CAB could no longer justify its existence. Acting on its own, the Board began to allow more latitude to airlines to choose destinations and establish fares.[10]

It was in this atmosphere that President Jimmy Carter took office in 1977. Deregulation appealed to Carter and he named Alfred E. Kahn as head of the CAB. A well-known professor of economics from Cornell University, Kahn's publications about federal regulation and the dynamics of business competition made him a respected figure in the corporate world as well as among academic colleagues. During his brief but dramatic tenure as head of the CAB, he relaxed fare structures, allowing the evolution of genuine bargain pricing such as the 'Peanuts Fare' of Texas International and 'Super Saver' promotions from American Airlines.

Other significant changes in traditional CAB attitudes included a relaxation of rules governing international carriers. US domestic routes were opened to foreign flag-carriers that made two stops in the United States. Kahn also encouraged American domestic carriers to enter the transatlantic market and cooperated with British authorities to accommodate Freddie Laker's low-fare Skytrain service on the important London–New York route. Finally in 1978 Congress passed the landmark Airline Deregulation Act, which eliminated restrictions on domestic routes and schedules and abolished federal controls over rates, although implementation was spread over a period of several years. In the event the CAB moved more quickly, so that by 1 January 1985, it had voted itself out of existence. Remaining elements of federal responsibility for airline service passed to the Department of Transportation (DOT).

Following the CAB's demise, the DOT became more involved in international agreements and became responsible for the final decision as to which US airlines would be granted the right to operate over international routes.

In the case of airline mergers, the government played a continuing role through the DOT and retained the right to determine whether a merger might contribute to reduced service and competition, thus compromising the best interests of the traveling public. The FAA, through the DOT, continued to hold responsibility for airline safety.[11] The DOT also acted to protect employees affected by mergers in terms of seniority lists for flight crews, as well as severance and relocation pay for other workers. Because many airlines abandoned services to smaller communities after deregulation, the DOT had responsibility for dealing with this problem. Through the Essential Air Service Program, it provided partial subsidies to carriers willing to maintain service on these smaller, less profitable routes.[12]

2 AIRCRAFT PROCUREMENT

With many European airlines functioning as nationalized operations, it was not unusual that their governments expected them to use locally produced aircraft. However pressure of competition, particularly on transoceanic routes, often led to the utilization of American equipment. In the US, airlines generally favoured American aircraft as they had better performance than European counterparts. Much of the airlines' success rested on the wings of the redoubtable Douglas DC-3, which entered domestic service in 1936 and became a global workhorse during World War II as the C-47. In addition, the US pushed rapidly ahead with development of a new series of long-range, four-engine passenger aircraft.

The notable features of postwar airliners first appeared in American airliners of the late 1930s and early 1940s. The competition of Sikorsky, Martin, and Boeing to sell flying boats, resulted in advanced structures that contributed to more cost-effective and profit-making airliners.[13] Pressurized passenger cabins also played a significant role. In 1937 the US Army tested a pressurized version of the Lockheed 12 airliner as part of its interest in high altitude bombers. In fact the commercial side of the business led the way with pressurized cabins, represented by the Boeing Model 307, which entered service in 1940.[14] In 1938 Douglas rolled out the experimental version of a new 52-passenger airliner called the DC-4E. The production version, called the DC-4, carried 42 passengers, and included many of its predecessor's features such as the tricycle landing gear. While the DC-4 remained unpressurized, the tricycle landing gear, well-equipped galley, and general spaciousness became synonymous with postwar travel. By the time the prototype became airborne in 1942, the US was at war and dozens of these modern planes went into military service as the C-54 transport. After 1945, it became standard equipment in America as well as abroad.

Similarly path-breaking was the Lockheed Constellation. Design work for this aircraft began in 1939 as a 40-passenger airliner based on requirements submitted by TWA. The plane had an improved pressurization system, four powerful engines, modern tricycle landing gear, carefully engineered seating, and several other advanced features. First flown in January 1943, it was also drafted for immediate military service – as the C-69 – although only 15 had been delivered by the end of the war. Military orders were converted to civilian contracts for the L-049 Constellation that could carry 43–49 passengers, depending on the route length and extra fuel requirements. TWA and Pan Am became the first customers; Pan Am introduced the Constellation into service on a New York–Bermuda trip early in 1946, quickly followed by a TWA service from the US to France. TWA also introduced the fast Constellations into its transcontinental services across the US, where its pressurized design contributed to a typical cruising speed of 313 mph, considerably faster than the 247 mph of the DC-4.[15]

In addition to outstanding aircraft, postwar American airlines benefited from the wartime experience of a qualified pool of air transport personnel. At the end of the war, the Air Transport Command (ATC) of the United States Air Force numbered 3090 planes and 313 000 men and women, including 104 000 civilians. The ATC had become the largest air transport service in the world, with routes that linked every continent on the globe. It was directed by Brigadier-General Cyrus R. Smith, who only four years earlier had been president of American Airlines. Although Smith reported to a senior officer, and military personnel dominated the ATC's roster of pilot and managerial positions, a considerable number of airline professionals from different companies entered service along with Smith. They organized schedules for military transports throughout the US, and created a network of pioneering routes around the globe. In many ways, they represented the most skilled group of air transport pilots, managers, and technicians available anywhere.[16] And in 1945, along with military veterans of the US Air Force, they were preparing to re-enter American civilian life. This legacy of experience is an important reason for the significant growth of America's postwar airline industry. Moreover, hundreds of war-surplus military transports like the C-47 and C-54 were available for civil service as the DC-3 and DC-4, planes which US personnel already knew inside out as the result of wartime duties.

During C.R. Smith's postwar tenure at American Airlines, its fleet not only included a number of Douglas products, but also a variety of others. Bill Littlewood, American's chief engineer, especially wanted a plane to succeed the unpressurized DC-4 on long transcontinental routes, where the pressurized Constellations in TWA's fleet were definitely setting the pace. The DC-3 seemed the earliest candidate for replacement. Smith trusted Littlewood, whose attention fastened on a Consolidated-Vultee proposal known as the

CV-110. Littlewood kept close tabs on the twin-engined aircraft's evaluation, making several suggestions that led to design changes. One of these involved the first foldable stairway built into an airliner as an integral piece of equipment. Larger airports had movable cabin stairs to move into place, but such equipment was nonexistent at many of the smaller airports where the twin-engine CV-110 would be calling. Littlewood's encouragement for the integral stairway made the plane much more versatile. Meanwhile Consolidated-Vultee was bought by General Dynamics and became its new Convair division. The CV-110 became known as the CV-240, with the last two digits denoting the number of seats in the plane. Littlewood apparently had a hand in finalizing the plane's designation as well, although in the end it was performance, not proprietary feeling, that led American Airlines to order the plane. The CV-240 was one of the earliest twin-engined designs to have a pressurized cabin, unlike its rival, the Martin 202. Littlewood and Smith both liked the idea of having a plane capable of flying over rough weather to keep schedules intact, especially on certain shorter routes. In 1946, American Airlines signed an US$18 million contract to launch the CV-240 the genesis of which showed the influence of potential customers on an aircraft's design.[17]

The desire for better equipment also prompted Smith and American Airlines to push the development of a large, four-engined pressurized transport to replace the DC-4. As usual, the driving force was competition. On the prestigious transcontinental route, TWA's Constellations easily overtook the slower DC-4s of American Airlines and United Air Lines. Although American and United were rivals over many domestic routes, they both needed an aircraft to contest TWA's fast Constellation. Competition made cautious bedfellows, as C.R. Smith and United's William A. Patterson agreed to join forces in urging Douglas to proceed with the pressurized DC-6. The early DC-6 models not only matched the Constellations in speed, but also led the way in air-conditioned passenger cabins. Part of this was due to C.R. Smith's stubborn adherence to shiny, bare-metal airliners, an American Airlines tradition. Although tests had proven that painting the top of the airliner's fuselage would reflect solar rays and cool the interior, Smith stubbornly refused to follow United's lead in this practice. He preferred to keep shiny metal as an American Airlines hallmark, adding air-conditioning to achieve passenger comfort.

While the CV-240s populated American Airlines' shorter routes, the DC-6 and remaining DC-4s equipped the premier, long distance services. However, improved versions of TWA's Constellations maintained an edge in speed and could also fly nonstop on eastbound journeys. Smith chafed at this advantage of his rival and he encouraged Douglas into building the bigger and faster DC-7. The larger plane would allow expansion of coach-class traffic and compete more effectively with new non-scheduled operators.

The DC-7 would take over on the important coast-to-coast routes, thereby releasing older equipment to compete in coach-class markets elsewhere. The Douglas engineers wavered. They felt that jets were just over the horizon and saw little need for another new piston-engined airliner. However Smith persisted, hinting that he might abandon his long association with Douglas and order a plane from someone else. Grudgingly, Douglas capitulated and the DC-7 became operational in 1953, the first airliner capable of flying nonstop on westbound routes against prevailing winds. American put it into service amidst elaborate publicity.

Loyalty to one manufacturer never became a ruling factor for American Airlines. When they decided that they needed turboprop aircraft in the mid-1950s, Lockheed proposed its L-188 Electra, based on the C–130 military transport. The final Electra configuration, with a low-wing arrangement, convinced American to order it off the drawing board in 1955. It became operational in 1959, although a series of structural failures eventually shortened its life-span with American.[18] When the first Douglas DC-8 jet airliners entered service in 1959 with United and Delta, American opted instead for Boeing 707s, but looking for a bigger, faster jet, American also ordered the Convair 990 in 1961. The Convair order reflected Smith's interest in aircraft that were faster than American's rivals, although problems caused the airline to discard the 990 before long. Nor was American reluctant to order foreign equipment. While United and Eastern served as launch customers for the short-range Boeing 727, American opted for the British BAC One-Eleven twin-jet, eventually becoming the largest single operator of the type in the United States.[19]

Meanwhile at Pan American Airways, Juan Trippe led the way in adopting the jet airliner. Charles Lindbergh, who served as a Pan American consultant for many years, was an enthusiastic supporter of gas turbines and urged Trippe to work with American manufacturers to develop a jet transport aircraft. But the high fuel consumption and maintenance costs of jet engines, plus their comparatively brief operational lives made them initially unacceptable for commercial application. Most airlines felt that turboprops were the next logical step and should be applied to existing airliners like the DC-6 and Constellation. Britain had stolen the march on its American competitors by developing the Vickers Viscount with Rolls-Royce turboprop engines, but the British also led the way in pure jet transports, putting the de Havilland Comet 1 into service during 1952. The same year, Pan Am placed orders for the more advanced Comet III, to be delivered in 1956.

A series of accidents, traced to structural flaws, grounded the Comets in 1954 and by the time the British had solved the problem and prepared to put modified Comets into service, American aircraft builders had surged ahead. About the same time that Trippe ordered the Comet, improved American jet engines were in production; several years of operating jet engines as well as building jet bombers, like the Boeing B-52, put US manufacturers in a more favorable position. In the process of developing a large jet-propelled military tanker for aerial refueling, Boeing were able to adapt it for a civil role as the ubiquitous 707 airliner. With his original order for Comets on hold, Trippe seized the opportunity of an improved American design and entered negotiations with Boeing, while at the same time holding talks with Douglas about their proposed jet, the DC-8. In similiar fashion Trippe conducted long conferences with the engine builders Pratt & Whitney, constantly hinting about a better deal with Rolls-Royce. Finally the Pan Am boss closed a deal with Boeing in 1955, took delivery of the first 707 in 1957 and introduced them on the premier New York–London route in 1958. It was a typically Byzantine performance for Trippe, prodding and hectoring all parties until he stitched together the best deal for Pan American. In the 1960s, he used similar tactics in negotiations with Boeing over the 747 jumbo jet, which entered service in 1970.[20]

3   ROUTES

3.1   Network development

Although both Europe and America had about the same physical size, geographic differences and population distribution created different air route patterns. Moreover political factors played a key role in the evolution of different networks. 'Whereas the European network grew up under the jurisdiction of many sovereign states, the US pattern has been regulated in its growth by a single authority.'[21] The European network comprised a series of radial patterns, with routes fanning out from the various national capitals, while the US had a variety of patterns, radial, linear, and grids, with some circular patterns as well. The network of American Airlines was essentially linear, connecting major cities along the Atlantic coast with those on the Pacific. Eastern Air Lines, serving the heartland east of the Mississippi and cities along the eastern seaboard, developed a grid pattern. This apparently patchwork system evolved under the powerful but benevolent bureaucracy of the CAB. In contrast to the political and prestige functions of Europe's flag-carriers, the CAB imposed a degree of operational logic and economic sense, presiding over an era of impressive air transport achievements.[22] Route expansion, mergers and deregulation rearranged these patterns, especially when US domestic airlines added intercontinental services in the 1980s.

In the northeast quadrant of the US, the density of population and closely spaced cities created radial patterns similar to Europe. For decades, the most heavily traveled air corridor was the New York–Chicago route. Also in the northeast, rail lines represented competition to air routes. At the same time, the northeast as well as the less populated areas of the West gave rise to 'a phenomenon almost unknown in Europe' for several decades after the war, the use of executive aircraft owned and operated by individual entrepreneurs and corporations. In the 1950s, such corporate aircraft accounted for as much as 60 per cent of business flights in the US.[23]

In any case, air travel became the principal means of postwar public transportation. For decades, the main rival of the airliner had been railroads and the Pullman service which offered passengers overnight sleeping berths and other amenities. By 1953 domestic airlines had already overtaken railroad travel in Pullman cars by a margin of three billion annual passenger miles. By the end of that year, airlines had also emerged as the prime mover for American travelers making trips of over 200 miles. Rail and bus passenger traffic continued to fade in succeeding years, due to increased competition from major airlines and the dozens of commuter airlines that appeared. During the late 1950s and 1960s, the expansion of the Interstate Highway system put further pressure on rail and bus lines. In many parts of the country, particularly the trans-Mississippi west, commuter airways and trunk line routes were often the only public transport systems available. The airlines also challenged the premier role of ocean liners. In 1958, over a million passengers bound for Europe traveled by air, surpassing steamship patrons for the first time. The introduction of jets made transoceanic flights even more appealing; within a decade, air traffic across the Atlantic soared to several million passengers annually. By 1970, when overseas travelers numbered five million, only 3 per cent boarded steamships.[24]

None of this happened smoothly. The experience of Capital Airlines demonstrates the problems faced by smaller airlines in their efforts to survive. Even stalwarts like Pan American faced increasing competition in the postwar era, not just from foreign flag-carriers, but also from US domestic carriers that the CAB had permitted to expand into international markets.

With the return of peacetime conditions, US airlines expected a strong market revival as consumer demand generated industrial activity and business travel. The scheduled airlines had 421 planes in 1945, 674 in 1946. Traffic met expectations, rising from about 6.5 million in 1945 to 12.3 million in 1946. The future seemed bright, but 1947 proved a disappointing year. Part of the reason stemmed from fare increases as airlines struggled to meet the costs of new equipment and route expansion, added to which, a series of highly publicized airline accidents dampened demand. In order to generate revenue and passenger loyalty, American Airline introduced family fares in 1948, providing 50 per cent discounts to family members, as long as one family member paid the full one-way fare and everyone traveled on Monday, Tuesday, or Wednesday. Other airlines tried different strategies to entice passengers. Western Airlines advertised a cut-price 'no-meal tariff', but the experiment lasted less than a year. The major airlines also had to contend with stiff competition from non-scheduled carriers which had used war-surplus equipment, eschewed the subsidies that would have meant fulfilling the CAB requirement to operate unprofitable routes and offered flights at rock-bottom prices. Many people who started these companies had wartime experience in the ATC. One such operator, Stanley Weiss, organized North American Airlines, and charged $99 for coast-to-coast trips and $160 for a round-trip ticket. Most of the *non-scheds* as they were known, quickly sank out of sight but Weiss's plucky airline and a few others forced the major airlines to respond.[25]

Up to this point, the major airlines offered first-class service to all passengers. Airlines had competed with the railway Pullman service since the 1920s, and the passengers on those trains expected a high level of attention and amenities. Airlines still catered to that expectation in the postwar years. Under competitive pressures from non-sched operators and slumping ticket sales, industry executives reluctantly gravitated towards the idea of *coach* fares. First off the mark was the small trunk carrier, Capital Airlines, who used high-density seating on its 60-passenger DC-4s to introduce the Nighthawk service between New York and Chicago. Beginning in November 1948, Capital announced coach fares of US$29.60 one-way on flights during the night, compared to the daytime fare of US$44.01. Even though rail coach fares were lower, passengers saved more than 12 hours in travel time. Capital's gamble proved to be popular, and coach class fares not only spread to other airlines, but also appeared on daytime flights as well. The trend to lower fares and economical service clearly came from competitive pressures, especially from outsiders eager to enter the airline business.

Capital's president, James H. Carmichael, had entered aviation in the 1920s as an enthusiastic youngster. Like many postwar airline figures, he had grown up with the industry. Having earned a private pilots license in 1926, when he was 19 years old, he had been a barnstormer, crop duster, instructor, and sometime airline pilot, before he joined Central Airlines in 1934, eventually moving into a series of management positions for Pennsylvania-Central Airlines (PCA). PCA became a vigorous regional airline in the mideast and southeast regions of the United States. In 1947 Carmichael became its president and changed its name to Capital Airlines, placing the headquarters in Washington DC. During the postwar era, Capital flew its DC-3 and DC-4 aircraft on longer routes, such as New York to Chicago and southern segments to Atlanta and New Orleans, as well as short-haul routes. The airline ranked fifth by the 1950s, and the CAB awarded Capital longer, choicer routes in order to prevent domination by the Big Four. Capital struggled to increase traffic on these longer, more profitable routes with innovative marketing and the acquisition of faster airplanes hoping that it might win CAB blessing to break out of its confinement in the east and to compete with the Big Four on a national basis.

In 1955, Capital made the stunning announcement that it had purchased 60 Vickers Viscount turboprop airliners. It introduced the first of the British aircraft on the intensely competitive New York–Chicago run, and the company's advertising trumpeted the arrival of 'Rolls-Royce Jets' on US routes. Despite the overstated boast, the Viscount's smooth flying qualities and higher speed proved to be effective in attracting passengers from the noisier and slower piston-engined aircraft flown by rivals. Thus encouraged, Capital ordered 15 more Viscounts and Carmichael even started negotiations to buy 14 de Havilland Comet jets.

Unfortunately Capital tried to go too far, too fast. The company's short-haul, high-cost route structure was simply not compatible with its fast, expensive equipment. Massive debt payments cut deeply into its revenues, forcing Carmichael to give up plans for the Comets and to scramble for cash while extending the Viscount mortgage several times. Meanwhile, the Lockheed Electra turboprop had entered service with other carriers and the CAB delayed awarding new routes to Capital. Some aircraft accidents in 1958 and 1959 hurt the carrier's image further as losses mounted. Finally in 1961 it was taken over by United Air Lines in the industry's largest merger up to that date. Although its identity had gone, Capital's legacy of lively competition and its use of advanced jet-powered aircraft pushed its larger, more cautious rivals into modernizing their operations.[26]

On overseas routes, Pan American continued as the principal postwar US carrier. Guided by Trippe, it enjoyed primacy for several decades. During the 1920s and 1930s, the airline enjoyed a tacit endorsement by US postal and military authorities, including the Department of State. With this sort of patronage, Pan Am came closest to becoming the American flag-carrier, with status similar to its European counterparts. Pan Am was the chosen instrument used to push American aviation routes into Latin America and to counter European, and in particular German, influence in that continent. Up to the beginning of World War II, Trippe campaigned tirelessly to keep Pan Am's near-monopolistic control of American air routes in the Caribbean, the Pacific, and across the Atlantic.[27] At the same time, Pan Am's success in developing radio communications, long-range navigation, and a logistical infrastructure, put it in the forefront of aeronautical progress. Moreover, Trippe's quest for effective, long-range airliners played a key role in the development of major flying boats built by Sikorsky, Martin, and Boeing. Trippe never displayed much loyalty toward manufacturers; indeed, from Pan Am's earliest years, he enjoyed playing them off against each other, in order to get the best deal for his airline. During World War II, Pan Am's experience with intercontinental routes helped it play a major role in global air transport, although government policy indicated an intention to create more opportunities for competitors in the future. With the blessing of the CAB, rivals chipped away at Pan Am's position during the 1940s and 1950s. Through route awards, subsidies, and mail contracts, Northwest Airlines became a contender on western Pacific routes, United made inroads into the central Pacific area with a Hawaiian destination, Braniff expanded through Central and South America and TWA not only competed with Pan Am on the prestigious north Atlantic, but also almost everywhere else around the globe. On the other hand while Pan Am made repeated efforts to secure domestic routes, the CAB and Congress rejected the idea until 1980, on the grounds that linking domestic markets to Pan Am's formidable international position would simply give it too much power and influence over the American air travel market.[28]

Early in the postwar era, air cargo emerged as a significant source of revenue. Nearly all passenger airlines after the war offered some sort of package service and several dedicated large aircraft to air cargo. But most airlines focused on the more familiar market of passenger travel. Several all-cargo outfits appeared just after World War II, trading on the experience of the Air Transport Command and the availability of war-surplus transports for immediate commercial work. With such ATC experience and a family fortune to back him, Earl Slick launched Slick Airways in 1946. The company became America's largest all-cargo line by the early 1950s, but passenger airlines then rediscovered the cargo business and Slick suffered from competition. Meanwhile, the Flying Tiger Line developed as an all-cargo rival and reaped considerable profits as an international carrier during the Korean war. Flying Tiger also developed a passenger charter division and during the Vietnam war its passenger and cargo operations enjoyed lucrative military contracts. By 1980, Flying Tiger reigned as the principal operator in the global air cargo industry, with a fleet that included 26 stretched DC-8 cargo planes, 13 Boeing 747 freighters, and an additional trio of 747s for passenger operations.

Like Slick, Flying Tiger began to encounter competition from the major airlines, many of whom now operated their own dedicated cargo jets. One major customer, United Parcel Service, started its own airline division, while Federal Express, organized in 1971, became a formidable rival. FedEx began as a small-package airline and operated modified Dassault Falcon 20 executive jets to carry out overnight deliveries as a specialized operation. After a rocky start it succeeded in developing this niche market into an amazingly successful venture. Deregulation proved to be a benefit, allowing it to introduce larger aircraft like the Boeing 727, permitting the company to compete for larger cargoes while continuing to integrate its small-package service to smaller cities abandoned by the larger operators. In 1989, FedEx acquired Flying Tiger, making it a global leader.[29]

A number of intrastate airlines emerged in the postwar era, two of them achieved notable success. They succeeded due to the unique nature of

American geography, with individual states like California and Texas having long stage lengths between major urban concentrations. Pacific Southwest Airlines (PSA) began in 1949 as a cut-rate Califorinian carrier between San Diego and San Francisco. This busy corridor gave PSA the opportunity to include Los Angeles as well as Sacramento during the 1960s. In terms of passengers boarded, PSA became one of the nation's leading operators. Following deregulation in 1978, it became an interstate carrier to several neighboring states, but fare wars and a heavy debt load led to PSA's acquisition by USAir in 1987. Nonetheless, PSA's early success prompted others to follow its low-fare formula.

By the early 1970s, Texas accounted for three of America's largest cities: Houston, Dallas-Fort Worth, and San Antonio. With two or three hundred miles between each of them, their triangular geographic pattern made for a promising intrastate airline system. A crucial political factor was the inauguration of the new Dallas-Fort Worth airport (DFW) in 1970, accompanied by FAA decisions that required major airlines to relocate there. A group of investors organized Southwest Airlines to operate out of the recently vacated Love Field in Dallas, an airport preferred by local airline passengers due to its closer location to the city center. By operating as an intrastate carrier, not subject to CAB authority, Southwest did not have to follow the major airlines to DFW. Southwest not only won a favorable situation in Dallas, but also in Houston where it operated from the convenient Hobby Airport rather than the distant Houston Intercontinental. Southwest's helpful staff, aggressive marketing and determined efforts to slash costs, kept customers loyal and its low fares were a major attraction. The airline served no meals, but offered packaged snacks, soft drinks, and reasonably priced wine and liquor. Its entire fleet of aircraft consisted of Boeing 737s. After deregulation, it also became an interstate carrier, but chose its markets cautiously and expanded first to the Pacific coast, then into the Midwest. By 1990, Southwest recorded 9.9 billion revenue passenger-miles and revenues of just over US$1 billion, giving it major airline status. Under the shrewd management of Herbert Kelleher, it continued to enlarge its network, reaching the Atlantic coast and southeast markets while following the same strategy of controlled growth. In the early 1990s it was still a consistent moneymaker, while other major airlines were struggling to stay out of the red.[30]

3.2 Airport Environment

Before World War II major airports in the US were characteristically owned and operated by the municipalities they served, although special commissions and independent airport authorities were also established. For example, the Port Authority of New York and New Jersey, a bistate organization, held responsibility for Kennedy International (JFK), LaGuardia, Newark, and

Teterboro, plus two smaller airports, including the Wall Street Heliport. A number of smaller airports were owned by corporations or individuals, who operated them as businesses. Two major airports in the Washington DC region functioned as federal entities, Washington National, within the federal district, and Dulles International, located in the Virginia countryside. Regardless of ownership, all airports in the US had to meet relevant standards set by the FAA. These requirements also covered the airlines that leased gates and services from the local operating authority.

During World War II, dozens of military airfields were constructed across the country and many cities acquired them after the war for conversion to civilian airfields. They required extensive renovation and improvement to attract both passengers and air service. Significant government support for flight facilities began with the Federal Airport Act of 1947. Congress allocated US$500 million for airports in the 48 states, plus $20 million for Alaska and Hawaii, the funds to be allocated over a seven-year period. These annual appropriations helped hundreds of airports and the program became an extended funding item for decades, totaling as much as US$84 million for a single year. The basic concept was for federal funds to cover up to 50 per cent of construction costs or improvements to publicly owned airfields. Additional money came from the states (usually a quarter of anticipated costs), with the remainder coming from local tax funds and bonds. The growth of air travel made such expenditures an ongoing feature of many airport authorities. Larger aircraft, requiring longer runways and expanded passenger terminals, kept major city airports in a state of constant upheaval. The total number of airfields in America swelled from 2000 in 1938 to over 10 000 in 1968. By 1993 over 18 000 airports existed, of which 70 per cent were private and two-thirds had unpaved runways, reflecting the use of over 200 000 general aviation aircraft in American life. Virtually all the paved and lighted airfields were publicly owned facilities; these included 670 certified by the FAA for air carrier operations and 417 with FAA towers and scheduled air service.[31]

Because cities such as New York and Boston possessed urban rail systems, air passengers could use them for arrivals and departures. But it was not always easy. During the 1980s New York's JFK Express required passengers to ride on a bus that shuttled between air terminals and the rail station. In Washington DC, National Airport required a similarly awkward procedure. Moreover Metro passengers arriving at National Airport found themselves dumped at a subway station perched above a parking lot. The station was open to the elements, freezing travelers in winter and soaking them whenever it rained. Railway transit systems elsewhere worked better, but travelers accustomed to the convenience of integrated airport rail links in Europe continued to find major American airports sadly lacking in comparison.

At Atlanta's Hartsfield airport, the basic design had been completed by 1976, and construction bids were about to go out, when the rapid rail transit system suddenly became a problem. The metropolitan area rapid transit, or MARTA, had planned to run a line out to the airport and locate the station adjacent to the airport's landside facilities. A reversal in planning incorporated the station in the main airport terminal complex, but required redrawing the preliminary designs, and rail service would not start until 1988. Even later, when American cities built billion-dollar transit lines, the airport link rarely received a high priority. The Dallas Area Rapid Transit, opened in June 1996, received enthusiastic acclaim in local news stories, which explained that plans called for various extensions over the next 15 years, one of which was scheduled to reach DFW airport sometime in the twenty-first century.

After the advent of high-speed trains in Japan and France, grandiose plans for similar services were conceived in several regions around the US. In the mid-1990s Texas became interested in a high-speed rail link between Dallas-Fort Worth, Houston and San Antonio. Plans called for hourly frequencies and journey times of only two hours between city centers. But the scheme required wholesale replacement of old track plus the purchase of long, straight stretches of land to eliminate curves, while pipelines and power lines had to be relocated. Not surprisingly the popular Southwest airline vigorously opposed the project; however thanks mainly to escalating cost projections, the scheme had disappeared by 1995 and the only American rail system that represented serious competition for the airlines remained Amtrak's service between Washington and New York.[32]

4 PERFORMANCE

4.1 Operating and Financial

In 1945 total revenue passenger-miles for certifed American airlines were 3.8 million. By 1960 it had surpassed 38 million and the figures continued to rise at a furious rate, totaling 113 million in 1968 and 519 million in 1994. Operating revenues rose more slowly, touching US$2 billion in 1960, US$12 billion in 1975 and US$87 billion in 1994.[33]

This whirlwind development created a far more complex world for airline managers and consequently the background of the airline industry's top executives became highly specialized, with increasing focus on economic abilities. They often came from outside the industry, with a professional career that had less to do with airplanes and air transport, and more to do with accounting and finances.

In the formative years of the airline industry, men like C.R. Smith, Tom Braniff and Bill Patterson had become aviation executives as a result of early careers in accounting and banking. But they entered the air transport business as young men and literally grew up with the airlines. For some time after World War II, personnel with Air Force careers, especially in the ATC, continued a pattern of aeronautically knowledgeable people filling airline management positions. By the 1950s, conditions had changed as the airlines expanded and simple war-surplus aircraft gave way to advanced planes with complex systems. Airline operations required meticulous cost analysis for highly intricate flight schedules as well as maintenance, repair, and financing for continuous funding cycles of new equipment.

C.R. Smith felt that finance represented a weak spot throughout the industry. Accordingly, when American Airlines decided to appoint a new vice-president and treasurer in 1950, he reached outside the airline fraternity to pick William J. Hogan who had been at Firestone Tire and Rubber and at H.J. Heinz, a food company. While Smith remained firmly in control of the airline, the financial expert Hogan played an increasingly significant role in its corporate affairs; the airline business was maturing from a swashbuckling adventure to a finance driven industry. In later years, when airline executives were appointed to key positions in top management, they tended to come not so much from the operational side, as from the financial and legal departmernts When C.R. Smith first retired in 1967, his replacement came from the airline's office of general counsel. Following a difficult period for the airlines, Smith again headed American Airlines in 1973–4, and was replaced by Albert Casey, a person with no airline experience, but strong credentials in transport and finance.

In 1979, Casey gave way to Robert Crandall whose credentials included experience at Eastman Kodak and Hallmark Cards. Crandall had then joined TWA as vice-president and controller, after which he worked at the Bloomingdales department store in New York City. In 1973 he joined American Airlines, working his way up through vice-presidencies in finance and marketing, becoming president and CEO in 1980, and finally chairman in 1985. In an industry forced to watch finances with increasing care, Crandall's tenure at American was notably successful. Crandall slashed costs and competed fiercely with rivals wherever American Airlines was challenged; if anyone represented the cost-conscious and competitive post-deregulation airline industry of the 1980s, it was Crandall.[34]

Financing new airliners represented a particularly daunting task for operators. Before placing orders for new aircraft, airlines had to do some careful economic forecasting and reach shrewd decisions in terms of having the right mix of equipment for duty on long, medium, and short haul segments. Airlines contemplating new or expanded transoceanic services had to commit extraordinary sums for wide-body aircraft. These decisions had to be made in the face of comparatively long-term deliveries; waiting time could be up to three years, or even longer if the manufacturers had a backlog of orders. Once delivered, the airlines could only hope that an economic downturn would not occur, cutting into the revenue stream needed to pay off the cost of the aircraft. On the other hand the airlines tried not to delay aircraft purchasing decisions for too long, fearing that they would lose market share to a rival who had gambled more successfully on the travel market.

Consequently, a comparatively large number of US planes were owned by someone else and operated under lease. Leasing circumvented some of the pitfalls involved in buying too soon or too late. For smaller, or marginally profitable airlines, it enhanced financial flexibility and in the era of post-deregulation takeovers, proved to be a useful buffer, since a carrier with fewer tangible assets was less appetizing to corporate raiders. While leasing has always been around, the practice spread in the 1960s as the cost of acquiring jet equipment involved higher commitments of cash reserves and credit lines. In the 1970s the development of new routes caused more operators to use leased equipment. By the 1990s, leasing had become a way of life, with major firms like GPA Group in Ireland supplying both European and American customers.[35] Interviewed in 1994 the president of McDonnell Douglas Corporation, Harry C. Stonecipher, commented that 'leasing companies that have come into vogue over the past years are going to be stronger and stronger in the future'. The airlines seemed to care less and less about owning planes. 'I think enough people have been bitten by the cycle now, that they look at it more as being a marketing company – marketing seats.'[36]

4.2 Labor and productivity

In terms of productivity, European airlines have lagged consistently behind their US counterparts. A variety of factors appear to account for this, such as more efficient and trouble-free aircraft, geography that maximizes the profitable use of large aircraft on long stage-lengths between cities, and a labor force which was not artifically inflated. European flag-carriers were frequently expected to serve uneconomical routes for prestige purposes while US airlines were not. As Peter Lyth shows in a comparison with BA, Delta Airlines in 1977 carried twice the number of passengers with the same size fleet and about half the staff. However the introduction of competitive European aircraft such as the regional ATR designs and the large Airbus airliners has clearly challenged the Americans in aircraft design. Continuing advantages in productivity on the American side seem to depend on controlling labor and fuel costs, and utilizing smart computer software for yield management in allocating the right number of airline seats to meet the competition in fares.[37]

4.3 Marketing

Prestigious air routes in the US were often supported by colorful promotional campaigns. In contrast to European styles, early American marketing seemed to be far more flamboyant. In 1956 Continental began operations on its Los Angeles–Denver–Chicago route, using DC-7s and calling it the Gold Carpet Service. On landing, Continental's planes were waved in by uniformed ground crews who wore gold helmets and trotted around in parade-ground formation. At airport terminals and travel agencies, the airline assigned groups of representatives fitted out with gold space suits, finned helmets, and a rocket back-pack that chirped signals from outer space.

At a more serious level, many airlines turned to experts in consumer psychology for advice about techniques to allay the fear of flying and attract more passengers. Consultants like the Institute for Motivational Research suggested that airlines should encourage wives to fly as passengers because that would reduce their concerns about husbands who made numerous business trips and result in more frequent travel. Stewardesses took lessons on presenting a calm demeanor to soothe nervous passengers and pilots received elocution lessons designed to reassure anxious travelers over the cabin loudspeaker system. Airlines also began to deal with other issues as flying became more popular. For airlines in the conservative South, where religious habits opposed the consumption of alcohol, the question of serving drinks to passengers became a hotly debated issue, although competition for passengers eventually led them to offer alcoholic beverages. Another feature of Southern life, public segregation of black and white passengers, changed slowly during the postwar years. Although during World War II black military personnel had begun to fly, achieving a sort of *de facto* desegregation on many airlines, postwar air terminals in the South retained segregated facilities until the civil rights legislation of the 1960s began to erase the barriers in the public sector.[38]

During the 1950s the introduction of new aircraft became an occasion for an airline news conference and entertainment. When the DC-7 was introduced, American Airlines launched a promotion scheme that featured Royal Coachman flights, with arrivals and departures heralded by a British actor in authentic coachman's clothing tooting an eighteenth-century coach-horn. By the 1970s and 1980s however, the democratization of flying meant less emphasis was placed on such snob appeal, and frugality became the watchword, especially after deregulation. One upstart airline, Texas International, boasted a low-cost 'Peanut fares', with stylized depictions of peanut-shaped airliners winging cheerful passengers across the US. The major carriers followed suit. American Airlines adopted similar tactics, introducing Super Saver fares in 1977. Such tickets offered deep discounts on round trips while restrictions meant that they were nonrefundable, had to be purchased well ahead of time and required a minimum stay. American also attempted to instill 'airline brand loyalty' by introducing a frequent-flyer program. Soon marketed as the AAdvantage plan, the scheme rewarded travelers with specified mileage credits on American Airline (AA) flights; with enough accumulation, the credits could later be traded in for free tickets.[39]

Some marketing ploys stressed frequent service and guaranteed seats. In 1961, this was Eastern Airlines' gambit along the heavily-traveled Boston–New York–Washington corridor, where American Airlines reigned supreme. Requiring no reservations, Eastern's Air-Shuttle featured departures at one-hour intervals, with six or seven planes committed to each departure. The Shuttle had less to do with competing with the railroad than with Eastern's desire to gain new customers in the lucrative northeast air travel market and it soon displaced American as the leader on this route. One year after the Shuttle's start, Alan Boyd, chairman of the CAB, described it as 'the greatest thing that has happened in air transportation in years'. Indeed during the next two decades, Eastern's Air-Shuttle became an institution for east coast travellers.[40]

Nothing in Europe matched the colorful promotions of Southwest Airlines. During the early 1970s, Southwest's stewardesses and female counter staff worked in eye-catching red hot-pants or orange vinyl mini-skirts, with white, over-the-calf vinyl boots. Feminists were appalled, but male travelers kept Southwest flying at full capacity. Other enticements included fares up to one-third less than competitors and automatic ticket machines that accepted credit cards.[41]

During the 1990s, startup airlines became a force in many markets, often leading to reduced fares and more productivity within the industry. Between 1990 and 1995, the Department of Transportation received 46 applications from such new airlines. Not every one of them succeeded in becoming airborne, but several emerged as very strong performers. Western Pacific Airlines (WPA) began in 1995 from a hub in Colorado Springs, south of Denver and by the end of the year, it listed 15 destinations from coast-to-coast, using leased Boeing 737s. WPA benefited from Colorado Springs airport's remodeled terminal facilities, especially as passengers were frustrated with the higher fares and irritating problems associated with the new Denver International Airport. Its biggest appeal however stemmed from ticketless reservations, stubbornly low prices, and imaginative marketing. Some aircraft became Logo Jets – flying billboards for corporate advertisers, with their names spelled out in bold letters from nose to tail. Each advertiser paid US$150 000 for the custom paint job and about US$88 000 for a year's exposure. WPA also sold *Mystery Fare* tickets of US$59 to any city in its system, although buyers did not know the destination until the plane's departure![42]

5 STRATEGY

5.1 Collaboration and competition

Running an airline in Europe often meant pooling and collaboration with other transport modes. Generally speaking, these strategies did not become a feature of major US carriers – indeed pooling would have violated American antitrust law – although independent charters made their appearance throughout the postwar years. However US companies experimented with other types of collaboration, for example over computerized reservation services (CRS), the costs of which were often shared by several airlines.

For two decades after the war, Pan American prospered, with an expanding world economy that increased business trips, stimulated leisure travel and benefitted from low fuel prices. Flush with cash, but concerned about its total reliance on overseas operations, Pan Am became one of the first airlines in the US to start a diversification program. During the 1960s it conducted a variety of management services for the Department of Defense (including a missile tracking range in the South Atlantic), operated one of the largest hotel chains in the world, and marketed the Falcon corporate jet, built by Dassault in France. In a grandiose project typical of the egocentric Juan Trippe, the company constructed the largest office building ever built on prestigious Park Avenue in New York City, making the site its world headquarters.[43]

The rising number of passengers, variety of airlines, and cascading lists of potential timetable variations left travelers and travel agencies in a constant state of anxiety. By the early 1960s, nearly all US airlines had some form of automated system in place to reduce the workload, but the capacity for storage of data remained inadequate and the continuing requirement of numerous manual operations made the process tedious. An electromechanical device would represent a major step forward, but nobody in the airline industry seemed to have an idea about which direction to take. One day, during a chance meeting with Tom Watson of IBM, C.R. Smith brought up the problem and the two men discussed various solutions. Eventually, IBM received a contract to develop something known as the Semi-Automated Business Reservations Environment, or SABRE. When American Airlines finally had all the equipment installed in 1962, SABRE proved a phenomenal success and cut its reservation process time from 45 minutes to three seconds. The system became a major marketing bonus for American Airlines and eventually spawned a series of competitors. Later versions of SABRE allowed travel agents to book hotel rooms, reserve a car, request special meals, and arrange medical assistance. Each time the travel agent used SABRE to sell a ticket on another airline, American Airlines received a fee. During the 1980s, this meant sales of US$400 million from SABRE and US$107 million profit.[44]

Deregulation in 1978 led to large investments in ancillary ventures. United Airlines' service to Mexico, beginning in 1980, seemed modest enough, but the parent company also began an unusually aggressive expansion into other areas. During the early 1980s United assumed leadership of the Apollo CRS. The company trumpeted its goal of becoming a totally integrated travel giant, acquiring Westin International Hotels and the Hertz rental business. Although Pan American had also developed a foreign hotel chain, the United scheme brought together domestic and foreign operations for air travel, lodging, car rental, and inclusive computerized reservations in a single travel package. These dramatic steps were only the beginning of rapidly expanded international service, underscoring the new orientation of an airline that had been domestically focused for decades.[45]

5.3 Deregulation and global alliances

The effects of deregulation after 1978 were mixed and observers of the air transport scene have recorded strong negative as well as positive opinions.

Hub and spoke networks were rapidly developed as the prevailing pattern on US routes. They had existed prior to deregulation, but airline managements now adopted them with zeal and passengers had to accept the new arrangement as a way of life. Airlines chose strategically located airports as hubs to serve as transfer points for passengers traveling from one city to another in the hub's region. The hubs also served as collection points, where an airline collected passengers departing or arriving from outside the region or from somewhere overseas. Airlines then proceeded to schedule departures and arrivals of planes in concentrated 'banks' of flights timed throughout the day. This meant transfers of passengers in the minimum time from incoming to outgoing flights. Hub and spoke systems had disadvantages too. Weather problems along the spokes, or at the hub itself, created havoc with a carrier's operations when aircraft were grounded, delayed, or rerouted. When delays occurred, passengers overwhelmed terminal facilities. Lost luggage also became a bigger frustration and very soon disgruntled passengers referred to the whole experience as the 'hub-and-choke' process.

Deregulation gave full rein to the expansionary aspirations of many airlines. An early wave of mergers had occurred during the early 1950s, ending the lives of many smaller operations. A second wave swept the industry in late 1960s and early 1970s, giving rise to new regionals like Allegheny and the expansion of old regionals like Delta. During the late 1970s and early 1980s, expansion and merger caused a tidal wave of change. Allegheny renamed itself USAir and pushed its market across new territory. Pan American bought National Airlines in 1980, finally acquiring a domestic market. Several airlines offered low-fare, no-frill service, such as People Express. Southwest built a sound network as a regional, then expanded further. Braniff added dozens of new cities in the US and stretched itself into new operations across the Pacific as well as the Atlantic; it also became one of the earliest casualties of deregulation. All the airlines experienced problems in the early 1980s, wounded by a sluggish economy, spiraling fuel costs, debt load from rapid expansion and fare wars. In the middle of all this, the Professional Air Traffic Controllers Organization (PATCO) walked out on strike during the summer of 1981. Domestic air traffic became snarled and for Braniff, already stretched thin, the loss of income was crucial and the company collapsed in 1982.[46]

Braniff's sudden death pales in terms of drama when compared to the saga of Frank Lorenzo and Continenal Airlines. In 1986 Lorenzo won control of Eastern Airlines after a bruising battle; the new airline holding company pitting one of the country's most arrogant labor organizations (at Eastern) against an equally notorious union-buster (Lorenzo). Escalating union troubles exacerbated the long-standing problems with creditors and Lorenzo was forced to sell much of Eastern's key assets, but to no avail. He then abruptly sold his interests in the holding company; while Continental staggered on through a series of presidents and bankruptcy proceedings, Eastern came apart in 1991, ending the life of one of the country's greatest pioneer airlines.[47]

United Air Lines consolidated its position in the US and made a series of bold moves into overseas routes. From Hawaii it offered services to Japan in 1983 and then, having bought long-range Boeing 747–400 airliners in 1989, to Australia. Next it launched schedules to Hong Kong, and then, in 1990, direct flights from Chicago and Washington to Frankfurt. With Pan American's collapse, United bought the old *chosen instrument*'s Heathrow operations in 1991, creating its first European hub. Later it picked off Pan Am's premier operation in Latin America, emerging as a global giant. By 1992, however, United faced the same sobering statistics as the rest of the airline industry and it was forced to sell its travel subsidiaries. Political emergencies, fuel costs, interest rates, and a stuttering international economy bit deeply into total revenues. Like everyone else, United cancelled orders and deferred delivery of some 400 planes, although its fleet at the end of 1992 numbered more than 500 aircraft, compared to 300 in the early 1980s. For Pan American deregulation had been a disaster. Although it had eventually acquired some domestic routes, stiff competition and rising fuel costs pushed the once-proud international giant to the wall. It sold off assets until there was nothing left to save. Delta, the conservative airline from Atlanta, emerged as the victor after picking up rights to dozens of Pan Am routes all over Europe, Africa, the Middle East, and as far away as India. Pan Am hung on desperately to its Caribbean and South American markets but it was dependent on contingency funds from Delta and eventually the newcomer was unwilling to continue the payments and the last vestiges of Pan Am collapsed in December 1991.

The growing troubles of the industry led to reports that the competitive structure of the American airline industry was becoming compromised. By 1991, the *Big Three* – American, United and Delta – controlled over half the market. The shifting alignment of US flag-carriers made the North Atlantic a major battleground in the struggle for international market share. The North Atlantic was the world's largest market in the early 1990s, accounting for some 28.7 million scheduled passengers per year. By 1992, American and Delta each expected to gain just over 10 per cent, United nearly 8 per cent. Two years earlier none of these three airlines had competed as major players in the Atlantic market.[48] The increasing power of the Big Three in both domestic and foreign markets put pressure on the surviving US airlines, leading them to negotiate special arrangements with foreign flag-carriers. In contrast to the selective marketing agreements of earlier years, the new alliances involved partial ownership of American operators by foreign airlines. This brought with it comprehensive coordination of schedules, equipment, and marketing. For the traditionally independent airlines in the United States, it meant a revolution in the business.

Leading the way in this new wave of investment were airlines from the so-called second tier of carriers, such as Northwest, Continental, and USAir. In 1989, Northwest signed a deal with KLM, giving the Dutch carrier a significant equity stake. Another European, SAS, became the largest single shareholder in the finances of Continental Airlines during the summer of 1990, when Frank Lorenzo announced he was stepping down as its head. One of the most controversial deals involved the troubled USAir and British Airways. Many analysts believed the new arrangements represented a harbinger of future airline operations. Studies of the US airline market noted that two-thirds of Americans heading for Europe began their travels from somewhere east of the Mississippi River. Theoretically therefore, USAir ought to have been tapping a big share of that market, but the company simply did not have the resources to shoulder its way into transatlantic competition. The agreement with British Airways was intended to feed passengers into USAir's gateway cities, where they would have access to BA's international network. In 1993, BA and USAir signed a deal, modified to placate critics, in which the British invested US$300 million in its American partner. Arrivals, departures, and ticketing for international passengers were integrated, offering a convenient travel product.[49]

Deregulation was probably not as bad as detractors claimed. The number of carriers, and travel choices, escalated from 36 scheduled airlines in 1978 to 123 in 1984. An inevitable shake-out accompanied this remarkable change and during the 1980s many hopeful new entrants failed to make the grade and several famous names disappeared. By the 1990s however, the number of entrants began to increase again as entrepreneurs took advantage of a surplus of used aircraft at attractive prices, stable fuel costs, and a skilled labor pool. Supporters of deregulation also pointed out that nearly 100 per cent of airline travelers enjoyed a choice of carriers in the decade following deregulation, compared with about two-thirds in 1978. There were concerns that a few big airlines might dominate major routes on an individual basis, resulting in higher ticket prices on some well-traveled corridors. But the major airlines continued to compete with each other in principal markets, and the appearance of smaller, low-cost airlines played a key role in keeping fares competitive. In some cases air fares dipped so low that bus and rail service found themselves hard-pressed by the airlines. A study by the Brookings Institution in 1993 concluded that airline passengers in the US were saving US$17.7 billion per year as the result of deregulation. Considered statistically, the years since the last year of government regulation (in 1977) represented impressive growth. A total of 240 million passengers boarded US airliners in 1977, compared to 490 million in 1993.[50]

Many veteran travelers on US airways might look back with nostalgia to an earlier era when terminals were less crowded, seating more spacious, meals fresher and cabin attendants more attentive. But the hordes of travelers in the late twentieth century reflect a unique level of success in the history of air transport.

NOTES

1. Air Transport Association, *The Airline Handbook* (Washington, DC: Air Transport Association, 1995), pp. 19–21, hereafter as ATA *Handbook*; Air Transport Association, *Air Transport 1995/Annual Report*, Washington DC, 1995, passim.
2. 'Major Airline Profiles', *Aviation Week and Space Technology*, 144, 8 January 1996, p. 239.
3. 'Postal aspects are covered in William Leary', *Aerial Pioneers: The U.S. Air Mail Service, 1918–1927*, Washington, 1985. For a summary of airmail and other institutional developments, see Roger E. Bilstein, *Flight Patterns: Trends of Aeronautical Development in the United States, 1918–1929*, Athens Ga., 1983, pp. 29–56, 127–44.
4. The evolution of federal legislation is covered in Nick A. Komons, *Bonfires to Beacons: Federal Aviation Policy under the Air Commerce Act, 1926–1938*, Washington DC, 1978, and John R.M. Wilson, *Turbulence Aloft: the Civil Aeronautics Administration amid Wars and Rumors of Wars, 1938–1953*, Washington DC, 1979.
5. See, for example, Ronald Miller and David Sawers, *The Technical Development of Modern Aviation* (New York, 1970), especially pp. 257–65, 277–79.
6. See, for example, the authoritative work by R.E.G. Davies, *Airlines of the United States since 1914*, Washington, 1982, as well as the FAA histories cited above.

Prewar mail subsidies are a rather different issue, as is tacit government support of Pan Am. Even in the case of Pan Am, however, operators like Braniff competed on certain pre-war Latin American routes, and competitors cropped up everywhere. After 1945, Pan Am's international position was freely challenged by TWA and others. For a concise history of US airlines, see Carl Solberg, *Conquest of the Skies: a History of Commercial Aviation in America*, Boston, 1979); on Pan Am, see Robert Daley, *An American Saga: Juan Trippe and His Pan Am Empire*, New York, 1980.

7.   Three additional FAA histories, all published by the Government Printing Office, cover these postwar years: Stuart I. Rochester, *Takeoff at Mid-Century: Federal Aviation Policy in the Eisenhower Years, 1953–1961*, 1976; Richard J. Kent, Jr., *Safe, Separated and Soaring: A History of Federal Civil Aviation Policy, 1961–1972*, 1980, and Edmund Preston, *Troubled Passage: The Federal Aviation Administration during the Nixon-Ford Term, 1973–1977*, 1987.

8.   Safety figures from ATA *Handbook*, p. 41.

9.   Mark Hosenball, 'Trouble in the Skies', *Newsweek*, July 29, 1996, pp. 34–6; Evan Thomas, et.al., 'Death on Flight 800', ibid., pp. 26–33. At the time of writing, the reason for Flight 800's crash remains unexplained. ValuJet restarted services in the autumn of 1996.

10.   Rochester, *Takeoff*, especially pp. 2, 9–32, 189–219; Preston, *Troubled Passage*, especially pp. 5–23, 130–3, 250–7; William A. Jordan, 'Civil Aeronautics Board', in Donald R. Whitnah, ed., *Government Agencies*, Westport, Conn., 1983, pp. 61–8.

11.   On deregulation, see Robin Higham, Alfred E. Kahn, in William Leary, ed., *The Airline Industry: Encyclopedia of American Business History and Industry*, New York, 1992, pp. 248–51; Anthony E. Brown, *The Politics of Airline Deregulation*, Knoxville, 1987; ATA *Handbook*, pp. 13–14.

12.   On the Essential Air Service Program, see R.E.G. Davies and O.E. Quastler, *Commuter Airlines of the United States*, Washington DC, 1995, pp. 116–21, 155–8.

13.   On the technological legacies of flying boats, see Richard K. Smith, 'The Intercontinental Airliner and the Essence of Airplane Performance', *Technology and Culture*, 24, July 1983, pp. 32–47.

14.   On the evolution of the Stratoliner and associated technology, see Douglas Ingells, *747: Story of the Boeing Super Jet*, Fallbrook, Ca. 1970, pp. 73–84, also Miller and Sawers, *Technical Development*, pp. 140–1.

15.   For a summary of airliner development, see Roger E. Bilstein, *Flight in America: from the Wrights to the Astronauts*, rev. edn, Baltimore, 1994, pp. 85–96, 167–76.

16.   The role of C.R. Smith is covered in Robert J. Serling, *Eagle: the History of American Airlines*, New York, 1985, pp. 160–80. On the ATC, see Roger Launius, *Anything, Anywhere, Anytime: an Illustrated History of the Military Airlift Command, 1941–1991*, Scott AFB, Illinois, 1991, pp. 55, 279.

17.   For American Airlines in this era, see Serling, *Eagle*, pp. 183–204; Davies, *Airlines of the United States*, pp. 324–34. For individual airliners, see Kenneth Munson, *Airliners since 1946*, New York, 1975.

18.   Serling, *Eagle*, pp. 277–82, 303–33, 341–42, 362–3; Munson, *Airliners*, pp. 151–2; Davies, *Airlines of the United States*, pp. 513–23.

19.   Serling, *Eagle*, pp. 236–73; Davies, *Airlines of the United States*, pp. 348–52; Douglas Ingells, *The McDonnell Douglas Story*, Fallbrook, Ca., 1979, pp. 115–27.

20.   For the advent of jet technology, Miller and Sawers, *Technical Development*, pp. 153–210. Jet developments are discussed in Davies, *Airlines of the US*, pp. 508–

32. For a commentary on Trippe, see Wesley Phillips Newton, 'Juan T. Trippe', in Leary, ed., *The Airline Industry*, pp. 464–76.

21. Kenneth R. Sealy, *The Geography of Air Transport*, London, 1968; p. 140.

22. Sealy, *Geography*, pp. 155–6.

23. Sealy, *Geography*; pp. 139–44. For a discussion of the corporate aviation phenomenon, see Bilstein, *Flight in America*, pp. 195–203, 239–44

24. Rochester, *Takeoff*, pp. 3–4; Bilstein, *Flight in America*, pp. 233, 238. On the evolution of commuter airline networks, see Davies and Quastler, *Commuter Airlines*; pp. 3–133.

25. Civil Aeronautics Board, *Handbook of Airline Statistics*: 1963 edn, Washington DC, 1964; p. 484. The story of the non-scheds is summarized by Davies, *Airlines of the US.*, pp. 447–65.

26. R.E.G. Davies, *Rebels and Reformers of the Airways*, Washington, 1987, pp. 75–9; Lloyd Cornet, Jr., James H. Carmichael, in Leary, ed., *The Airline Industry*, pp. 95–8.

27. Newton, *Trippe*, in Leary, ed., *The Airline Industry*, pp. 464–76; Marilyn Bender and Selig Altschul, *The Chosen Instrument*, New York, 1982, passim.

28. Robert Daley, *An American Saga: Juan Trippe and his Pan Am Empire*, New York: Random House, 1980, pp. 335–46, 370–86; Davies, *Airliners of the United States*, pp. 358–87; Wesley Phillips Newton, 'Pan American Airways', in Leary, ed., *Airline Industry*, 347–8.

29. John C. Bishop, 'Delivery of Goods by Air: the United States Air-Cargo Industry, 1945–1955', in *Essays in Business and Economic History*, Los Angeles, 1989; Robert A. Sigafoos and Roger R. Easson, *Absolutely Positively Overnight: The Unofficial Corporate History of Federal Express*, Memphis, Tenn., 1988.

30. Davies, *Airlines of the United States*, pp. 548–52; Myron Smith, Jr., 'Pacific Southwest Airlines', in Leary, ed., *The Airline Industry*, pp. 342–3; George E. Hopkins, 'Southwest Airlines', ibid., pp. 451–2; Roger Bilstein and Jay Miller, *Aviation in Texas*, Austin, Texas, 1985, pp. 231, 234.

31. Federal Aviation Administration, *FAA Historical Fact Book: a Chronology, 1926–1971* Washington DC: Government Printing Office, 1974, pp. 53–54, 277; Wilson, *Turbulence Aloft*, pp. 178–92. The total expended by 1970 was $1 165 200 000 (Wilson, p. 191). See also Bilstein, *Flight in America*, pp. 232–4. For an encyclopedic coverage of airports, including administration, see John Stroud, *Airports of the World*, London, 1980, passim.

32. Stroud, *Airports*, passim; Betsy Braden and Paul Hogan, *A Dream Takes Flight: Hartsfield Atlanta International Airport*, Athens Ga., p. 190; George A. Brown, *The Airline Passenger's Guerrilla Handbook*, Washington DC, 1989, p. 131; DART Opens, *Houston Chronicle*, 15 June 1996; Davies and Quastler, *Commuter Airlines*, pp. 386–7, passim.

33. Numbers compiled from selected copies of Air Transport Association, *Air Transport . . .* Annual Report and Aerospace Industries Association, *Aerospace Facts and Figures* Washington, annually, passim.

34. Serling, *Eagle*, pp. 217–19, 238–40, 437–39; Lloyd Cornet, Jr., Robert J. Crandall, in Leary, ed., *Airline Industry*, pp. 126–9.

35. ATA, *Handbook*, pp. 29–30; 'Crisis Worsens for GPA Group', *Aviation Week and Space Technology*, 141, 10 May 1993, p. 32; James Ott, 'New Market Reshapes Embattled Big Three', *Aviation Week and Space Technology*, 141, 10 May 1993, pp. 31–2, hereafter *AWST*.

36. Bruce Smith, et.al., 'New Transport Orders will Bolster Industry', 143, *AWST*, 13 March 1995, pp. 42–4; Standard and Poor, *Standard and Poor's Register of*

Corporations, Directors, and Executives, vol. 1, New York: McGraw-Hill, 1996, p. 1309.

37. European productivity comparisons are implicit in the above chapters; see especially Peter Lyth's survey of British Airways in this volume. Typical US press assessments include Christopher Power and Aaron Bernstein, et.al., 'The Frenzied Skies', *Business Week*, 19 December 1988, pp. 70–3, 76, 80; Agis Salpukas, 'Changing Course at American Air Lines', Business Section, *New York Times*, 25 April 1993, pp. 1, 6.

38. Roger Bilstein, 'Air Travel and the Traveling Public: the American Experience, 1920–1970', in William F. Trimble, ed., *From Airships to Airbus: the History of Civil and Commercial Aviation*, vol. II , Washington, 1995, pp. 102–4.

39. Serling, *Eagle*, 100, pp. 144–9; Cornet, Crandall, in Leary, ed., *Airline Industry*, pp. 127, 129.

40. Davies, *Airlines of the United States*, pp. 541–4; W. David Lewis, *Eastern Air Lines*, in Leary, ed., *Airline Industry*, pp. 164, 167. In 1989 financier Donald Trump bought the Shuttle operation, although it soon passed to USAir. By the 1990s, travel times and fares approximated that of rail; the choice of plane or train became a matter of personal preference.

41. Bilstein and Miller, *Aviation in Texas*, pp. 231, 241.

42. David Hughes, 'Air Canada CEO Mulls Startup DC-9 Airline', *AWST*, 4 December 1995, p. 34; William B. Scott, 'Rapid Growth Tests Westpac's Low-Fare Formula', ibid., pp. 37–8.

43. Wesley Phillips Newton, 'Pan American Airways', in Leary, ed., *Airline Industry*, p. 347.

44. Serling, *Eagle*, pp. 347–8, 443–4; Christopher Power et.al., 'The Frenzied Skies', *Business Week*, 19 December 1988, p. 76.

45. Robert van der Linden, 'United Air Lines', in Leary, ed., *Airline Industry*, pp. 479–80.

46. ATA *Handbook*, 15, passim; Brown, *Politics of Airline Deregulation*, passim; Paul Stephen Dempsey, *Flying Blind: The Failure of Airline Deregulation*, Washington, 1990, passim; 'A Failed Experiment? Deregulation's Sad Legacy', *Consumer Reports* 56, July 1991, p. 465; Bilstein, *Flight in America*, pp. 291–2, 341–2.

47. Aaron Bernstein, *Grounded: Frank Lorenzo and the Destruction of Eastern Air Lines* New York, 1990; Davies, *Rebels and Reformers*, pp. 143–52.

48. Information and observations drawn from industry publications used to organize an annual essay, 'Aerospace', prepared by the author for *Collier's Encyclopedia Yearbook* during the years 1978–93, New York: Macmillan, 1978–93; Bilstein, *Flight in America*, pp. 344–5.

49. Richard Weintraub, Rebuilding USAir, *Washington Post*, 22 March 1993; pp. 19–21; Bilstein, *Flight in America*, p. 346.

50. ATA *Handbook*, pp. 15–16; David Hughes, 'Air Canada CEO Mulls Startup DC-9 Airline', *AWST* 14, 4 December 1995, p. 34.

Bibliography

AIR FRANCE

Carlier, C., ed. 'La construction aéronautique, le transport aérien, à l'aube du XXIème siècle'. Centre d'Histoire de l'Industrie Aéronautique et Spatiale, University of Paris I, 1989.

Chadeau, E., ed. 'Histoire de l'aviation civile'. Vincennes, Comite Latécoère and Service Historique de l'Armée de l'Air, 1994.

Chadeau, E., ed. 'Airbus, un succès industriel européen'. Paris, Institut d'Histoire de l'Industrie, 1995.

Chadeau, E. 'Le rêve et la puissance, l'avion et son siècle'. Paris, Fayard, 1996.

Dacharry, M. 'Géographie du transport aérien'. Paris, LITEC, 1981.

Esperou, R. 'Histoire d'Air France'. La Guerche, Editions Ouest France, 1986.

Funel, P., ed. 'Le transport aérien français'. Paris, Documentation française, 1982.

Guarino, J-G. 'La politique économique des entreprises de transport aérien, le cas Air France, environnement et choix'. Doctoral thesis, Nice University, 1977.

Hamelin, P., ed. 'Transports 1993, professions en devenir, enjeux et déréglementation'. Paris, Ecole nationale des Ponts et Chaussées, 1992.

Le Duc, M., ed. 'Services publics de reseau et Europe'. Paris, Documentation française, 1995.

Maoui, G., Neiertz, N., ed. 'Entre ciel et terre', History of Paris Airports. Paris, Le Cherche Midi, 1995.

Marais, J-G. and Simi, F. 'L'aviation commerciale'. Paris, Presses universitaires de France, 1964.

Merlin, P. 'Géographie, économie et planification des transports'. Paris, Presses universitaires de France, 1991.

Merlin, P. 'Les transports en France'. Paris, Documentation française, 1994.

Naveau, J. 'L'Europe et le transport aérien'. Brussels, Bruylant, 1983.

Neiertz, N. 'La coordination des transports en France de 1918 à nos jours'. Unpublished doctoral thesis, Paris IV- Sorbonne University, 1995.

Neiertz, N. 'Etude d'un mode d'interconnexion de reséaux: le cas d'Aéroports de Paris'. *Transports urbains*, no. 84 (July-September 1994), pp. 15–22.

Pavaux, J. 'L'économie du transport aérien, la concurrence impraticable'. Paris, Economica, 1984.

Pavaux, J., ed. 'Les complémentarites train-avion en Europe'. Paris, Air Transport Institute, 1991.

Saint-Yves, M. 'L'intermodalité air-fer, une nécessité pour l'Europe'. *Aviation International*, no. 995 (1 December 1989), pp. 32–4.

Spira, N., ed. 'Le transport aérien'. *Les Cahiers français*, no. 176 (May–June 1976), pp. 1–70.

Stoffaës, C., ed. 'Transport aérien, libéralisme et déréglementation'. Paris, Ecole nationale des Ponts et Chaussées, 1987.

Thomson, J.M. 'Road, rail and air competition for passengers in Europe'. *Aeronautical Journal*, London (April 1978), pp. 139–47.

Varlet, J. 'L'interconnexion des réseaux de transport en Europe'. Paris, Air Transport Institute, 1992.

BRITISH AIRWAYS

Birkhead, Eric. 'The Financial Failure of British Air Transport Companies 1919–1924' *Journal of Transport History*, vol. 4, no. 3, 1960; 133–45.

Bray, Winston. *The History of BOAC, 1939–1974*, London, BOAC, 1975.

British Air Transport in the Seventies, Report of the Committee of Inquiry into Civil Air Transport (the Edwards Report), London, HMSO, May 1969.

Campbell-Smith, Duncan. *The British Airways Story: Struggle for Take-Off*, London, Coronet, 1986.

Cronshaw, M. and Thompson, D. Competitive advantage in European aviation – or whatever happened to BCal? *Fiscal Studies*, 12, 1991, pp. 44–66.

Dobson, Alan. *Peaceful Air Warfare*, Oxford, Clarendon, 1991.

Doganis, Rigas. *The Airport Business*, London, Routledge, 1992.

Ellison, A.P. and Stafford, E.M. *The Dynamics of the Civil Aviation Industry*, Farnborough, 1974.

Hayward, Keith. *Government and British Civil Aerospace*, Manchester, 1983.

Higham, Robin. *Britain's Imperial Air Routes, 1918 to 1939*, London, 1960.

Higham, Robin. British Airways Ltd., 1935–1940, *Journal of Transport History*, Vol. 4, No.2, 1959; 113–123.

Knight, Geoffrey, *Concorde: the Inside Story*, London, 1976.

Lyth, Peter J. 'The changing role of government in British civil air transport, 1919–1940', in Robert Millward and John Singleton (eds), *The Political Economy of Nationalisation in Britain, 1920–1950*, Cambridge University Press, 1995.

Lyth, Peter J. 'A Multiplicity of Instruments: 'the 1946 decision to create a separate British European Airline', *Journal of Transport History*, vol. 12, no.2, 1990, pp. 1–24.

Lyth, Peter J. 'Experiencing turbulence', in James McConville (ed.), *Regulation and Deregulation in National and International Transport*, London, Cassell, 1997.

Miller, Ronald and Sawers, David. *The Technical Development of Modern Aviation*, London, 1968.

Nicolson. D.L, British Airways in the Eighties, *Journal of the Institute of Transport*, (May 1974).

Pryke, Richard. *The Nationalised Industries: Policies and Performance since 1968*, Oxford, 1981.

Shibata, Kyohei. *Privatisation of British Airways: Its Management and Politics 1982–1987*, EUI Working Paper EPU No. 93/9, European University Institute, Florence, 1994.

Thompson, A.W.J. and Hunter, L.C. *The Nationalised Transport Industries*, 1973.

LUFTHANSA

Bongers, Hans M. *Es lag in der Luft. Erinnerungen aus fünf Jahrzehnten Luftverkehr*, Düsseldorf, Econ, 1971.

Bongers, Hans M. *Deutscher Luftverkehr. Entwicklung, Politik, Wirtschaft, Organisation. Versuch einer Analyse der Lufthansa*, Bad Godesberg, Kirschbaum, 1967.

Dienel, Hans-Liudger. 'Das wahre Wirtschaftswunder. Flugzeugproduktion und innerdeutscher Flugverkehr im West-Ost-Vergleich, 1955–80', in Johannes Bähr und Dietmar Petzina (eds), *Innovationsverhalten und Entscheidungsstrukturen. Vergleichende Studien zur wirtschaftlichen Entwicklung im geteilten Deutschland, 1945–1990*, Berlin, Akademie Verlag, 1996, pp. 341–72.

Reul, Georg. *Planung und Gründung der Deutschen Lufthansa AG, 1949–1955*, Köln, Botermann & Botermann, 1992.

Treibel, Werner. *Geschichte der deutschen Verkehrsflughäfen. Eine Dokumentation von 1909 bis 1989*, Bonn, Bernard und Graefe, 1992.

Weigelt, Kurt, *Von der alten zur neuen Lufthansa*, Bad Homburg, published privately, 1966.

KLM

van Bakelen, F.A. (ed.), *Teksten vervoerrecht: Luchtrecht*, Zwolle, Tjeenk Willink, 1983.

Bouwens, Bram and Dierikx, Marc, *Op de drempel van de lucht. Tachtig jaar Schiphol* The Hague, SdU, 1996.

Davies, R.E.G. *A History of the World's Airlines*, Oxford University Press, 1967.

Dierikx, Marc. *Bevlogen Jaren. Nederlandse burgerluchtvaart tussen de wereldoorlogen*, Houten, Unieboek, 1986.

Dierikx, Marc. *Begrensde Horizonten. De Nederlandse burgerluchtvaartpolitiek in het interbellum*, Zwolle, Tjeenk Willink, 1988.

Haanappel, Peter, Petsikas, George, Rosales, Rex, and Thaker, Jitendra (eds), *EEC Air Transport Policy and Regulation and their Implications for North America*, Deventer, Kluwer, 1990.

Hellema D.A., Wiebes, C., Zeeman, B. (eds), *Jaarboek Buitenlandse Zaken: Derde Jaarboek voor de geschiedenis van de Nederlandse buitenlandse politiek*, The Hague, SdU, 1997.

Maas, P.F. (ed.) *Parlementaire Geschiedenis van Nederland na 1945. Deel III: Het kabinet Drees-Van Schaik (1948–1950). Band B: Anti-communisme, rechtsherstel en infrastructurele opbouw*, Nijmegen, Gerard Noodt Instituut, 1992.

Slot, P.J. and Dagtoglou, P.D. *Toward a Community Air Transport Policy. The Legal Dimension*, Deventer, Kluwer, 1989.

Smit, G.I., Wunderink, R.C.J., and Hoogland, I. *KLM in beeld. 75 jaar vormgeving en promotie*, Naarden, V+K/Inmerc, 1994.

de Vries, Leonard. *Vlucht KL–50*, Amsterdam, Meyer Pers, 1969.

Wassenbergh, H.A, *Aspects of Air Law and Civil Air Policy in the Seventies*, The Hague, Nijhoff, 1970.

ALITALIA

D'Avanzo, G. *I lupi dell'aria: l'Italia e gli italiani in un secolo di evoluzione del trasporto aereo*, Rome, 1992; *Vent'anni Alitalia*, *Freccia Alata*, special edition, 1967.

Camera dei deputati, *Problemi delle gestioni aeroportuali*, Rome, Servizio commissioni parlamentari, 1972.

Camera dei deputati, *Situazione dell'aviazione civile in Italia. Indagine conoscitiva della X Commissione permanente*, Rome, 1976.

Struttura di mercato e regolamentazione del trasporto aereo, F. Padoa Schioppa Kostoris, Bologna, il Mulino, 1995.

LOT

Czownicki, Jerzy. *Ekonomiczna Efektywnośċ Samolotów i lotnisk komunikacyjnych*, Warsaw, Szkoła Główna Planowania i Statystyki, 1976.

Czownicki, Jerzy Dariusz Kalińki, Elżbieta Marciszewska, *Transport Lotniczy w Gospodarce Rynkowej*, Warsaw, Szkoła Główna Handlowa, 1992.

Filipczyk, Joanna. *Ponadnarodowy System Transportowy w Warunkach Integracji Europejskiej*. Dissertation, Academy of Economics Katowice, private printing, 1995.

Januszkiewicz, Włodzimierz. *Transport i Spedycja w Handlu Międzynarodowym*, Warsaw, 1986.

Kneifel, J.L. *Fluggesellschaften und Luftverkehrssysteme der sozialistischen Staaten: UdSSR, Polen, USW.* Nördlingen, Verlag F. Steinmeier, 1980.

Mikulski, Mieczysław, Andrzej Glass, *Polski Transport Lotniczy*, Warsaw, Wydawnictwo Komunikacji i łązności, 1980.

Morawski, Wojciech, Krysztof Baduch, Bogumiła Krzyżewska, Romana Plejewska, Zofia Podgrodzka, Krystyna Zabórska, *Prognoza Potrzeb przewozowych do roku 1009 I 2000*, Warsaw, Ośrodek Badawczy Ekonomiki Transportu, 1974.

Prucha, Václav. (ed.) *The System of Centrally Planned Economies in Central-Eastern and South-Eastern Europe after World War II and the Causes of its Decay*, Prague, Vysoká Skola Economická v Praze, 1994.

Osiński, Jerzy. *Sytuacja Ekonomiczna Transportu Lotniczego 1960–1970*, Warsaw, Branżowy Ośrodek Informacji Technicznej i Ekonomicznej Lotniczwa Cywilnego, 1972.

Osiński, Jan. Henryk Żwirko. *Biuletyn Informacyjny Lotnictwa Cywilnego. 50 Lat PLL LOT. Wydanie Specjalne 20/80*, Warsaw, Branżowy Ośrodek Informacji Technicznej i Ekonomicznej Lotniczwa Cywilnego, 1980.

Sławiłska, Jadwiga. *Działalność Lotnisk Komunikacyjnych w Polsce w Latach 1991–1995*, Warsaw, Główny Inspektorat Lotnictwa Cywilnego, Ośrodek Analiz I Informacji, 1996.

Sobcki, Eugeniusz (ed.). *Podstawowe Informacje Dotyczące Lotnictwa Cywilnego*, Warsaw, Główny Inspektorat Lotnictwa Cywilnego, Ośrodek Analiz I Informacji, 1994.

Sobecki, Eugeniusz, Jadwiga Sławińska. *Rys Historyczny Rozwoju Władzy Lotniczej w Polsce, 1919–1994* Warsaw, Główny Inspektorat Lotnictwa Cywilnego, 1995.

Szumielewicz, Kazimierz. *Polskie Linie Lotnicze S.A.*, Warsaw, PLL LOT Zakładowy Ośrodek Informacji Technicznej i Ekonomicznej, 1996.

Żylicz, Marek. *Międzynarodowy obrót lotniczy*, Warsaw, Wydawnictwa Komunikacji i łączności, 1972.

Żyliez, Marek. *International air transport law*, Dordrecht, Nijhoff, 1992.

USA

Bender, Marilyn and Altschul, Selig. *The Chosen Instrument*, New York, Simon & Schuster, 1982.

Bernstein, Aaron. *Grounded: Frank Lorenzo and the Destruction of Eastern Air Lines*, New York, Simon & Schuster, 1990.

Bilstein, Roger E. *Flight Patterns: Trends of Aeronautical Development in the United States, 1918–1929*, Athens, University of Georgia Press, 1983.

Bilstein, Roger E. *Flight in America: From the Wrights to the Astronauts*, rev. edn, Baltimore, Johns Hopkins University Press, 1994.

Bilstein, Roger E. 'Air Travel and the Traveling Public: the American Experience, 1920–1970', in William F. Trimble, ed., *From Airships to Airbus: the History of Civil and Commercial Aviation*, vol. II, Washington, Smithsonian Institution Press, 1995.

Bishop, John C. 'Delivery of Goods by Air: The United States Air-Cargo Industry, 1945–1955', in *Essays in Business and Economic History*, Los Angeles, Economic and Business Historical Society, 1989.

Daley, Robert. *An American Saga: Juan Trippe and His Pan Am Empire*, New York, Random House, 1980.

Davies, R.E.G. *Airlines of the United States since 1914*, Washington, Smithsonian Institution Press, 1982.

Davies, R.E.G. *Rebels and Reformers of the Airways*, Washington, Smithsonian Institution Press, 1987.

Ingells, Douglas. *747: Story of the Boeing Super Jet*, Fallbrook, Calif., Aero Publishers, 1970.

Kent, Richard J. Jr., *Safe, Separated and Soaring: A History of Federal Civil Aviation Policy, 1961–1972*, Washington DC, 1980.

Komons, Nick A. *Bonfires to Beacons: Federal Aviation Policy under the Air Commerce Act, 1926–1938*, Washington, DC, Government Printing Office, 1978.

Leary, William. *Aerial Pioneers: The U.S. Air Mail Service, 1918–1927*, Washington, Smithsonian Institution Press, 1985.

Lewis, W. David. 'Eastern Air Lines', in William Leary, ed. *The Airline Industry*. Encyclopedia of American Business History and Industry, New York, Facts on File, 1992.

Preston, Edmund. *Troubled Passage: the Federal Aviation Administration during the Nixon-Ford Term, 1973–1977*, Washington DC, 1987.

Rochester, Stuart I. *Takeoff at Mid-Century: Federal Aviation Policy in the Eisenhower Years, 1953–1961*, Washington DC, 1976.

Serling, Robert J. *Eagle: the History of American Airlines*, New York, St. Martin's Press, 1985.

Sigafoos, Robert A. and Easson, Roger R. *Absolutely Positively Overnight: the Unofficial Corporate History of Federal Express*, Memphis, St. Luke's Press, 1988.

Smith, Richard K. 'The Intercontinental Airliner and the Essence of Airplane Performance', *Technology and Culture* 24, July 1983, pp. 32–47.

Solberg, Carl. *Conquest of the Skies: a History of Commercial Aviation in America*, Boston, Little, Brown, 1979.

Wilson, John R.M. *Turbulence Aloft: the Civil Aeronautics Administration amid Wars and Rumors of Wars, 1938–1953*, Washington, DC, Government Printing Office, 1979.

Glossary

Cabotage	Right of airline from country A to carry revenue passengers between two points in country B, (often a colony or former colony of country A)
Capacity or available tonne-kilometres (CTK/ATK)	Measure of airline output. ATK is the payload capacity multiplied by the stage distance
Chosen instrument	National airline, 'chosen' for receipt of government support
Code sharing	Market practice in which two airlines share the same two-letter code used to identify carriers in a computer reservation system (CRS)
Combi	Combined passenger/cargo aircraft
Computer reservation system (CRS)	System owned by airlines, used by travel agents, for displaying flight information and reserving seats
Double designation	Two scheduled airlines from the same country allowed to fly an international route
1st Freedom	Right to fly over another country without landing
2nd Freedom	Right to land in another country for technical reasons only, e.g. refuelling
3rd Freedom	Right to carry revenue passengers from your country (A) to another country (B)
4th Freedom	Right to carry revenue passengers from another country (B) to your country (A)
5th Freedom	Right of airline from country A to carry revenue passengers from country B to country C (or D)
6th Freedom	Right of airline from country A to carry revenue passengers from country B to country C, via its base in country A
Frequent Flyer programme	Marketing scheme designed to enhance loyalty to a particular airline
Gateway	An airport in western Europe or the east coast of the United States from which intercontinental flights begin
Grandfather rights	Long-standing landing and take-off rights – slots – held by an airline at a particular airport
Hub and spoke	Route system by which an airline can channel passenger traffic through a strategically located 'hub' airport
Load factor	Percentage of aircraft capacity filled with fare-paying passengers or freight

Non-scheduled service	Non-regular or charter services
Pool	An international air route between two (usually European) cities, the capacity on which, and the revenue from which, is shared between the two national airlines according to a pre-arranged formula
Scheduled service	Regular services at generally published tariff rates
Slot	Take-off or landing time at an airport reserved to an airline
Turbofan	Jet engine with a front fan in which a portion of the airflow bypasses the combustion chamber
Turbojet	A pure jet engine, in which thrust is derived solely from the jet exhaust
Turboprop	Propeller-turbine aircraft engine
Utilization	Aircraft flying hours, based on revenue earning hours only
Widebody	Aircraft with more than one aisle in the passenger cabin, e.g. Boeing 747, 767; Lockheed L-1011 Tristar, McDonnell Douglas DC-10, Airbus A300, A310
Yield management	Revenue-maximizing fare-fixing system run by a CRS

Index

260